Computer Simulation Applications

Computer Simulation Applications

**DISCRETE-EVENT SIMULATION FOR SYNTHESIS
AND ANALYSIS OF COMPLEX SYSTEMS**

Julian Reitman

**Norden Division
United Aircraft Corporation
Norwalk, Connecticut**

ROBERT E. KRIEGER PUBLISHING COMPANY
MALABAR, FLORIDA
1981

Original Edition 1971
Reprint Edition 1981 w/ new preface

Printed and Published by
ROBERT E. KRIEGER PUBLISHING COMPANY, INC.
KRIEGER BUILDING, KRIEGER DRIVE
MALABAR, FLORIDA 32950

Printed in the United States of America

Library of Congress Cataloging in Publication Data

Reitman, Julian.
    Computer Simulation Applications.

    Reprint of the edition published by Wiley-Interscience, New York, in series: Wiley series on systems engineering and analysis.
    Includes bibliographies and index.
    1. Digital computer simulation.   2. System design.
I. Title.
[T57.62.R45 1981]   001.4'24   80-26972
ISBN 0-89874-310-9

To Nora, Marc, and Randa
for whom simulation
has been the real thing

SYSTEMS ENGINEERING AND ANALYSIS SERIES

In a society which is producing more people, more materials, more things, and more information than ever before, systems engineering is indispensable in meeting the challenge of complexity. This series of books is an attempt to bring together in a complementary as well as unified fashion the many specialities of the subject, such as modeling and simulation, computing, control, probability and statistics, optimization, reliability, and economics, and to emphasize the interrelationship between them.

The aim is to make the series as comprehensive as possible without dwelling on the myriad details of each specialty and at the same time to provide a broad basic framework on which to build these details. The design of these books will be fundamental in nature to meet the needs of students and engineers and to insure they remain of lasting interest and importance.

Preface to Reprint

When the opportunity arose to revise and reprint this volume, I asked a basic question: does this volume still have something to say that is not covered elsewhere? At the time I was preparing a paper on interactive graphics and discrete event simulation. So I had a current look at the literature. My conclusion is that the features of this volume, the emphasis on fundamentals of simulation, the recognition that complex systems are not easily modeled, the use of detailed simulation applications, and the emphasis on a variety of techniques, all these are still relevant. Several features of this volume continue to provide valuable insights to learning the art of simulation. Among these are the treatment of the drive to the office application from a number of different simulation approaches, the treatment of the logical structures of the application areas, and the emphasis on man-machine interaction. However, the opportunity to see how the state-of-the-art has evolved cannot be overlooked, so a few comments are offered.

Changes in the State-of-the-Art

The choice of discrete event simulation language is still from GPSS, SIMSCRIPT, SIMULA and GASP or their direct derivatives. The latest versions do not indicate extensive changes. Those changes that have occurred are evolutionary. In GPSS there is now a compiler version, GPSS-H, developed by Jim Hendriksen that has a speed improvement ranging from three to twenty times. This significantly reduces the computer cost. In addition, GPSS-H has interactive debugging and some enhancements including those originally introduced as part of the GPSS/Norden extension.

SIMSCRIPT II.5 is well established on a variety of computers through the efforts of a single organization CACI. Pritsker and Assoc. have produced a variety of tools based on GASP and its derivatives. All these languages are widely used.

Changes That Haven't Happened

There was considerable optimism that simulation would become easier to learn and that computer terminals would provide an interactive tutorial for use when constructing models. So far neither has happened. The capability to present results via film or video tape in a user oriented fashion is actually further from usage today that was the case ten years ago. We may be on the threshold of great changes, but they are still not here.

Changes In Practice

The use of simulation languages has grown and become routine in many areas: military scenarios, assembly lines, horizontal and vertical conveyors, telephone, and transportation systems. However, FORTRAN is still very much in use.

There has also been an emphasis on statistical and theoretical aspect of simulation. This emphasis has suggested the possibility of highly accurate results through the statistical inference from the model results. The unfortunate aspect of this emphasis has been to create the illusion that because the statistics are well behaved the results are valid. This has served to detract from the major and difficult problem in simulation, namely, the accuracy of the assumptions. It is difficult to ensure that the assumptions represent the problem accurately. Therefore the accuracy of the results must remain suspect until there is a confirmation that the logic structure does implement the assumptions and that these assumptions mirror the problem.

Some models have survived for a long time, providing numerous different users with a tool. One example is CASEE (Comprehensive Aircraft Support Effectiveness Evaluation), a model developed for the Navy by S. Seidler and used by numerous Navy and industry facilities. CASEE is a classic example of a model evolving over a period of years to satisfy a number of different users.

CASEE is a system analysis tool to determine aircraft operational readiness and mission availability based on the following interrelationships:

- Aircraft system supportability concepts
- Aircraft system and subsystem reliability and maintainability
- Organizational and intermediate maintenance manning levels
- Spare parts stock levels and support environment

CASEE can and has been used to assist in answering such questions as: Given an aircraft type with specified reliability and maintainability characteristics, what will be the operational readiness and the capability to carry out its scheduled mission demands in a given support and operational environment? Or, "What are the benefits of improving the reliability of a given aircraft subsystem as compared to incorporating built-in-test capability or varying the initially out-fitted spares complement?"

The CASEE model is a large **GPSS** discrete-event computer simulation model. It is capable of simulating up to 127 individual aircraft of as many as 10 different type/model/series designations and as many as 10 different organizational units. Thus, anything from a single aircraft to an entire carrier air wing can be simulated at one time. The model can be used with land-based as well as with carrier-based aircraft.

The CASEE data base permits use with either currently operational or projected aircraft. The output statistics of CASEE describe the results of the simulation run for a specified simulated time period under a specific set of model inputs. Data formats for both input and output are compatible with those of the Navy Aviation 3M (Maintenance and Material Management) System. The current version of CASEE has been logically and numerically validated against real-world naval aircraft operations and support.

Where We Are

Over the last few years an active community of simulation practitioners has emerged. These individuals have been conducting meetings, publishing papers, and establishing a simulation literature. The proceedings of the Winter Simulation Conference, the Annual Simulation Symposium and to a lesser extent the Summer Simulation Conference provide numerous illustrations of what is being done and the various approaches utilized. Overall, the activity in simulation is of considerable extent and increasing.

Julian Reitman

January 1981

Preface

This book gives system designers, engineers, and analysts a new computer-aided design tool—discrete-event simulation. Such a method for representing complex systems in a computer language is useful in system design and analysis. It is specifically suggested for those systems in which relationships between key variables cannot be expressed analytically or the major system attributes are characterized by stochastic processes or probability distributions. Resource management, command and control-system analysis, information and traffic-flow problems, chemical-plant design, reliability and maintainability analysis, and weapons-system cost-effectiveness trade-offs are examples of system problems that can benefit from the application of discrete-event simulation.

This book should give the reader the confidence to solve complex system problems by using a digital computer and one of the available discrete-event simulation languages with a minimum of support from mathematicians and programmers. Access to a computer and prior experience in running problems on computers in any language are helpful. The General Purpose Simulation System (GPSS) and SIMSCRIPT are the two most widely used languages. Except for one example, which is treated in SIMSCRIPT II, the simulations used as examples are in the GPSS language.

Readers already knowledgeable in GPSS or SIMSCRIPT may choose to skip the initial treatment of these languages in Chapters 2 and 7. Those who find these chapters too advanced should consult one of the elementary books on simulation languages.

The degree of success expected from simulation depends primarily on the accuracy of problem definition. Fortunately this is one of the first and most important products of the use of simulation languages.

The purpose of this book is to instill in the reader confidence in the application of simulation and to provide an understanding of how to approach difficult problems. The variety of applications cover a sufficiently broad range that the reader will feel that each new problem has a relationship to something described in one or another of the applications. Discrete-event simulation is not something to be applied to trivial problems. The applications concentrate on those complex problems that require this powerful tool.

This book is divided into three parts. Part I begins with an introductory section covering the background and development of discrete-event simulation, the illustration of a specific simulation example using GPSS V; some brief background on the peculiarities of simulation and computer languages; a review of how random numbers are used and obtained; a comparison of different approaches to using computer languages for simulation; a second GPSS V look at the simulation example with the broader perspective developed from the additional material; and another version of the example using SIMSCRIPT II.

Part II is the specific review of five different applications. Limitations of space and interest reduce the coverage of each application to an understanding of the approach followed. Any of the applications could be expanded to an entire volume. This level of detail is unnecessary. The reader must learn by doing. The five applications cover scheduling and cost analysis of passenger railroad system operation, resource allocation in production scheduling, prediction of effectiveness for a weapons system, simulation of a computer system, and an evaluation of alternative automobile traffic control through a series of intersections.

Part III is devoted to relating the applications to each other and to placing the role of simulation into perspective. In particular, it relates the obvious successes of simulation with the less obvious difficulties overcome. The human factor in computer-aided system design is not overlooked, since there is so much more to be achieved toward the effective use of computers.

An extensive bibliography is included at the end of each chapter. As far as possible, references are made to available books, periodicals, and conference proceedings. Documents from the Defense Documentation Center are not included, because they are not readily available in libraries or bookstores, or from professional societies, and are preempted for use by defense industries. As a result, the application area without adequate references is the analysis of weapons systems.

References to pseudorandum number generation are extensive. Usually this is not the concern of the system analyst, but there are so many different approaches to generating and evaluating pseudorandom numbers that the reader may wish to consult the literature in order to be knowledgeable on this arcane subject.

Undergraduate and graduate students should be able to use this material in a variety of engineering areas: systems analysis, computer science, applied mathematics, and statistics. Until now, discrete-event simulation has mainly been taught in industry, since the subject is still rarely treated in a comprehensive manner in university curricula.

The problems are structured to provoke thought and understanding of concepts rather than to help in finding answers. Toward this end the problems are not designed to teach GPSS or SIMSCRIPT, inasmuch as there are specific books devoted to teaching simulation languages. Access to a computer will greatly extend the understanding of the possibilities of discrete event simulation.

Many people have helped in the development of this book. On a general level, I am indebted to Geoffrey Gordon for introducing me to GPSS, and Harold Chestnut, Ed Hall, and L. T. E. Tommy Thompson for their views of the role of discrete event simulation. I also thank those who helped me with specific chapters: Chapter 7, Phil Kiviat; Chapter 4, Harry Felder; Chapter 11, John Maneschi; and Chapter 12, Al Blum. Many thanks, too, to those who offered encouragement: Dick Baxter, John Bult, Dave Eig, Gene Ehrlich, Ivan Flores, Howard Halpern, Harold Hixson, Dick Horton, Sherman Hunter, Jerry Katzke, John Lovkay, Arnold Ockene, Frank Preston, Bernard Radack, Ted Rosenberg, Sandy Seidler, Jon Shapiro, and Burt Smith. Finally, I extend my appreciation to Shirley Lawrence and Sharon Luciani for typing the manuscript.

JULIAN REITMAN

Norwalk, Connecticut
March 1971

Contents

Part I

BACKGROUND

The term simulation in the sense used in this book describes a strange aggregate of elements. Simulation is a practical, application-oriented procedure. In order to use it, however, one must construct an abstraction of the problem, transfer the problem to a foreign device, the computer, and then obtain indications pertaining only to the representation of the system. Therefore there must be strong forces advocating simulation. The goal of Part I is, first, to delineate these forces and, second, to provide a perspective on the simulation technology—not just the present state of the art, but also some background on how it has been reached. Once this background has been established, we can investigate specific applications.

The powerful forces advocating simulation start with problem definition, follow through during system synthesis and analysis, and continue into operational evaluation. There are alternative approaches which may be quicker, cheaper, or easier if they work. Therefore we must learn when to use which tool, how to use computers, and how to express the problem clearly. Since the precision of answers obtained from simulation is different from an analytical result, we must understand when to utilize relative rather than absolute results. Hence, little attention is paid here to the theoretical aspect of pseudorandom number generation or to elegant distinctions among simulation languages.

The material is organized to familiarize the novice with the field and prepare him to simulate specific complex problems. The challenge is not to get the computer to provide a useful result, but rather to learn to achieve useful results quickly, with adequate accuracy, and within available funds, and for the method to be accepted by the customer.

1

1

Discrete-Event Simulation—A New Tool for System Designers, Engineers, and Analysts

The modern system designer or analyst is frequently confronted with problems which, upon study, are extremely difficult or impossible to solve using conventional analytical techniques. In general he has relied on intuition to solve them. A better approach must be used, however, especially since these problems are not exotic, only highly complex. Figure 1.1 is an attempt to compare the rapidly increasing complexity of one type of implemented system—long haul transportation.

Long haul transportation is one example of complex system design problems which range over a broad spectrum. At one end are the strategic, global, or economic problems. An example would be the repercussions of an income tax hike on the gross national product or on the force structure of the U.S. armed forces—very complex and usually not the concern of system designers. At the other end of the spectrum are the defined analytical problems, such as designing the control system of an aircraft—problems whose complexity can be restricted and readily partitioned into manageable pieces. Between these two extremes, as shown in Figure 1.2, are the tactical type of problems faced by the system designer.

The following examples indicate the wide range of these problems:

- Scheduling the items processed by a job shop so as to provide a long-term, high-level output and to satisfy immediate demand but use limited resources.
- Determining the number of new higher-speed passenger cars needed to reequip a railroad.
- Estimating the computer system throughput for a mix of program tasks.
- Contrasting different policies for inventory level, shipping schedule, and available resources with overall cost and performance of an enterprise.

3

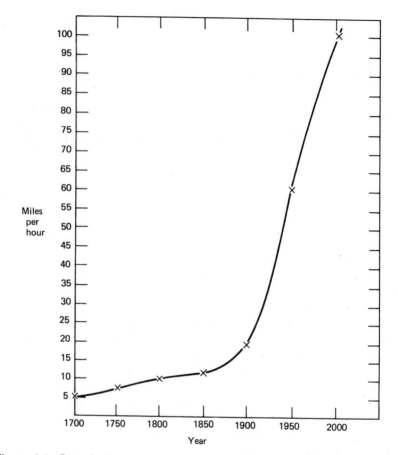

Figure 1.1 Long haul transportation speed as a measure of system complexity.

Figure 1.2 Spectrum of problems.

- Anticipating the number and mix of toll booth attendants and automatic gates needed to improve traffic flow.
- Comparing the life-cycle costs of initially using reliable and expensive parts to replacing and repairing less dependable and less expensive parts.
- Predicting the effect of increased accuracy on overall performance of a weapons system.
- Evaluating alternative configurations of playground equipments for elementary school-age children.

These few examples illustrate the type of problems requiring new techniques for solution. In the past such problems have been included in queuing theory, system behavior prediction, job shop analysis, systems effectiveness, and management information. Some analytical techniques have been effective, but frequently they have been too laborious to apply or have reduced the problem to an oversimplified shadow of its full self. A tool for solving these problems is now available through use of digital computers. Before discussing how to use the computer let us characterize typical problems.

Problem Characteristics

Three major characteristics of system design problems[1-9] are complexity, poor definition, and mathematical unpleasantness. The presence of any one of these is cause for trouble in the system design process. For example the problem of determining the minimum number of high-speed cars to maintain a specified level of service after reequipping a railroad is easy to define, but difficult to solve. All facets of the problem are interrelated. The number of cars depends on the schedule; the schedule, in turn, depends on car speed and terminal turnaround time. Then if the schedule is fast and frequent, a greater number of passengers may be attracted by the service—if the fares are not raised. Whichever way the problem is turned in an attempt to find some straightforward analytical expressions, the number of variables and expressions relating the different facets of the problem cannot be reduced to expressions solvable by conventional means. There are interrelated factors involving too many possible combinations of actions and situations, and an overwhelming number of possible states needed to fully describe the system.

In addition to the lack of analytical relationships representing the problem, there is frequently a lack of input data. For example what would happen if scheduled train service were increased? Would more passengers be attracted? And when would the additional passengers request service? An empirical relationship for passenger demand could be represented as a probability distribution of the number of passengers requesting service for each 10-minute interval during the day. This is certainly not a straightforward expression, and is further complicated by the fact that all days are not alike.

The determination of railroad fleet size should be a well-defined problem. Investigation, however, reveals large gaps in our understanding of the problem. One aspect is the choice between running an express train to a limited number of stations or stopping at every station. Now modify the problem to intermix some of the trains, expresses that make a few stops and locals that skip a few stations. The rules governing this choice are restrictions on train headway, passenger demand, and car storage yards. The problem may be stated easily. The actual definition of the problem, however, is expressed only in generalities that offer little help to the problem solver who must keep the trains from colliding and make sure that all the cars do not end up in a storage yard farthest from the next day's passengers.

Paradoxically there are many mathematical solutions lacking problems to which they can be applied. Unfortunately the mathematical tools for solving problems of a combinational nature have not yet been developed. If the railroad would only run point to point—Boston to New York—and the passengers appear with equal frequency during the 24-hour day, numerous techniques could be used to provide adequate, if not optimal, solutions. But in the real world there are stops in Providence, New Haven, Bridgeport, and so on, and passenger demand is high in the morning and evening and very low during the rest of the day.

The problems the system designer, engineer, or analyst must solve involve the performance of a system—that mixture of capital and labor organized to meet specified objectives within specified constraints, either minimum cost for specified performance or maximum performance for a specified cost.[10-12] Examples of these systems or subsystems are found in the design of computers, streets, right-of-ways, runways, machine tools, communication lines, military hardware, telephone switchboards, supermarkets, toll plazas, and so on. The measure of use made of the system—computer throughput, traffic flow, passengers carried, items machined, messages transmitted, missiles on target, calls placed, customers satisfied, motorists on their way—provides some of the criteria on how well a system meets its problems of services, delays, processes, failures, confusions, misses, losses, and so on, and at what cost. Notice that the tool we are looking for is a general technique, one capable of helping us study not just the obvious processes but the underlying interactions of the systems they represent. With the three system elements of "what is to be done," "how is it to be done," and "what is the measure of success," the fundamental basis of the problem may be described. People are going to be moved on schedule by ground transportation using high-speed trains over an existing right-of-way. A way must be found to compare different systems according to services provided at what costs.

Computer Logic Models and Simulation

A representation of a problem in terms of its logical and mathematical rules is a simulation of a system. Simulation in one or another form has been used for years as a technique of evaluating system performance. There have been scale models, mathematical models, dynamic analogies, and a wide variety of techniques to represent continuous behavior of a system. What is new is the development of a number of higher-level computer languages that provide a means of analysis for system problems that up to now have not been satisfactorily expressed by the above techniques.

The representation of the system—the rules and relationships that describe it—is defined as the model. In a sense the model becomes the algorithm, or statement of the problem. The use of the model under specific conditions is defined as simulation. The running of the model on the digital computer is the computer simulation of the system.

One purpose of simulation is to predict the system behavior. Once the model of the system has been developed and *confidence* has been established that a model is a valid representation of the real world, then a series of parametric simulations can be run to gain understanding of system behavior.

Uses of Simulation

The development of the model and the use of simulation can give the system designer something no other tool in his repertory can give—the feeling, insight, and opportunity to operate and manipulate a system plus a measure of insurance—while the system is still a paper concept.

The designer of a complex system traditionally has been limited. Rarely does he have the opportunity to redesign a system; even more rarely can he design the system from the start the way he would like. Usually there are restraints, rules, previous commitments, limited funds, and a tight schedule. In the face of these difficulties, and when conventional mathematical tools fail, the only technique available, other than intuition, is to use computer models.

The characteristics of complex systems that challenge the designer also conspire to make his task difficult and fill his path with pitfalls. Complex systems are unique. The system designer is deprived of already possessing that most necessary commodity—experience with the system. A good look at our complex systems—urban roads, railroads, even our computer systems—shows that the checks and balances of the existing system have stymied new developments or escalated the problem to the point where only the most highly trained individuals can work on these systems. When improvements must be made, they are achievable only slowly, in an evolutionary step by step manner. System designers can not limit their efforts to the improvements obtained by cut and try. They need to gather insight while the system design is flexible, before capital equipment is ordered and schedules established; and while many cases may be examined cheaply.

Simulation has many roles in improving the system designer's degree of success. At each stage of system design simulation serves different functions:

- Aid in problem definition.
- Help in relating numerous factors with their influences on the design.
- Insight into the sensitivity of the design to wide ranges of parameters.
- Support for selecting the final design from among the alternatives uncovered.
- Guidance in predicting system performance.

Problem Definition

The system designer obtains useful information even before the simulation is run merely from the disciplined efforts to construct a model. The definition of complex problems is subject to a wide range of interpretations, especially when several analysts of varying backgrounds are collaborating on the system design. Fortunately the computer accepts only clear rules. If, for example, when analyzing use of a road complex we want to simulate whether or not a particular vehicle will make a right turn at an intersection, certain rules would govern. One rule might be never on Sunday, another, only when the road straight ahead is "congested"; a third, derived from historical statistics, an even choice. Any one or all of these rules could be included in the model. Whether this degree of detail is necessary or desirable depends on the use to be made of that simulation. The computer model itself can be made to represent the real-world problem as it exists or as it is anticipated to be. Once the model is structured it becomes the definition of the problem, the *lingua franca* of system design—the interpretation around which the various members of the design team can gather when they discuss the problem. After all, the computer interprets the problem strictly according to rules. In summary constructing a computer model forces the system designer to state clearly and explicitly his understanding of the system.

Isolate Areas Where Most Good Can Be Done

Once a coarse model has been established of the overall system concept, it becomes a working tool. True, adequate data are usually not available and insight into system behavior is weak, but now tests can be made. Using speculative data ranging from optimistic to pessimistic, the design team can determine where the glaring holes are; what needs to be done if system performance is to meet specifications; and, most important, from the sensitivity test which additional data are needed and which are *not* needed.

Systematically the model can help determine the most sensitive and critical areas of the system. It is comforting to know, even when there are no available data, that possible data values for one function will not be the system determining factor—one more area that can wait until later. Some areas, however, cannot wait; there data must be obtained. The representation of the system is dynamic, changing as required by the model to reflect the greater detail needed to represent the system. Unfortunately models rarely shrink; the demands on them grow and grow. In this early stage simulation is serving to guide the design process by showing where the design is sufficiently strong that detailing can be safely delayed as opposed to where it is weak and needs immediate modification.

Choosing Among Alternate Designs

Complex systems permit a wide variety of implementations. In the past the designer selected those that intuitively seemed best and evaluated them. One risk he faced was the frequent request to investigate another alternative design. Since complex system designs were involved, a whole series of domino effects came into play. Service, performance, costs, and development schedules were affected, and the designer could not evaluate in a reasonable time frame all the additional possibilities. Herein lies one of the great advantages of computer-based simulation. Once a model is developed with a reasonably flexible structure, then it can be quickly and cheaply varied to include new wrinkles. Frequently only input data need be changed. It makes for friendly relations to be able to use the same model to evaluate additional alternatives.

The system designer has advantages when he is able to respond to yet another concept later in the design process, as well as when he is able to evaluate a larger number of alternatives. Once the set of rules, scenario, and model structure have been developed, all alternatives are evaluated on a common basis. The comparison is not to an absolute standard; but one alternative compared with another. It is helpful to know which system design will do 10 percent more work than any suggested alternative. This comparison must be based on the same work load, for the same time interval and sequence of demand. Where this is impractical the alternatives can be compared instead to an ideal system.

Expected System Performance

Before the system design is set, there must be assurance that the proposed system will meet specifications. This, too, may be obtained through simulation. The simulation of the proposed system represents an opportunity to put together all of the complex factors and discover how they interact to meet the system requirements. To illustrate, consider the problem of message garbling on a single circuit because two widely separated transmitters both acted as though they had line control. Prior to the simulation, opinions of several experienced engineers were solicited. Their opinions were either optimistic or pessimistic. The optimists felt that garbling would occur so rarely no error control would be needed, while the pessimists felt that the number of errors would be so great there would be almost no successful traffic. The simulation, operating on historical data, provided an answer midway between the two groups. This was almost a trivial case, but a useful one; and the needed insight could be obtained readily via simulation.

Using Simulation Requires Some Skills

At this point it is well to state an item of caution. Simulation can greatly improve the designer's effectiveness, but it cannot perform miracles. A system under analysis needs data, the establishment of system constraints, and, most of all, adequate effort to perform these tasks. While the GIGO principle (garbage in, garbage out) is well accepted, something can be learned from "garbage in." The input data, while not verified, can be made to represent the range of possible values. Insight is gained in spite of the lack of input data. To achieve the goal for the particular analysis requires the definition of the potential system and its constraints. The point is that developing these items requires considerable effort *and* the help of simulation.

The system designer needs the opportunity to learn how to use simulation as a tool of system design. New tools are used only after they have been proven and the user has gained experience. Therefore how does one go about becoming proficient in simulation? It would be nice if simulation were well-mannered, readily implemented, and readily subject to scientific reason. Simulation at this stage in its development, however, is primarily an art, demanding constant effort on the part of the user to understand the problem, goals, and extent of design effort required—to obtain, in short, the system designer's perspective. Under these conditions the effort to use simulation will provide meaningful assistance in system design. Fortunately it is not difficult to develop proficiency in the practice of simulation.

Simulation is not an end in itself nor should it be the specific tool of a small group. It should be used primarily by those faced with the problems of system design and should not require more than limited assistance from computer specialists—those middlemen between the designers—with their problems and the computer with its answers. From the designer's viewpoint simulation consists of a problem, a computer, and a language. These enable *him* to express that problem on the computer and obtain the needed results in a presentable manner with efficiency and speed.

Simulation Using Computers

Systems have been and are being designed, implemented, and made to work without a simulation. Designers have tried to use many techniques to aid in improving system design. The toy soldiers used by ancients to prepare for and study battles could be considered one of the earliest examples of the use of models. In a sense these were special-purpose simulations. Basically the desire has been to use one bounded, controllable medium to represent one less bounded and uncontrollable. From toy soldiers to sand tables to towing tanks, one medium after another has been used, until the latest medium, the computer.

Modern simulation can be said to have begun with the opportunity to represent mechanical systems in terms of electrical ones. True Vannevar Bush[13-16] had used mechanical elements to represent different mechanical systems in the 1920s, but not until electrical systems were used in modeling did the modern art of simulation become a practical technique. Power-system simulators were built in the 1930s as was acoustical analogy representation of microphones and loudspeakers.[17-20] These efforts produced useful but relatively inflexible representations of a system in different medium. Electrical simulation systems overcame the lead of mechanical systems because they could more easily be used to study a variety of systems with greater ease in changing parameter values. The breakthrough, which made simulation broadly feasible, was to substitute the multipurpose computer for the special purpose electrical system.

Continuous Simulation

A variety of systems can be represented by a mathematical model in the form of a set of differential equations. For the cases where a solution to the differential equations is known, the system model could be studied and the design evaluated. The equations of aerodynamic motion of an aircraft, for example, could be investigated to determine whether the proposed control system would work or how it could be improved. Similarly the suspension of an automobile could be analyzed as it traveled over different road surfaces. From this set of equations the designer would obtain data so that he could anticipate and understand various suspension problems without having to build and test a large number of suspension systems. The representation of the differential equations of a system by a computer is the simulation of system behavior in a continuous manner.[21]

Today the study of continuous system behavior can use a variety of analog and digital computers.[22-25] The computer can be used to analyze suspension system excursion from its steady-state condition as a result of a pothole. The transient phenomena explored concern the time required to return to normal, amount of overshoot, and maximum and minimum deviations experienced by a suspension system. The goal for continuous simulation is to predict the behavior of the system under a series of disturbing conditions.

The analog computer solves differential equations by the use of electrical integrators. These are combined with adders, multipliers, and sometimes nonlinear elements to produce time-varying electrical signals which are a direct representation of the solution for specific inputs.

The digital computer can solve the same differential equations. Instead of using electrical integrators, however, a method of numerical integration is combined with addition, multiplication, and logic. These are driven by the numerical representation of a time-varying signal. The key to the economics of using digital computers for solving these equations is the selection of the time interval for integration. Once this increment is selected it becomes the equal interval of simulated time when all calculations are repeated. Too short an increment requires too many calculations; and too long an increment results in poor accuracy. Ideally the length of the interval should vary according to the accuracy required. This concept ties in with a changing need for detail in response to different inputs. The obvious development, then, is to be able to control the increment of time in a simulation according to the problem requirements.

Discrete Event Simulation

Suppose the reason for conducting the study were different. Instead of the detailed behavior of the car suspension, our interest is in the behavior of several cars on a highway. The areas of study could cover such things as highway entrances, one car passing another, speed limits, weather, traffic density and, possibly, the suspension system since potholes may affect speed. The problem is complex in a different way. It is a system study of a bounded subset of the global transportation system. Now the problem defies any set of differential equations and cannot be represented in continuous form. No longer is the study limited to one vehicle. Vehicles must be considered as they interact with one another. Individual cars must be followed and statistical results considered. The elements of importance could be occasions when a car accelerates into traffic, changes speed, passes another car, exits, decelerates to avoid a pothole, or the various factors which prevent maintaining a constant speed.

A problem of this nature does not require a computer to continuously simulate the activity of the vehicles. Instead the simulation looks ahead to each occurrence of a change-of-state in the system and then evaluates the situation. Discrete event simulation uses the computer to indicate the system state for each occasion when something of interest occurs. The interval between such occasions is of no interest and therefore does not require computation. Discrete event computer simulation skips from one event at one time to the simulated time for the next event. The computer capacity to simulate the problem is concentrated into only those occasions the problem dictates. There is no waste of computer capacity to scan situations which have not changed. The vehicle moving at constant speed will arrive at its destination at a predicted time. Heavy traffic density, however, will provide numerous interactions with other vehicles and prevent constant speed. Discrete event simulation selects the times for each change of speed and skips the intervals between.

Characteristics of Problems Suitable
for Discrete Event Simulation

Large-scale, complex system analysis and design do not lend themselves to a conventional approach. There are no overall equations to be used. There is no clear place to start nor is there a clear ending. A system behaves, however, according to a set of relationships. This is the precise point of attack. Even though initially the rules for the interrelationships may not be consistent, they form the basis for model structure.

Logic, Rules, and Equations

A model must start somewhere. Since there are numerous choices, it is difficult to develop a model without going through a process of setting down a hypothesis and testing its validity. A means of implementing this process is the flow chart. For example what are the average, best, and worst times required to travel from A to X? The flow chart of Figure 1.3 shows a basis for a model. The two main factors for the model are:

1. The traffic desiring to go from A to X with the competing traffic already on the highway.
2. The highway geography with the number and distance between exits.

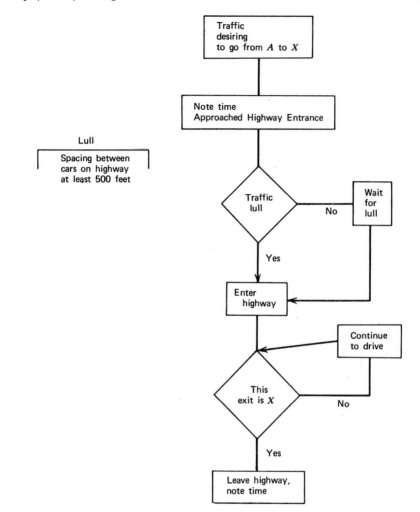

Figure 1.3   Flow chart representing travel from *A* to *X*.

To determine the required times, a model must be structured to follow each individual car as it passes the intermediate exits and kept a record of the time taken to travel from *A* to *X*. Finally the result and data should be organized into a tabular listing or graphical presentation.

The flow chart of Figure 1.3 may be adequate for the purpose of the model. More likely it does not fully answer the questions since there are additional factors which could be significant. The flow chart may have exposed a need for greater detail. Where initially the definition of a lull in traffic was a spacing between cars of at least 500 feet, a fuller description might include the following:

- Acceleration capability—horsepower-to-weight ratio below 0.025 keep 500 feet, decreasing ratio to 0.1 linearly reduce distance to 250 feet, ratios above 0.1 keep 250 feet.
- Speed of oncoming traffic—90 percent of speed limit or below retain 500 feet, 90 to 100 percent of speed limit increase distance linearly, above 100 percent of speed limit no increase in distance.

The flow chart under these new conditions might be modified to where the section determining the spacing defining a lull appears as shown in Figure 1.4.

The flow chart has established the relationships of the model. Going from one point on the flow chart to the next requires the application of rules. At each step the rules must reduce to a single choice for the next step. The rules need not be simple—entire sets of complex

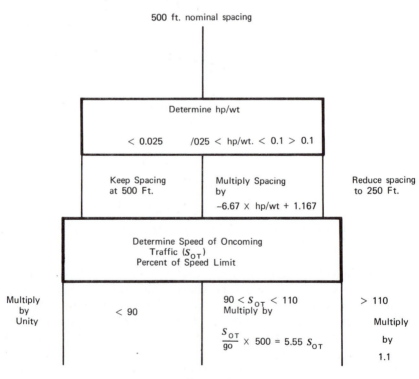

Figure 1.4

relationships can be used. At different times different subsets of rules may be selected. When a variety of meanings is possible, however, there must be further rules defining the choice of each possibility. In many cases statistical relationships can be utilized to provide a choice on a probabilistic basis. Many of the choices will be dictated by numerous straightforward analytical expressions, such as computation of the time and distance required to accelerate to 60 mph. The fundamental point is the ability to establish rules using any available data, not to be restricted to certain forms. Actual historical data may be used in either raw or smoothed form.

Historical Data

In this example simulation is directed toward designing a highway system of entrances, exits, and traffic. Each car represents an item of interest with its entrance, route, and exit. A number of events occur and many more could have. These interactions are not definite, but depend on circumstances. The decision concerning the occurrence of a particular event is made by use of equations, logic, or probability. Probabilities can be based on historical experience or an idealized premise. Cars can be grouped according to a bell-shaped normal distribution or according to road records for clear weather on a straight and level four-lane highway. The simulation should handle either equally well. When only normal distributions are used the need for simulation is considerably reduced since analytical methods may suffice. When only well-behaved analytical equations are involved, they may be solved more quickly and cheaply than the effort to establish and use a system simulation.

Role of the Computer

For the last few years we have progressed from communication with the computer on its own terms (computer machine language) to using the computer to convert some easily understood instructions to machine language on a one-for-one basis (assembler) to the present-day practice in which one gross instruction becomes hundreds or even thousands of machine language instructions (compiler).

Improved computer languages have permitted the solution of increasingly complex problems. In addition the computer itself attends to many of the details of determining which problem is next, for up to how long, and who is charged for the computer time. Under these circumstances the way human beings communicate with the computer must emphasize convenience, speed in obtaining solution, limited training for standard problems, and high levels of training only for certain complex problems. Higher-order computer languages have become the order of the day and for good reason—the actual output per total dollar spent is way up. Analysts whose problems lent themselves to FORTRAN, ALGOL, and COBOL were able to solve complex equations and involved data processing manipulations for fractions of previous labor cost. Cost of computing has come down rapidly. The addition of compilation costs to the cost of processing the problem has been gracefully accepted. Almost all data processing today is based on some form of higher-order special programming language.

Where in all this world of plenty was the system designer? Before looking at the possible solutions, consider his rather unique requirements.

System Designer's Language Requirements

Frequently the system designer uses simulation when problem definition is vague and insight is needed. Under these circumstances his desire to use the computer makes quick response the paramount requirement.

During early stages of problem definition the system designer may range far and wide in his sensitivity analyses. The computer must be able to reflect the latest change in problem definition and respond to that change *before the next one is introduced.*

- Response is fastest when the system designer uses the computer himself. Having to use intermediaries takes time and is frequently a source of misinterpretation.
- The degree of model detail represents an area for deliberate choice by the system designer. Greater detail increases the computer time required, the cost of time and labor of the simulation, and the amount of information needed. Less detail improves the running time. The system designer should be the judge of what details are necessary and which can be left out.
- Simulation can conform to the idiosyncrasies of the real world at the required level of detail. The step-by-step process of representing real-world conditions one at a time and having the simulation tie the interactions together helps to gain acceptance of the computer simulation. Usually this process turns out to be faster and less

painful than convincing someone that an idealized form of the problem is an equivalent representation.

- Acquisition of data is never easy. There usually are too many inappropriate data or too few useful data. When it is desirable to use raw data the cost of data acquisition and processing is reduced. When the designer can influence the form in which data are delivered, he can reduce the total amount to be processed and can come closer to obtaining only what is pertinent.
- Computer time is one cost the system designer can afford because at any stage of system effort the simulation computer time cost is usually the least significant element in the entire process. Computer time costs rarely compare to the costs of design team manpower, system hardware, or redesign. Computer systems are the exceptions since their simulation can use quantities of computer time.
- The system designer can best recognize whether the model is an accurate representation of the system. Because he knows the system and what is expected from it, he can readily judge whether the initial simulation results are realistic. If they are he proceeds with confidence; if they are not he must modify the approach and try again.
- Complex problems which can profit from simulation do not remain the same. They are dynamic. A model that represented yesterday's version of the problem is equivalent to yesterday's newspaper. The problem must be represented as it is currently, with the understanding that change is normal and can be accommodated. The conclusions, however, must be relatively insensitive to minor problem changes.
- The system designer uses the computer only part of the time he is working on the problem. Therefore his experience with computers is not on a familiar basis. The simulation itself should take care of routine details, housekeeping, and model documentation. The host of computer ills possible in modern complex computer systems should not impede the designer. His is a special use of the machine, and his limited set of requirements should be met with simplicity and ease.
- While a model is changing it is advantageous to have a trail that can be retraced. Although this requirement may seem trivial, there is a strong need for documentation of structure and assumptions in each successive version of the model. If these early versions are lost, the same mistakes can and probably will be repeated. Some of the responsibility for documentation of the model must be inherent in the simulation language.

- Until our universities train system designers in the use of simulation techniques, these designers will have to be trained on the job. Short courses have great value, but complex systems are not solved in short courses. Ideally the learning process should be accomplished in stages—some of the tools first, more as confidence grows, and finally the full complement. There will be useful service even during the earlier stages, and we must expect that some designers will never have the time to develop the full set of skills.

- The system designer does not operate by himself. Usually he is only one member of a team and his ability to communicate with his teammates is critical. If only one individual can understand the model and the simulation results, it is unlikely that the model will be accepted and shared by all. The output and description of the model, therefore, should be in some form other than mountains of computer printout. Graphical output is necessary. Designers have been taught to use graphs, histograms, and plots; the output must be in similar form.

- Simulation must be a portable tool. The conditions under which systems are designed are subject to considerable transient disturbance. The team may meet at one location one day and at a different site on another day. If simulation is required at the first location, it is required at the other. Either our communications network will improve to the extent that the home base computer can be interrogated from anywhere, or the model must be portable so that it can be taken to some other computer installation where results can be obtained.

- The simulation language should accommodate the system designer. Representation of a problem in the language of simulation takes effort and is dependent on the form of the simulation language used. Some languages make it easier for a system designer to structure and describe a problem. The ideal language permits conditions of the real world to be represented to the degree desired, with a minimum of artificiality and without forcing the actual problem to conform to the available simulation language.

- System designers' tools must be reliable. This is true of software as well as hardware. The representation of the problem by the designer should be the representation by the simulation. Since the system designer is forced to learn from an instruction book, that instruction book must be clear.

When these requirements are added up, it is apparent that there is no ideal simulation language. FORTRAN does not begin to fulfill these special requirements. It lacks the quick response, flexibility, built-in housekeeping, and logical rule structure required of a simulation language. Simulation languages have been written in FORTRAN, but the system designer who is not concerned with FORTRAN then becomes enmeshed in it. Similar statements can also be made concerning the inadequacy of other algebraic languages for the logic and rule structure of discrete event simulation.

Some Discrete Event Simulation Languages

Life would be more reasonable for the system designer if all the language criteria just discussed were fulfilled by a particular simulation language. There are many languages available, but none meets all the criteria. From the system designer's viewpoint some are more adequate than others. The languages will be covered in detail later, but at this point the basic distinctions will be noted.

System design problems are different. Discrete event simulation languages devised for a single problem require each designer to develop a private simulation technique. Use of private techniques make the process of using the computer for simulation awkward, slow, expensive, and dependent on individual experience. Therefore a different approach, a general language, must be used if simulation is to be a widely used system design tool.

There are two basic structures for general purpose simulation languages represented by General Purpose Simulation System (GPSS) and SIMSCRIPT. The GPSS is highly structured, relatively easy for the nonprogrammer to learn, and includes many convenience features. SIMSCRIPT is less structured, requires greater programming competence, and demands more effort by the user to obtain equivalent convenience features. GPSS was developed by Geoffrey Gordon and released in its primitive form in 1961 by IBM.[26] Several versions have appeared since, the latest being a version for the IBM 360 system, GPSS V.[27-31] SIMSCRIPT was developed at RAND Corporation by H. H. Markowitz, B. Hausner, and H. W. Karr and released by RAND in 1962.[32-34] A radically different version, SIMSCRIPT II, was released by RAND in 1968.[35] Among languages of the less-structured form are SIMULA,[36-38] GASP,[39] MILITRAN,[40] and OPS-3.[41]

Since these languages have been introduced, they have been subjected to evolutionary change through several versions. There are advantages to the system designer from this process. Other designers have used the language, found shortcomings, and suggested changes. The language gains in reliability through evolution. Moreover, since usually a single group does the updating and improving rather than each user, the language evolves with many different inputs and from a variety of uses.

Developing a simulation language is a long and arduous undertaking. The task is just as arduous for the documentor. Nothing is quite as frustrating for the system designer as documentation in the form of an instruction book which says one thing should happen and simulation results which show something else. Widely used languages have had their innards well explored, and there is a better chance that the simulation will do what the instruction manual says it will. Yet it is impossible to develop a simulation language that is free of program bugs. The hope is that someone else has already found them and that they are or have been corrected. Obviously widely used and well supported languages should have fewer bugs remaining.

Discrete and Continuous Simulation

It is easy to visualize the need for a single simulation structure for the analysis of both car suspension and interactions among cars in a highway system. One approach is to simulate the continuous problem in a general-purpose analytical language—FORTRAN, ALGOL—or a special-purpose continuous simulation language—MIDAS, PACTOLUS, CSMP[43]—and subordinate the continuous simulation in the framework and overall structure of the discrete event language by trying together the two simulation systems. Other approaches are to use either the analog-digital hybrid computer complex or a multiprocessor computer complex with continuous simulation on the analog or special digital computer under the control of the master digital computer, which is also doing discrete event simulation. The burden on the master digital computer becomes a time-shared programming problem with command and control intermixed with the discrete event simulation paced in time by the slowest element. Hybrid facilities of this type do exist, but so far have been used extensively for continuous simulation.

Uses of Simulation Languages

Simulation languages have a number of distinct functions in the design of complex systems. While it is true that systems have been implemented without the benefit of simulation or even computers, there are advantages gained in using simulation. Problem definition is clearer, tie-in to future use is better analyzed, uncertainties of the design should be less, and all this almost invariably at a saving in overall cost. On the other side there is the reluctance to study and restudy rather than "cut metal" the suspicion of a broad set of interdisciplinary activities, and not least the reticence to use a new and "unproven" tool.

Uses of Simulation Languages—Problem Definition

The initial gain from simulation is obtained even before the simulation is actually run. There are a number of people working on the design at one time, and it is very useful to the entire group to have a clear definition of the problem. Even when the definition is only an agreement to disagree, the boundaries of the disputed area are identified, quantified, and seen in focus. This is not a trivial achievement, as designers who have tried to participate in large-scale system work will attest. It is difficult to get the design team to visualize before actual experience that a simulation language can become the best vehicle in which to express the problem definition.

The language structure has an important role in determining the clarity of the problem definition. If each language expression is readily understood by the design team, the statement of the problem is useful. When translators are needed or when the structure of the language is obscure to any but the most knowledgeable, the problem definition is not useful. It may seem strange, but the simulation language is put to one of its most important uses before the problem meets the computer.

The major expectations from simulation are as an aid in synthesis and analysis of complex systems. The choice of simulation over conventional methods goes back to the inability to make the real world conform to classical mathematical techniques. Simulation provides a convenient means for the designers to participate in the design process by using the computer in an iterative process. It is hoped that through a series of steps, the design process will converge toward an adequate system; and on occasion the use of simulation will spark sufficient insight to tie the problem neatly into a simple mathematical relationship.

Users of Simulation Language—Real-World Perspectives

Designers using simulation can accept real-world rules. Sometimes this approach shows almost painlessly that things could not possibly be as black as painted. Its real advantage lies is not having to accept an idealized version of the problem. Nothing in nature guarantees that complex problems are consistent or structured; or for that matter that the purpose of each member of the design team is the same. Under these conditions it is helpful to accept the problem in its ugly reality and convert it to a computer representation with its peculiarities relatively intact.

Uses of Simulation Languages—Input Data

The capability of conforming to the real world presents a severe problem. Where are the data for use in the model? Complex problems tend to fall into two categories: those with too many data and those with too few. The two-many-data case calls for using all the methods of statistical analysis to gain the sense of the data without the deluge of detail. Yet simulation languages permit introduction of actual quantities of data in raw or processed form or in combination. Because raw data can be used, historical data can be introduced when available. The designer should not hesitate to try records that could be meaningful, even though in numerous instances real-world data can be worthless because modifying restrictions are not apparent. The important fact is that when the real world does provide useful historical data, simulation can accept those data; when there are not enough historical data, simulation can mix statistical distributions with historical data.

In the too-few-data case the system designer has a different set of problems. New systems change the status quo. When the system designer lacks real data he must devise pseudodata. By relying on probabilistic simulation he can develop data or pseudodata for analysis of the problem. Guiding rules obtained from historical reports, knowledgeable sources, or even intuition can be used to provide sets of data that are not mathematically idealized or isolated to a small subset of the environment.

Uses of Simulation Languages—Relative Solutions

Analytical and mathematical relationships have led us to expect a specific answer or set of answers. Ambiguity in the problem relationships has led to statistical techniques in the expectation that a relationship holds when enough individual events occur.

The system designer operates in yet a different environment. Rather than seek a single optimum solution, he investigates alternate solutions, searching for a possible system which appears most adequate. The answer does not require all possible situations to be thoroughly explored. Many of the possible solutions can be ruled out on the basis of simple analysis. The need is for meaningful comparisons among the remaining alternatives. Here simulation has a clear advantage. Alternative system designs can be evaluated against sets of requirements—traffic density, mission profiles, or a combination of factors. The results are quantitative, consistent with established criteria, and indicative of possible system behavior. The results can be compared to an ideal case or compared to each other on a relative basis. Care must, of course, be exercised before extrapolating to all conditions.

Uses of Simulation Languages—Sequence of Steps

The use of simulation proceeds along well-defined steps. First, the overall possibilities are investigated with a coarse overview model. Then the evolutionary process of system design takes over. Somehow the validity of the initial or derivations model becomes established. Then the simulation becomes more and more detailed in those areas determined by the sensitivity analysis. Eventually the model is composed of both detailed and coarse parts. The coarse parts remain when the range of possible data indicates insufficient effect on the system design.

Statistical significance sometimes becomes a factor in simulation; how often radio communication becomes unsatisfactory due to sun spot activity could be one such instance. When analyzing the behavior of a specific system, comparative results of alternative systems are preferred to straight statistical significance because of the great reduction in computer time required for adequate results.

Simulation gives the system designer a tool that will conform to his requirements during system synthesis and analysis. He can use data in circumstances considered meaningful and compare possible solutions for the most attractive prospects. When detailed questions such as the worst-case condition are raised, he can study the system configuration closely enough to provide detailed analyses. As the system design progresses, the characteristics of his use of simulation change to reflect increased interest in the behavior of some aspects of the system and decreased interest in others. At each stage, simulation has provided extra insurance that the system is understood and what is expected. The flexibility and growth of problem understanding are mirrored in the changing uses of simulation as the system evolves from concept to implementation.

PROBLEMS

1. Some problems are solvable by analytic methods and others by simulation. Discuss the advantages and disadvantages of each approach for the following systems:
    Select one route over another for a particular auto trip.
    Compare different modes of transportation for the trip.
    Compare different schedules for the introduction of new vehicles in a transportation system.
    Determine the life-cycle costs and performance for a weapons system.
    Schedule limited machinery and personnel in a production shop.
    Determine the probability of a false radar return from clutter, noise, or a combination of the two.
    Compare the performance and cost of high-speed and low-speed communications lines for computer remote terminals.
    Predict childrens' choice of playground activities. This may be difficult, but estimates could be based on group preferences.
    Contrast computer installation performance based on turnaround, cost, and implementation.
    Lay out the schedule for classroom assignment.
2. Apply the following elements to the above systems:
    System boundaries and constraints.
    Criteria to evaluate each system.
    Sources for input data.
    Approach to validate results.
    Applicability of results to changes in input data.
    Utility of approach for changes in system description.
3. The scale of the problem has an influence on the method of solution, consider how the following could be solved and with what approach to mechanization:
    What is the expected performance for elevator systems used in: six-story building, sixty-story office building. How would elevator systems for building sizes and functions in between be analyzed?
    What is the quality of an automobile ride over a road with potholes, numerous traffic lights, and in both light and very heavy traffic?
    What methods should be used to schedule a charter airline, one

with predominately long-range, low-density flights, and one with numerous short flights over heavily populated routes?

4. Develop a schedule for a restricted right-of-way transportation system with the following characteristics:

n stops.

varying passenger demand during the day.

multi unit vehicles.

vehicle storage yards at m locations.

Establish criteria for system performance.

Determine how these criteria would be measured, alternative systems evaluated, and the difficulty of performing the analysis.

BIBLIOGRAPHY

1. *Systems Engineering,* H. H. Goode and R. E. Machol, McGraw-Hill, New York, 1959.
2. *The Design of Engineering Systems,* W. Gosling, Wiley, New York, 1962.
3. Systems Engineering, H. A. Affel, Jr., *Intern. Sci. Tech.,* 18–26 (November 1964).
4. *Systems Engineering Tools,* H. Chestnut, Wiley, New York, 1965.
5. *Systems Engineering Methods,* H. Chestnut, Wiley, New York, 1967.
6. *Techniques of System Engineering,* S. M. Shinners, McGraw-Hill, New York, 1967.
7. *A Methodology for Systems Engineering,* A. D. Hall, Van Nostrand, New York, 1962.
8. When and Why Is Systems Engineering Used? J. Reitman, *Proc. 1964 Syst. Eng. Conf.,* 65–74.
9. A New Look at Systems Engineering, R. A. Frosch, *IEEE Spectrum,* 24–28 (September 1969).
10. *Systems Analysis: A Computer Approach to Decision Models,* C. McMillan and R. F. Gonzalez, Irwin, Homewood, Ill., 1965.
11. *Systems Analysis for Effective Planning,* B. H. Rudwick, Wiley, New York, 1969.
12. *Systems and Simulation,* D. N. Chorafas, Academic, New York, 1965.
13. A Continuous Integraph, V. Bush, F. D. Gage, and H. R. Stewart, *J. Franklin Inst.,* **203,** 63–84 (1927).
14. The Differential Analyzer, V. Bush, *J. Franklin Inst.,* **212,** 447–488 (1931).
15. Structural Analysis by Electric Circuit Analogues, V. Bush, *J. Franklin Inst.,* **217,** 289–329, (1934).
16. Integraph Solution of Differential Equations, V. Bush and H. L. Hazen, *J. Franklin Inst.,* **204,** 575–615 (1927).
17. MIT Network Analyzer, H. L. Hazen, O. R. Scherig, and M. F. Gardner, *Trans. AIEE,* **49,** 3:1102–1113 (1930).
18. *Elements of Acoustical Engineering,* H. F. Olson, Van Nostrand, New York, 1940.
19. A New Analogy Between Mechanical and Electrical Systems, F. A. Firestone, *J. Acoust. Soc. Am.,* **4,** 249–267 (1933).

20. The Mobility Method of Computing the Vibration of Linear Mechanical and Acoustical Systems: Mechanical-Electrical Analogies, F. A. Firestone, *J. Appl. Phys.,* **9,** 373–387 (1938).
21. Simulation—Its Place in System Design, H. H. Goode, *Proc. IRE,* **39,** 12:1501–1506 (1951).
22. *Analog Methods in Computation and Simulation,* W. W. Soroka, McGraw-Hill, New York, 1964.
23. *Electronic Analog Computers,* G. A. Korn and T. M. Korn, McGraw-Hill, New York, 1956.
24. *Introduction to Electronic Analog Computers,* J. N. Warfield, Prentice-Hall, Englewood Cliffs, N.J., 1959.
25. *Simulation,* J. McLeod (ed.), McGraw-Hill, New York, 1968.
26. A General Purpose Simulation Program, G. Gordon, *Proc. EJCC, Washington, D.C.,* 87–104, Macmillan, New York, 1961.
27. A General Purpose Digital Simulator and Examples of its Application: Part 1—Description of the Simulator, R. Effron and G. Gordon, *IBM Syst. J.,* **3,** 1:21–34 (1964).
28. GPSS III—An Expanded General Purpose Simulator, H. Herscovitch and T. Schneider, *IBM Syst. J.,* **4,** 3:174–183 (1965).
29. GPSS V Introductory User's Manual, IBM Corporation Form No. SH20-0866, 1970.
30. GPSS V User's Manual, IBM Corporation Form No. SH20-0851-0, 1970.
31. GPSS/360—An Improved General Purpose Simulator, R. L. Gould, *IBM Syst. J.,* **8,** 1:1627 (1969).
32. *SIMSCRIPT—A Simulation Programming Language,* H. Markowitz, B. Hausner, and H. W. Karr, Prentice-Hall, Englewood Cliffs, N.J., 1963.
33. A Description of the SIMSCRIPT Language, B. Dimsdale, and H. M. Markowitz, *IBM Syst. J.,* **3,** 1:57–67 (1964).
34. SIMSCRIPT I.5, H. W. Karr, H. Kleine, and H. M. Markowitz, Consolidated Analysis Centers, CACI 65-INT-1, Santa Monica, Calif., 1965.
35. *The SIMSCRIPT II Programming Language,* P. J. Kiviat, R. Villanueva, and H. M. Markowitz, Prentice-Hall, Englewood Cliffs, N.J., 1968.
36. SIMULA—A Language for Programming and Description of Discrete Event Systems: Introduction and User's Manual, O. J. Dahl and K. Nygaard, Norwegian Computing Center, 1967.
37. SIMULA—An ALGOL-Based Simulation Language, O. J. Dahl and K. Nygaard, *Commun. Assoc. Comput. Mach.,* **9,** 671–678 (September 1966).
38. *Some Features of the SIMULA 67 Language,* O. J. Dahl, B. Myhrhaug, and K. Nygaard, *Proc. 2nd Conf. Appl. Simul.,* New York, 29–31, 1968.
39. *Simulation with GASP II: A FORTRAN Based Simulation Language,* A. A. B. Pritsker and P. J. Kiviat, Prentice-Hall, Englewood Cliffs, N.J., 1969.
40. MILITRAN Programming Manual, Systems Research Group Report ESD-TDR-64-320, 1964.
41. *On-Line Computation and Simulation: The OPS-3 System,* M. Greenberger, M. M. Jones, J. H. Morris Jr., and D. M. Ness, MIT Press, Cambridge, Mass., 1965.
42. A Survey of Digital Simulation: Digital-Analog Simulation Programs, R. N. Linebarger and R. D. Brennan, *Simulation,* **3,** 6:22–26 (1964).
43. Two Continuous System Modeling Programs, R. D. Brennan and M. Y. Silberberg, *IBM Sys. J.,* **6,** 4:242–266 (1967).

2

Characteristics of Simulation—A Tour Through One Simulation

An Example

A single problem, the simulation of the drive to the office each morning, will be used in this chapter to show the problem being formulated, describe the model development in GPSS V, and demonstrate simulation results. The drive-to-the-office model stresses those elements of simulation methodology that are generally useful within the framework of a common problem.[1-3] Only part of the general problem is covered, however, namely one specific trip to the office competing with other traffic. For comparison in Chapter 7 the same problem is modeled using SIMSCRIPT II. Chapter 6 again treats this problem with only advanced GPSS techniques.

What the Example Should Show

From the various possible purposes of the drive-to-the-office model, probably the most useful information would be to learn whether or not *changing the route* would reduce the trip duration. Other items could be departure time, driving strategy, vehicle characteristics, and abnormal weather or road conditions. These elements could be studied by running the simulation many times with variations and comparing the results, or through the artifice of each morning simulating each route with a separate vehicle.

Statistical Departure Time Considerations

Departure time is constrained within narrow bounds of the desired arrival time at the office. Numerous elements of the environment interact with the departure time and route. For example the trip duration may be affected by school and office starting time, public transportation facilities, routes and schedules, and changes in police shifts.

Other Changes

Alternatives to be explored include changing the route, improving vehicle characteristics, and modifying driving strategy. Of these a change of route would appear the most promising alternative to investigate, but actually this is what can never be done in reality. Only in a simulation may we change routes and find out which route would have been faster for a given trip.

It might be claimed that vehicle characteristics are the least important factor, since speed laws act as a constraint. This could be tested by the change in trip duration caused by very different vehicles—a small import versus the highest Detroit horsepower-to-weight ratio. Their acceleration characteristics would contribute the main differences to the trip duration.

Driving strategy might consider traveling longer distances at higher speed rather than seeking the shortest route. Other strategies might be to seek the route which is sequentially timed for fewer traffic lights, or the route with the lowest accident rate.

Simulation Factors

A systematic investigation using simulation of the above factors would permit direct numerical comparison between the average trip duration over one route with the average duration over another. A more meaningful comparison, however, might be obtained through an overall figure of merit. This might be expressed as a function of highest speed attained, total time stalled in traffic, amount of fuel consumed, and exposure to traffic lights. This last item might require further stratification into lights timed to speed the flow of traffic and those with unrelated random settings.

Results based on averages are useful, but what was the longest trip duration, what was the shortest, and why did each come about? Special insight can be gained from the extreme cases. How much had to go wrong to cause the longest delay may provide that extra understanding needed to justify strategy, route, or vehicle. When the worst case is fairly close to the average (a tightly bounded system) there is little to choose. The worst case that is far from the average could indicate "one of those things"—everything that could go wrong did. The distribution of trip durations is in itself a significant result from the simulation.

Abnormal Conditions

So far the analysis has been geared to help understand the normal situation. Additional areas remain to be accounted for if the simulation is to be adequately close to real life. Emergency vehicles may halt traffic at an intersection. During a snowstorm, one route may be preferable to another. When there is a traffic tie-up at an intersection, it may be necessary to choose an alternate route. This could require that the simulation consider traffic information relayed by radio. Other abnormal conditions to be considered are accidents and road repairs.

Defining the Model

Problem definition is the first result to evolve from simulation—a result long before the computer is used and a result which helps fix the boundaries of the problem and, in so doing, the model structure. The next step is to fill in the model with a minimum number of necessary details.

1. Route definitions.
2. Speed equations.
3. Rules governing the choice of alternate routes.
4. Degree of interaction with other traffic.
5. Influence of abnormal events.

For illustrative purposes some of the factors in the example will be developed to a greater degree than others. But this will remain, basically, a simple example. The coarse overview is shown in Figure 2.1, a flow chart of the basic outline of a minimal drive to the office. The following sections discuss the various factors that are required to define this example.

Input Data

Specific input data for the model are as follows:

1. Characteristics of each route segment: mileage, traffic lights, and average speed. Some of these data are available; where they are not estimates for each route segment would have to be used, especially for the average speed.

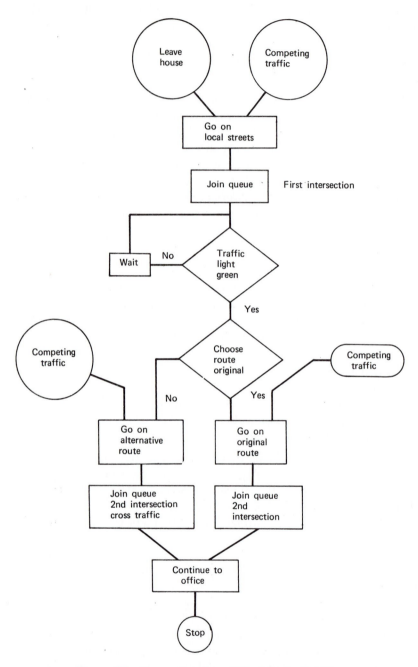

Figure 2.1   Coarse overview of the drive-to-the-office.

32

2. Delays at points of traffic confluence—these data are more the reason for the study than a set of obtainable input data, although sometimes they are available from intersection surveys. Where data are not available, delays may be estimated from the action of traffic signals. Only the proportion of time the traffic signal is red causes delays. For the case of traffic below the saturation density, the delays average half the duration of the red phase.

3. Vehicle characteristics: acceleration, handling, and passing performance—in the coarse initial stages of model development these factors might be combined into an average vehicle speed.

4. Weather factors—these can be obtained from a random selection of possible weather conditions, preprogrammed, historical storm data, or selected from a limited set of on-road surface conditions, dry, wet or snowy.

5. Synchronization of traffic lights—the passage through each intersection has to be controlled either from a separate timing sequence for all signals or individual sequences of so many seconds.

This set of diverse input data represents a typically heterogeneous group. That is what would be expected; that is the nature of complex systems. There are data that are based on historical records, some completely on the laws of chance, and a few on equations. The degree of certainty with which these are known at this stage is typically too many "guesstimates" and too few facts.

Model Structure Rules

Some of the rules governing the behavior of the driver are fixed and form the basic framework for the model. This set of rules, or algorithms, might include both general and specific rules for how the simulation is to be run this time as contrasted with a different set of rules to be used the next time. Examples of the rules governing the model are as follows:

1. Merging into the traffic stream—merger can occur under discipline of signal lights, right-of-way preference, and degree of risk.

2. Route segment selection—selection based on traffic density and queuing delay.

3. External influences—weather in the form of heavy rain may influence speed, possibly by forcing an arbitrary 30 percent speed reduction or arbitrarily restricting speed to under 20 mph.

Rules must be easy to change and adaptable to new and better information. This process is intimately tied to the use of the model for problem definition. Rules that are the underlying foundation of the model, however, can become difficult to change later on. Therefore it is well to consider how to *change* the rules even while outlining and structuring the model. For example add another traffic signal or change the number of intersections in a route segment.

Choices Within the Model

Three radically different types of factors govern the choice of possible alternatives: equations, rules of logic, and probability.

1. Equations—there are plenty of opportunities for simulation to use analytical relationships ranging from simple to complex. The use of analytical relationships is desirable, such as the computation of travel time based on distance and average speed. If the problem is adequately defined in these terms, simulation is unnecessary. Moreover analytical rules will reduce computer running time.

2. Logic—frequently the analytical equations may be reduced to a decision tree or a series of yes or no expressions. For example when there is no oncoming traffic, enter the traffic stream; or if there is no queue at the traffic signal and fhe signal is green, proceed.

3. Probability—all too often the facts are known for the behavior of the multitude, not the individual. The behavior of the individual can be selected from the range of possibilities by random selection from the set of possible behaviors. One specific departure time is selected from the distribution of the possible departure times. Of course if enough selections are made then each value should appear as often as its popularity warrants.

Care must be taken, however, *not* to account for all possible eventualities. Taking care of everything takes too long and costs too much. The advantage of simulation is to resolve new eventualities when and if they occur. Exceptions to the general behavior may be most interesting, but trying to anticipate all eventualities is unnecessary, delays the model start-up, radically increases the costs, and can be anticipated by the use of alarms for unexpected occurrences.

Random Processes

Many of the factors governing the drive to the office are deterministic. Other factors are more random but within well-established limits, such as the departure time from the house, driving speed, and competing traffic. The simulation governs the selection of these factors for each trip through the use of random numbers. Within each simulation there must be provision for obtaining a random number. Those aspects of the simulation which cannot be tied down to specific numbers or relationships are obtained through this technique.

In Figure 2.2 several common popularity distributions are shown. The normal distribution describes a large number of natural random processes. For all cars on the highway the distribution of the deviation from average speed could be normal. Another distribution is the uniform or rectangular. This could be used when there is equal probability that each distinct alternative value could be selected; for example to set the time for the initial switching of each traffic light, after which a fixed cycle governs.

A third possibility is the case where actual data are available from records of the daily number of cars passing through a particular point, as shown in Figure 2.2c. In this case traffic-flow data for each simulated data is selected from the processed historical data shown in the plot of Figure 2.2d by a random number draw. The raw historical data of Figure 2.2c were sorted into ascending order and the frequency or popularity of each value was calculated as a percentage or in terms of parts per thousand.

The random number generator selects one value from the multitude. It allows freedom to use data in any form. Mathematical convenience no longer is a criterion.

Timing Focus

It is not necessary for a simulation to follow a process continuously. Instead the discrete event simulation is limited to the times which represent the focus of interest. Those times when events occur which indicate change or continuance of courses of action. Furthermore the time interval is not constant, but varies as the events arise. There are three basic timing considerations:

· When to start the simulation.
· What is the minimum interval between events.
· When to stop the simulation.

If the simulation were started at 7:56 each morning for our nominal 8 a.m. departure, there would be no assurance that traffic flow had settled down to a steady-state condition by the departure time for our car. With insight gained from several simulations, the time required to initialize the system becomes clear. Once the initiating times are known,

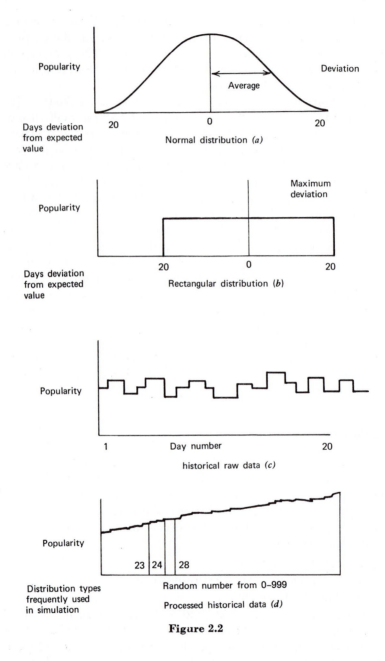

Figure 2.2

those factors which are time-related become obvious. After that each new simulation must start at the same time and with the same situations if it is to use the same scenario. If different situations were considered, then direct comparison of results cannot be used and a statistical analysis becomes necessary.

Each simulation requires an analysis to determine the minimum time interval between two events. The interval could be the time for a car to pass a traffic signal. In this case any two events closer together than one second would be considered to occur at the same instant. The selection of this interval helps to define the level of detail since any series of events which require less than one second would be considered as at least a second apart and, therefore, their detailed interactions would be ignored. The choice of one second would be far too long for a study of suspension-system behavior and far too short for the analysis of long-haul traffic on limited access highways. The type of problem dictates the interval, and the shortest one controls the separation between events in that simulation—nanoseconds, minutes, or months.

Stopping the simulation depends on one of two separate timing considerations. Either a particular sequence of events has been completed, our car reached the office one morning; or enough sequences of events have passed to acquire sufficient data, our car has completed trips on one hundred mornings. If it were considered necessary to cover the case where the trip might not be completed, the simulation could be terminated on the basis of time. In this case the time for a run might be one hundred days.

Degree of Model Detail

This model is a coarse representation of the drive to the office. It is not the detailed inch-by-inch representation of the problem with all its ramifications and interactions. The degree of detail should be adjustable, however, depending on the desired accuracy of the results and the sensitivity of the results to certain details. Care must be taken since simplifications may submerge the critical areas. However, detail for the sake of detail is expensive in time and progress. For the limited purpose of showing the effects of changing the route to the office, changing departure time may be deliberately eliminated. In this case the simulation is restricted to subset of the problem. This enables insight to be gained from the results before the scope of the simulation is enlarged.

When the model is enlarged in increments, the behavior of the system is analyzed under arbitrary control. First, variation in one factor is considered; then the scope of the model is enlarged to include additional factors. Later the process is further enlarged and some of the earlier detail is removed. Sometimes the simulation will be so successful that an analytical relationship will emerge, thereby eliminating the need for further simulation.

Model Generality

Complete problem definition is a distant goal when the system designer starts to develop the model. In fact the insight provided by the model will assist in refining the problem definition. This in turn leads to further refinement of the model. The key to a useful model is its ability to be modified as the system becomes more definite.

In the example route definition can be used to indicate model flexibility and generality. Initially the route segment may be considered as having a single intersection. Subsequently the insight gained will enable consideration of more realistic route segment definitions for n intersections. If the form and structure of the model were inflexible, it would be necessary to specifically structure the model for each new set of routes. This approach is too costly, both in the amount of design effort and in getting each new version of the simulation to work. Therefore the model structure must be flexible and general to accept the variety of potential input data.

Combined Models

The need to understand the system behavior demands that some model areas be deeply defined while other less-significant areas are only broadly covered. Therefore the model may be structured to varying levels of detail reflecting the design team's approach and knowledge. Moreover it is impractical for a single model to encompass the full scope and ultimate detail for the total problem. An approach to resolve this difficulty is to partition the problem into a series of submodels which eventually combine to form an overall simulation. The submodels provide a means of separating various influences which eventually will have to be combined. One of these submodels could be used to develop the competing traffic flow at the first intersection. In a fully partitioned model each source of traffic would separately introduce its traffic into the system. These would compete with our car to form the total traffic picture. Caution must be observed since partitioning assumes each source of traffic is independent of any other.

After data has been obtained from the detailed simulation, there are occasions when these results may be substituted in a broader model. Here our car would compete with an aggregate characteristic of the traffic and a distribution of delays rather than individual cars, thus simplifying those parts of the simulation which do not strongly contribute to the results.

Sometimes it may be necessary to examine one part of the system in greater depth. Consider the route segment composed of limited-access highway. Here the rules for one car passing another, ignored until now, may have to be added. A submodel representing passing may be developed separately as a subroutine, independently verified, and then introduced into the main model.

The drive to the office is a part of a much larger system. The results from this simulation should be available for inclusion in a larger simulation or set of analytical expressions. The form of output data must tie into another model so that a series of models can nest together into a much larger picture, permitting the analysis of larger systems. The technique of partitioning, however, may only be used where the submodels do not interact. Where there is interaction the problem may be partitioned for ease of model construction, but eventually it must be tied together and run as a single entity.

External Effects

The behavior of the system under abnormal conditions may be the purpose for the simulation. The focus of interest, then, is on the effect of abnormal conditions and the question of which route is most immune to emergency vehicles, accidents, detours, rain, and snowstorms.

The introduction of abnormal events into the simulation imposes an additional burden on the requirements for generality and flexibility since these events make no contribution to the results under normal conditions. For example when the weather has no influence, its factor is an additional term which is evaluated to zero or a unity multiplier of an entire expression. When the weather has an influence, it becomes a variable element in the model proportionate to the disruption caused.

Accuracy of Results

There are two objectives in simulating the drive to the office. The first is to compare the effects of specific changes on performance and the second to learn how long the drive will take. The first has been covered through the discussion of relative results. Now consider how simulation can provide results with absolute value.

Uncertainty is characteristic of the input data for simulation. Therefore to obtain a prediction of system performance the results must be based on numerous trials. Confidence in the results may vary on the basis of the number of independent trials. Primarily the results depend on the confidence level in the input data. When the accuracy and limitations of input data are known, then the simulation duration for a given degree of accuracy may be determined according to the number of trials required to satisfy the statistical relationships. Even so several series of results should be obtained with a different series of random numbers used each time, and these results compared. This will overcome some of the bias which may be introduced by the limited set of random numbers used.

Model Structure

The flow chart of Figure 2.1 shows a very simplified version of the drive to the office. Now the simplified version of the problem will be developed, analyzed, and simulated. To introduce the simulation example with an easy, clear, and straight-forward technique a higher-order simulation language, GPSS, will be used. The use of computer-assembly language for a model is not considered since this is limited to very experienced programmers and is slow and costly. FORTRAN similarly is not considered because of the difficulties of developing models for complex systems. The reasons for choosing a simulation language are covered in detail in Chapter 5.

Model Representations

The simplified, idealized drive to the office is shown schematically in Figure 2.3. The problem is intentionally simplified to emphasize the basic characteristics of the simulation technique, rather than the development of a realistic solution. The basic assumptions are as follows:

1. Our driver gets into the car every morning at the same time, 7:30 a.m.

2. The car leaves the driveway and merges into local street traffic without being delayed.

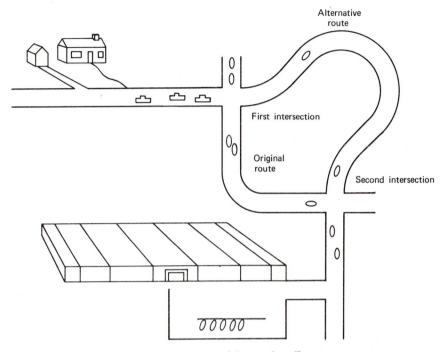

Figure 2.3 The drive to the office.

3. Average speed on local streets is considered to be 25 mph, unless subject to slowdown from traffic and abnormal conditions.

4. At the first major intersection there is a control signal and interaction with other traffic.

5. Competing cars will interact to provide degrees of traffic density. These cars come from separate sources. Their arrival rate at the intersection is variable and externally controlled.

6. At the first intersection there is a choice of two routes to the next major intersection. The choice is determined by traffic density, intersection delays, and abnormal conditions.

7. Average speed to the next intersection depends on the selection of route, either 35 mph for the original route or 55 mph on the alternate. Both are subject to slowdown from traffic and abnormal conditions.

8. The second intersection is considered in a similar manner except that there is no route selection.

9. Average speed to the office parking lot, 15 mph, is also subject to slowdown and abnormal conditions.

These nine assumptions provide a framework for the specific input data, logical relationships, and analytical equations required to simulate the drive to the office.   The major items are listed below:

1.   The average levels of competing traffic over the various parts of the trip are shown in Figure 2.4*a–c.*   Over the local segment there is a maximum of 10 vehicles generated every three minutes for a rate

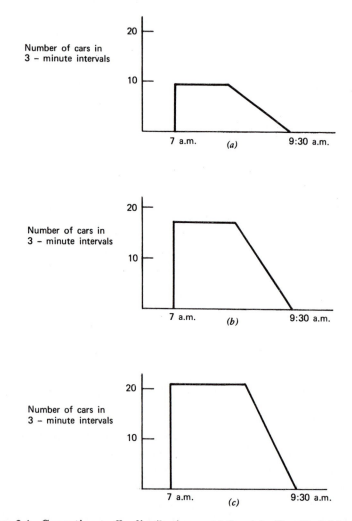

Figure 2.4   Competing traffic distributions.   (*a*) Local traffic; (*b*) Original route; (*c*) Alternate route.

of 200 vehicles per hour. The amount of traffic generated during each three-minute interval from 7 to 9:30 a.m. is used as a basis for each day's traffic. While there may still be traffic after 9:30, our interest should be over by that time. The actual traffic for each day is derived from the value for the 3-minute interval as modified by a random number.

2. The equations for the time to traverse each part of the trip are in this form: time = route distance/nominal speed × slowdown factor caused by traffic density × abnormal condition factor where either the slowdown factor or abnormal are unity for no influence.

3. At each intersection the rules are for the car to join the queue, if there is one. The queue discipline is first in, first out (FIFO).

4. The rules for the route selection at the first intersection are either to continue on the original route or take the alternative. Any one of the following factors is sufficient reason to choose the alternate:

(a) Signal condition at signal light—when the light is green or there is no queue, continue on the original route.

(b) Indication of traffic density—lighter traffic, stay on original route.

(c) Abnormal conditions—when the weather is poor, rainy or snowy, or there is an indication of abnormal situation—take alternate route.

At this stage the problem is defined, the assumptions stated, and the paths for flow of information established. The background is complete and we are ready to develop the model.

General-Purpose Simulation System (GPSS) Model

The reader will be aided by reading the beginning of an introductory manual to GPSS before continuing with the next section.[4] The example is described in terms of the block-diagram flow chart structure of GPSS; each block representing the statement of a function to be performed in the simulation. The GPSS language simulates the system with a limited set of information processing block types. Only 14 of the almost 60 types of blocks are required to define the basic actions and conditions that occur in the drive-to-the-office problem. The blocks are connected to form a path for the flow of information represented by transactions. In turn the transactions represent the active elements in the simulation vehicles. The GPSS program creates the transactions, representing ve-hicles, moves them through blocks, executes the action associated with the block, computes when the next event is scheduled to occur, and preserves the proper order for all transactions. Now the initial part of the model and the blocks needed to establish and process our transaction through the first intersection are covered item by item, in Figure 2.5.

Transactions representing all vehicles, both ours and competing traffic, are the first element of the model. These transactions are introduced through several GENERATE blocks. The transaction representing our car is generated by the first block of the model. GPSS operates on a specific block name, in this case GENERATE, and its associated fields. Each block type has different uses for the fields.

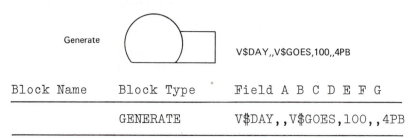

Block Name	Block Type	Field A B C D E F G
	GENERATE	V\$DAY,,V\$GOES,100,,4PB

Once each day a transaction is generated to represent the daily drive. The A field of the GENERATE block defines the interval between generation of successive transactions once a day for our vehicle from the same GENERATE block. In this case field A contains the symbolic name DAY for a variable statement, which is the equation for determining the number of seconds in a day. The A field could also have referenced the first variable statement if we chose to number it as V1: GENERATE V1,,V2,100,,4PB.

	Statement Name	Statement Type	
	DAY	VARIABLE	24*60*60
or	1	VARIABLE	24*60*60

One second has been selected as the minimum increment within the simulation. Any events which in the real world might occur within a fraction of a second are in the model truncated to the nearest integer, namely the same instant of simulated clock time.

The B field of the GENERATE block contains the term which provides variation in the interval between each successive transaction as would be the case if our next departure were approximately 24 hours later. Since the daily departure time is assumed to be the same the B field is intentionally left blank.

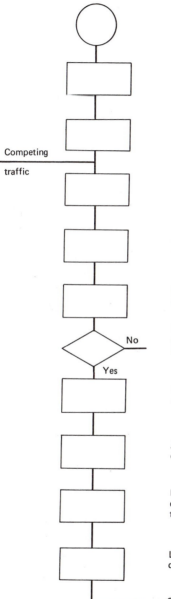

Generate a transaction to represent our car.

Our car needs a distinctive attribute – place the value 10 in parameter 1.

Calculate and spend the time to travel on the local streets.

Competing

traffic

Join the queue before the first intersection.

Record the current size of the queue in transaction parameter 2.

Restrict only one vehicle at a time to cross the intersection.

Traffic light condition green – yes – pass the intersection red – no – wait till light changes

No

Yes

Calculate and spend the time crossing the intersection.

Allow another vehicle to cross the intersection.

Find out and record if our car was alone. Add this value to what was already in parameter 2.

Leave the intersection and its queue.

Go on to next intersection.

Figure 2.5 Actions needed to reach and pass the first intersection.

It is desired to have the transaction leave the GENERATE block at the simulated time of 7:30 a.m. If the first transaction were generated at 7:30 a.m., then all the rest would follow at 24-hour intervals according to the A field. This offset to delay the first transaction until 7:30 a.m. is determined from the symbolic variable V$GOES in the C field of the GENERATE block.

	Statement Name	Statement Type	
	GOES	VARIABLE	V$FIRST+30*60
or	2	VARIABLE	V3+30*60
	FIRST	VARIABLE	7*60*60
or	3	VARIABLE	7*60*60

Variable GOES includes the equation of variable FIRST, the delay until 7 a.m., and adds an additional 30-minute delay. GPSS allows symbolic names for statements or numbers. GPSS will, during assembly, convert the symbolic names into statement numbers.

The total number of transactions that this block may generate is limited by the contents of the D field. In this case the 100 provides one transaction to represent trips for 100 days.

The E field, which is blank, may be used to establish a distinction among transactions with regard to priority. If two transactions were to request the same service at the same instant, the modeler could control which transaction receives the service by providing one transaction with a higher priority. In this simulation the concept of priority is ignored.

The remaining fields of the GENERATE block, F, G, H, and I define the type and maximum number of attributes each transaction may have. These attributes or labels are termed transaction parameters in GPSS. Parameters in GPSS V may be 8, 16, or 32 bits, byte, half or full word, 1, 2, or 4 bytes, and in addition a floating point value used in a special type of fullword parameter, coded as PB, PH, PF and PL, respectively. In this model, 4 byte parameters are used, the 4PB in field F. There are provisions for up to 255 parameters of each type in a single transaction. If G, H and I are intentionally left blank, however, only the byte parameter will be used. There is no need to specify these parameters in a particular order. As few parameters as possible should be used to conserve the amount of core storage required.

The actual placing of information into parameters will be accomplished by ASSIGN blocks. This permits transactions to be separated according to a particular property. At this point one transaction will have been generated by our single GENERATE block and the statements it references. At the appointed time that transaction automatically goes to the next block and will continue on either until some time is to be spent in the block or it is halted by logical conditions.

So far our model has only generated one transaction representing our vehicle. However, competing traffic using additional GENERATE blocks will be introduced. Therefore it is necessary to distinguish the transaction as being unique representing our car. This is accomplished by inserting into byte parameter one the number 10. A byte parameter could be used to contain any number between 0 and 127. The A field of the ASSIGN block selects the parameter number, one, and the B field describes the value to be inserted into the parameter, 10. If the transaction had additional types of parameters the C field would contain a PB to select the parameter type.

Block Type	Field AB
ASSIGN	1,10

Now that our transaction is uniquely identifiable it can enter a sequence of blocks representing the intersection where it will compete for service with other transactions from the other GENERATE blocks. The first block in the sequence, the ADVANCE block, retains the transaction for a period of time equivalent to the time to drive on local streets from the house to the first intersection.

Block name	Block Type	Field A
AAA	ADVANCE	V$LOCAL

Statement Name	Statement Type	
LOCAL	VARIABLE	1*60*60/25*FN$DENS1*XB$ABNR1
DENS1	FUNCTION	W$AAA,C4
0,1/24,1/250,10/5000,10		
	INITIAL	XB$ABNR1,1

This ADVANCE block is given the mnemonic AAA for later use. It is good practice to divide a GPSS model into subroutines. Since the mnemonics must begin with three letters out of a total of five characters the subroutines are lettered rather than numbered. For consistency all mnemonics on page A should begin with the letter A. How long each transaction remains in the ADVANCE block, the equivalent of the time to traverse the street, is determined by the equation LOCAL VARIABLE. This equation contains terms representing the distance over the local street, 1 mile; the speed, 25 mph; the slowdown factor for traffic congestion; and the slowdown factor from abnormal conditions. In GPSS terminology the equation becomes: 1 mile/25 mph converted to miles per second * DENS1, which is the graphical FUNCTION relating slowdown to congestion, * ABNR1, the value representing the current state of abnormal conditions.

FN$DENS1 is a continuous FUNCTION,C4, defined by four points with continuous interpolation for values selected between these points. The particular value to be selected is determined from the current number of vehicles on the local street. This is determined by W$AAA, the number of transactions currently spending time or waiting in BLOCK AAA. When the number of transactions represent from 0 to 25 vehicles on the street, then the value for the slowdown factor DENS1 is unity. Between 25 and 250 vehicles increases the slowdown factor linearly to a value of 10. Greater traffic congestion, however, does not increase the delay. Figure 2.6 is a graphical representation of this continuous GPSS function.

The second modifying factor is the value in a specific independent data-storage location, or data cell, SAVEVALUE ABNR1. This location is set to unity at the start of the simulation by the INITIAL

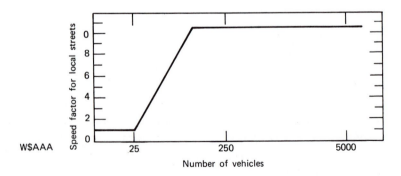

Figure 2.6 Speed reduced according to traffic density.

XB$ABNR1,1 statement. The XB is the symbol for a byte, 8 bit, data-storage location. In a structure similar to parameter types savevalues may contain byte, halfword, fullword or floating point data, XB, XH, XF, and XL, respectively. Later during the simulation the value in this data storage location may be changed by a transaction passing through the SAVEVALUE statement used to update the contents of this storage-location.

                                          QUEUE INTS1
                                    or   QUEUE 1

After transactions representing traffic have waited in the ADVANCE block the simulated time, they join the queue at the first intersection, INTS1, in order of arrival. The A field, INTS1, is the symbolic name for this queue. GPSS will convert INTS1 to queue 1. We could, instead, have numbered the queues ourselves.

The current number of vehicles in the queue as each vehicle joins the queue is of interest since it indicates whether or not the vehicle entered an empty intersection. This information is stored in byte parameter 2 of each entering transaction.

ASSIGN 2Q$INTS1

The control of the intersection is based on the traffic light, but the immediate element of control is maintained by restricting the use of the intersection to one vehicle at a time. A FACILITY POS1 is used to represent the area of the intersection. The SEIZE block allows one transaction representing one vehicle to be in the intersection area at a time. The intersection, FACILITY POS1, is in turn controlled by the traffic signal LGHT1. The first vehicle in the queue SEIZEs the POS1 facility and holds it for a time to cross the intersection if the light is green or waits when the light is red until it changes. These four blocks control one vehicle at a time to cross the intersection according to the condition of the traffic light.

Traffic light control of the intersection is performed by switching the condition of GATE LR LGHT1 from set to reset, stop to go, respectively. A separate timing sequence of LOGIC and ADVANCE blocks is described later as part of the routines on page D of the model, Figure 2.12. Initially the condition of the GATE is in the reset status LR. Not

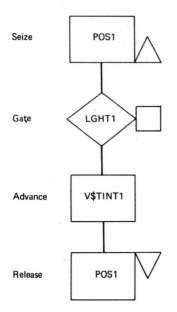

```
SEIZE    POS1
GATE LR  LGHT1
ADVANCE  V$TINT1
RELEASE  POS1
```

until some action occurs elsewhere in the model and the GATE is in the LR reset condition will a transaction be able to pass this block. The actual time for the transaction to cross the intersection when allowed by the condition of the GATE is set by the ADVANCE block, and determined by VARIABLE TINT1.

```
TINT1    VARIABLE    1/Q$INTS1*2+PB2/2*4-(PB2-2)/2*4
```

This equation is composed of two parts, but only one is used to determine the transit time for the intersection. If the queue is empty of other traffic, our vehicle establishes its value to unity so the first term becomes 2 seconds. The second part becomes zero since PB2 was set to the contents of the queue, one; and in a GPSS integer VARIABLE the fraction $\frac{1}{2}$ is truncated to zero, eliminating the second term. Similarly the third term remains zero until the queue has three or more vehicles in it. When the queue has vehicles in it, the first part becomes zero and the second and third parts remain constant at 4 seconds. After the car passes through the intersection the FACILITY POS1 is released by the RELEASE block.

The number of cars queued at the traffic light when each vehicle passes is recorded for future use by adding to the contents of byte parameter

2 the current contents of the queue, which must be at least 1 since our vehicle is still within the queue. This information is used to determine whether a car is affected by traffic moving with it. The last item to consider at the intersection is the transaction to leave the queue—the DEPART block.

DEPART INTS1

or DEPART 1

The next two blocks are TEST blocks to separate competing traffic and select the route.

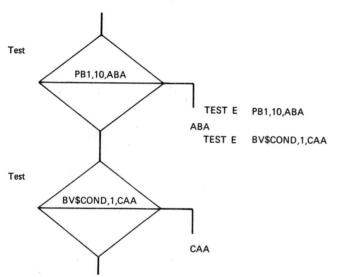

Test

PB1,10,ABA

TEST E   PB1,10,ABA

ABA

TEST E   BV$COND,1,CAA

Test

BV$COND,1,CAA

CAA

The first TEST block determines if byte parameter 1 of each entering transaction has the value 10. Those that do, and only one each day does, continue on to select the route to the next intersection. All other transactions are removed from the simulation by the TERMINATE block located by the mnemonic ABA.

The second TEST block selects the next route segment by directing the transaction to either BAA or CAA, the addresses of the starting points for the original and alternate route segment subroutines on pages B and C, respectively. When the test conditions are met the transaction continues on the next block, TRANSFER,BAA. When they are not met, however, the transaction goes to the block specified in the C field, CAA. That is there are one or more conditions requiring use of the alternate route. This test is performed using the binary variable

BV\$COND. This variable has the useful property of being evaluated
to either unity when satisfied, or zero when not satisfied.

Statement Name		Statement Type
COND	BVARIABLE	(Q\$INTS1'G'1)+(PB2'G'5)
		+(W\$AAA'G'10)
		+(XB\$ABNR1'G'3)

Any of the four terms must be true for the binary variable COND
to be evaluated to unity; namely competing traffic was still present
as our car passed the intersection, Q\$INTS1'G'1; or at least four
other cars were in the queue when our car was there, P2'G'5; or
during the trip on the local route segment the density exceeded 10
vehicles, W\$AAA'G'10; or the abnormal conditions became severe,
XB\$ABNR1'G'3. If none of these conditions are met the vehicle con-
tinues on the original route segment.

One of the advantages of GPSS is that we do not have to complete
the structure of the model at one time. Instead a set of blocks can
be implemented and debugged, then another subroutine and another.
Eventually they will all be combined. To illustrate the piecemeal construction of the model, the blocks and statements used so far can be
executed. All that must be added is to provide a block for the addresses
left hanging, BAA and CAA. Two TERMINATE blocks labeled BAA
and CAA, respectively, accomplish this. In addition there is the need
to state how long the simulation should continue.

$$\text{START} \qquad 100$$

The START statement will require 100 transactions to be counted at
a TERMINATE block with a value in the A field. The BAA TER-
MINATE block has the value one. Comment statements in GPSS are
indicated by an asterisk in Column 1. The listing of the model blocks,
variables, functions, comments, and control statements, appears below.

Not only can part of the model be built, it can also be debugged
and tested. GPSS provides a series of cross-reference lists to help the
conversion from symbolic notation to the absolute numerical notation
used to execute the model. For illustration each block mnemonic is
listed, as shown below, in alphabetical order in the cross-reference table.
Also listed are its sequential block number and all references to that
symbol with the card numbers of the statement referencing it.

```
BLOCK
NUMBER    *LOC    OPERATION   A,B,C,D,E,F,G,H,I            COMMENTS
                  SIMULATE
          *
          *                          VARIABLES
          *
          * VARIABLE STATEMENTS ARE USED TO REPRESENT ALL NUMERICAL CONSTANTS AND
          * EQUATIONS IN THE MODEL
          *
           DAY    VARIABLE    60*60*24    NUMBER OF SECONDS IN A DAY
          *
           FIRST VARIABLE     7*60*60     EARLIEST POSSIBLE DEPARTURE TIME IS
          * SEVEN A M. THIS STATEMENT CONVERTS SEVEN A M TO SECONDS
          *
           GOES   VARIABLE    V$FIRST+30*60      SEVEN THIRTY A M DEPARTURE TIME IS
          *                   VARIABLE FIRST   PLUS THIRTY MINUTES
          *
           LOCAL VARIABLE     1*60*60/25*FN$DENS1*XB$ABNR1   THE TIME REQUIRED TO
          *                   TRAVEL THE DISTANCE ON LOCAL STREETS
          *
           TINT1 VARIABLE     1/Q$INTS1*2+PB2/2*4-(PB2-2)/2*4            TWO SECONDS
          *                                    FOR NO TRAFFIC,FOUR SECONDS FOR TRAFFIC
          *
          *
           TIME   VARIABLE    C1386400/60/60*100+C1@3600/60 INTERNAL CLOCK TO
          *                                    DAILY TIME IN HOURS AND MINUTES
          *
           TRIPT VARIABLE     M1/60/60*100+M1@3600/60+1 TRIP TIME - HOURS & MINUTES
          *
          *
           COND   BVARIABLE   (Q$INTS1'G'1)+(PB2'G'5)+(W$AAA'G'10)+(XB$ABNR1'G'3)
          *
          *
          * OTHERS ARE AT THE INTERSECTION WITH OUR CAR
          * THE QUEUE AT THE INTERSECTION CONTAINED MORE THAN FOUR CARS
          * A TOTAL OF TEN CARS ARE ON THE LOCAL STREET AT THE SAME TIME
          * ABNORMAL CONDITIONS EXCEED A FACTOR OF THREE
          *          ANY ONE OF THE ABOVE CONDITIONS SELECTS ALTERNATIVE ROUTE
          *
          *
          *                          FUNCTIONS
          *
          * FACTORS IN THE SIMULATION WHICH CANNOT BE HANDLED WITH EQUATIONS ARE
          * PLOTTED AND HANDLED USING GRAPHICAL RELATIONSHIPS - FUNCTIONS
          *
          *
           DENS1 FUNCTION      W$AAA,C4    SPEED ON LOCAL STREETS
          0.1/25.1/250,10/5000,10
          * NUMBER OF CARS ON STREET                     TRAFFIC SLOWDOWN FACTOR
          *
           BNCH1 FUNCTION      P2,C4       SLOWDOWN CAUSED BY INTERSECTION TRAFFIC
          0.10/2.10/10,20/100,20
          * FIRST INTERSECTION QUEUES                    BUNCHING FACTOR
          *
          * INITIAL INPUT DATA - ABNORMAL CONDITIONS - HALFWORD SAVEVALUES
                   INITIAL     XB$ABNR1,1       NO ABNORMAL CONDITIONS
          *
```

In this model symbolic notation was employed for all entities used: FACILITIES, QUEUES, integer and boolean VARIABLES, LOGIC SWITCHES, FUNCTIONS, and SAVEVALUES.

GPSS provides a standard set of output statistics. A variety of optional statistics may be gathered by introducing additional statements into the model. Basically the results may be separated into those used mostly for debugging and those containing the desired results.

```
         *
         *  PAGE A DISTANCE TO FIRST INTERSECTION AND PROCEDURE TO GET THROUGH IT
         *
1                 GENERATE    V$DAY,,V$GOES,100,,4PB      START THE TRANSACTION
         *                                          TO REPRESENT OUR CAR EVERY MORNING
2                 ASSIGN      1,10                  LABEL OUR CAR WITH A TEN
3        AAA      ADVANCE     V$LOCAL         TIME SPENT ON LOCAL STREETS TO REACH
         *                                          FIRST INTERSECTION
4                 QUEUE       INTS1           JOIN QUEUE AT INTERSECTION ONE
5                 ASSIGN      2,Q$INTS1       FIND OUT SIZE OF QUEUE AT INTERSECTION
6                 SEIZE       POS1            GET INTO POSITION TO PASS INTERSECTION
7                 GATE LR     LGHT1           SIGNAL LIGHT CONDITION
8                 ADVANCE     V$TINT1         MOVE THROUGH FIRST CAR POSITION
9                 RELEASE     POS1            LET NEXT CAR INTO POSITION
10                ASSIGN      2+,Q$INTS1      NOTE IF CAR LEFT INTERSECTION ALONE
11                DEPART      INTS1           LEAVE INTERSECTION
12                TEST E      PB1,10,ABA         KEEP TRACK OF OUR CAR
         *                                          ONLY OUR CAR REMAINS IN THE MODEL
13                TEST E      BV$COND,0,CAA      CHECK CONDITIONS TO DETERMINE ROUTE
14                TRANSFER    ,BAA            CONTINUE ON NEXT SEGMENT
15       ABA      TERMINATE                  COMPETING TRAFFIC NOW REMOVED
         *
         *  PAGE B PORTION OF DRIVE OVER ORIGINAL ROUTE SEGMENT
         *
16       BAA      TERMINATE   1
         *
         *
         *  PAGE C PORTION OF DRIVE OVER ALTERNATIVE ROUTE SEGMENT
         *
17       CAA      TERMINATE
         *
                  START       100
                  END
```

The debugging results from running this portion of the model contain the count of how many transactions passed through each block and whether any transactions are still at any block. For this run there were 100 transactions GENERATED, and they all followed the same path through the model. Since the model logic is not yet implemented neither block ABA(15) nor CAA(17) has been entered. The clock time at the end, 8,580,747 seconds, corresponds to the first day starting at 7:30 a.m., 27,000 seconds, and continuing for 100 days, 86,400 seconds each, plus the trip duration, so far 146 seconds, the evaluation of V$LOCAL and V$TINT1, 144 and 2 seconds, respectively.

```
                                                    CROSS-REFERENCE
                                                    BLOCKS


      SYMBOL            NUMBER            REFERENCES

      AAA                 3                 32       48
      ABA                15                 75
      BAA                16                 78
      CAA                17                 77
```

There are standard statistics gathered by the GPSS simulation. The first set covers the FACILITY POS1, the intersection. The statistics provided are the average utilization which in this case appears as 0.000,

SYMBOL	NUMBER	REFERENCES					
POS1	1	69	72				

CROSS-REFERENCE
QUEUES

SYMBOL	NUMBER	REFERENCES					
INTS1	1	22	32	67	68	73	74

CROSS-REFERENCE
VARIABLES

SYMBOL	NUMBER	REFERENCES	
DAY	1	11	62
FIRST	2	13	16
GOES	3	16	62
LOCAL	4	19	65
TIME	6	26	
TINT1	5	22	71
TRIPT	7	29	

CROSS-REFERENCE
LOGIC SWITCHES

SYMBOL	NUMBER	REFERENCES
LGHT1	1	70

CROSS-REFERENCE
FUNCTIONS.

SYMBOL	NUMBER	REFERENCES	
BNCH1	2	52	
DENS1	1	19	48

CROSS-REFERENCE
BYTE SAVEVALUES

SYMBOL	NUMBER	REFERENCES		
ABNR1	1	19	32	57

CROSS-REFERENCE
BOOLEAN VARIABLES

SYMBOL	NUMBER	REFERENCES	
COND	1	32	77

**** ASSEMBLY TIME = .05 MINUTES ****

```
RELATIVE CLOCK          8580746   ABSOLUTE CLOCK         8580746
BLOCK COUNTS
BLOCK  CURRENT     TOTAL     BLOCK CURRENT     TOTAL     BLOCK CURRENT          TOTAL
   1       0        100        11       0       100
   2       0        100        12       0       100
   3       0        100        13       0       100
   4       0        100        14       0       100
   5       0        100        15       0         0
   6       0        100        16       0       100
   7       0        100        17       0         0
   8       0        100
   9       0        100
  10       0        100
```

which is 2/86400, the number of transactions to use the facility, the
average time each transaction spent using the facility, and if the facility
was in use at the time the statistics were collected, which transaction
had possession of the facility. Queue statistics are similarly gathered.
For queues there are additional statistics: the number of transactions
to pass through the queue in zero time, which optional table also has
queue statistics, and the current contents of the queue. The last standard output is the contents of the btye SAVEVALUE ABNR1. Its
value is still one which was introduced by the INITIAL statement.

Now that the initial subroutine has been completed, the next part
of the model can be developed. Before following our car onto its next
route segment, however, it is necessary to generate competing traffic.
The competing traffic for the local route segment originates at a second
GENERATE block.

Block Type	Field A B C D E F
GENERATE	V$DAY,,V$FIRST,100,,4PB

This block is the same as the GENERATE block for our vehicle
with the exception that the offset time to introduce the first transaction
is described by VARIABLE FIRST, which starts the competing traffic
at 7 a.m. instead of waiting until 7:30 a.m. as in VARIABLE GOES.
Only one transaction is generated each day. This serves to trigger the
generation of additional transactions representing individual competing
vehicles. The requirements for a strict reproducible scenario governing
the competing traffic is met by using a fixed number of transactions
to represent each day's traffic. Therefore the same number of random
number draws will occur during each simulated day, although not at
the same time each day. The SPLIT block duplicates the trigger transaction the number of times specified by the competing traffic function,

```
*********************************************
*                                           *
*              FACILITIES                   *
*                                           *
*********************************************
```

FACILITY	NUMBER ENTRIES	AVERAGE TIME/TRAN	-AVERAGE UTILIZATION DURING- TOTAL TIME	AVAIL. TIME	UNAVAIL. TIME	CURRENT STATUS	PERCENT AVAILABILITY	TRANSACTION NUMBER SEIZING PREEMPTING
PQS1	100	2.000	.000					

```
*********************************************
*                                           *
*                QUEUES                     *
*                                           *
*********************************************
```

QUEUE	MAXIMUM CONTENTS	AVERAGE CONTENTS	TOTAL ENTRIES	ZERO ENTRIES	PERCENT ZEROS	AVERAGE TIME/TRANS	$AVERAGE TIME/TRANS	TABLE NUMBER	CURRENT CONTENTS
INTS1	1	.000	100	0		2.000	2.000		

$AVERAGE TIME/TRANS = AVERAGE TIME/TRANS EXCLUDING ZERO ENTRIES

```
*********************************************
*                                           *
*            BYTE SAVEVALUES                *
*                                           *
*********************************************
```

NUMBER - CONTENTS   NUMBER - CONTENTS   NUMBER - CONTENTS   NUMBER - CONTENTS   NUMBER - CONTENTS
ABNR1   1
END

***** TOTAL RUN TIME (INCLUDING ASSEMBLY) = .12 MINUTES *****

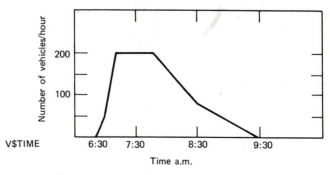

Figure 2.7 Local segment traffic density.

TRFC1, and sends the duplicate transactions to the block identified by the mnemonic AACA.

Block Mnemonic	Block Type	Field A B
AABA	SPLIT	FN$TRFC1,AACA

Therefore, these will emerge from the SPLIT block transaction representing the average number of vehicles for each 3-minute interval during the morning. Figure 2.7 is a plot of the results of FUNCTION TRFC1.

The competing traffic on the local segment starts at 6:45 and by the time our car departs at 7:30, it has reached a rate of 200 vehicles per hour. This rate continues until 8:00 a.m. and falls off to zero by 9:30 a.m. The time in this figure is described in normal hours and minutes. This requires a conversion from the simulation interval, 1 second.

    TIME VARIABLE   C1@86400/60/60*100+C1@3600/60

The current interval simulated time, C1, in units since the simulation began, is converted into hours and minutes by two terms. First, C1 is divided by the number of seconds in a day and only the remainder is retained, modulo division. This remainder is divided by the number of seconds in an hour to give the hour of the day, which in turn is multiplied by 100. The second term converts C1 into minutes past the hour. Then the two terms are added.

The actual FUNCTION, TRFC1, is arranged according to the traffic density during each three-minute interval.

    TRFC1   FUNCTION   V$TIME, C7
    0,0/645,0/700,2/715,10/800,10/830,4/930,0

Since in real life the traffic is not the same every morning, the simulation uses the random number generator to randomize the traffic generated every 3 minutes. The transactions representing competing traffic are directed from the SPLIT block to ADVANCE block AACA for this randomizing.

Block Mnemonic	Block Type	Field A B
AACA	ADVANCE	180,180

The ADVANCE block holds the transactions for anywhere from 0 to 6 minutes. The A field describes the average delay and B the modification, plus or minus, to the delay. Since these two numbers are the same, the range of delay will average out to a uniform delay of 0 to 360 clock units. After the delay the transactions transfer to the main part of the model to compete with our vehicle at block AAA.

The trigger transaction continues to cycle through the SPLIT block until the FUNCTION TRFC1 becomes zero at 9:30 each day when it is directed to the TERMINATE block. The FUNCTION acts as input data which could be used to provide competing traffic at different times of the day by changing its values, not the model.

```
AABA   SPLIT        FN$TRFC1,AACA
       ADVANCE      180
       TEST E       FN$TRFC1,0,AABA
       TERMINATE
```

Each segment of competing traffic requires a set of similar blocks and FUNCTIONS, those for the original route and alternative segments appearing on pages B and C, respectively.

The signal light control, page D, is the remaining part of the model concerning the local segment of the trip. Again a GENERATE block is used. This one is identical to the block for the competing traffic.

Block Type	Field A B C D E F
GENERATE	V$DAY,,V$FIRST,100,,4PB

The transaction from this block will cycle repeatedly simulating the action of the traffic control signal—alternately green and red. Since the model covers only the trip to the office it is only necessary that

the traffic signal work part of the day. This is arranged through the
following sequence of blocks:

```
        ASSIGN      1,96
DAA     LOGICS      LGHT1
        ADVANCE     60
        LOGICR      LGHT1
        ADVANCE     60
        LOOP        1PB,DAA
        TERMINATE   1
```

The ASSIGN block provides in byte parameter 1 the initial value
for the number of control signal cycles, 96. The GATE LGHT1 block,
in the LOCAL route segment of page A is controlled by the settings
of LOGIC LGHT switches. First, this GATE LGHT1 is set by
LOGICS LGHT1 block and held in that state for 60 seconds in the
ADVANCE block. Then, GATE LGHT1 is reset by the LOGICR
LGHT1 block and held in that condition for another 60 seconds. The
LOOP block decrements byte parameter one by one and directs the
transaction back to DAA the LOGICS block. When parameter is decremented to zero, after 96 cycles, the transaction continues on to the next
block TERMINATE to end the daily cycle. The one in the A field
of the TERMINATE block indicates that each termination will count
in determining the number of terminations to end the simulation.

All factors involved in the local segment have been covered. The
next segment may be either the sequence of blocks on page B starting
at BAA, covering the original route segment, or that on page C starting
at CAA, the alternative route segment. Both these block sequences
are similar to the page A subroutine of the local route segment starting
at AAA. Therefore only the major differences will be covered in detail;
the remainder may be interpreted with reference to page A, the local
route segment.

The first block of each routine is an ADVANCE block, either BAA
or CAA. The time spent in each block is determined from VARIABLE
statements ORGRT AND ALTRT, which are similar to the LOCAL
statement previously used. ORGRT does include one additional factor:
the representation of a slowdown when our car follows the original route
with a number of other cars as a result of delay at the traffic light. A
modifying FUNCTION BNCHO can reduce average speed up to a factor
of 2.

The next block in each sequence is an ASSIGN block that records
which route was used by placing in byte parameter 3 either 1 for the
original route segment or 2 for the alternate.

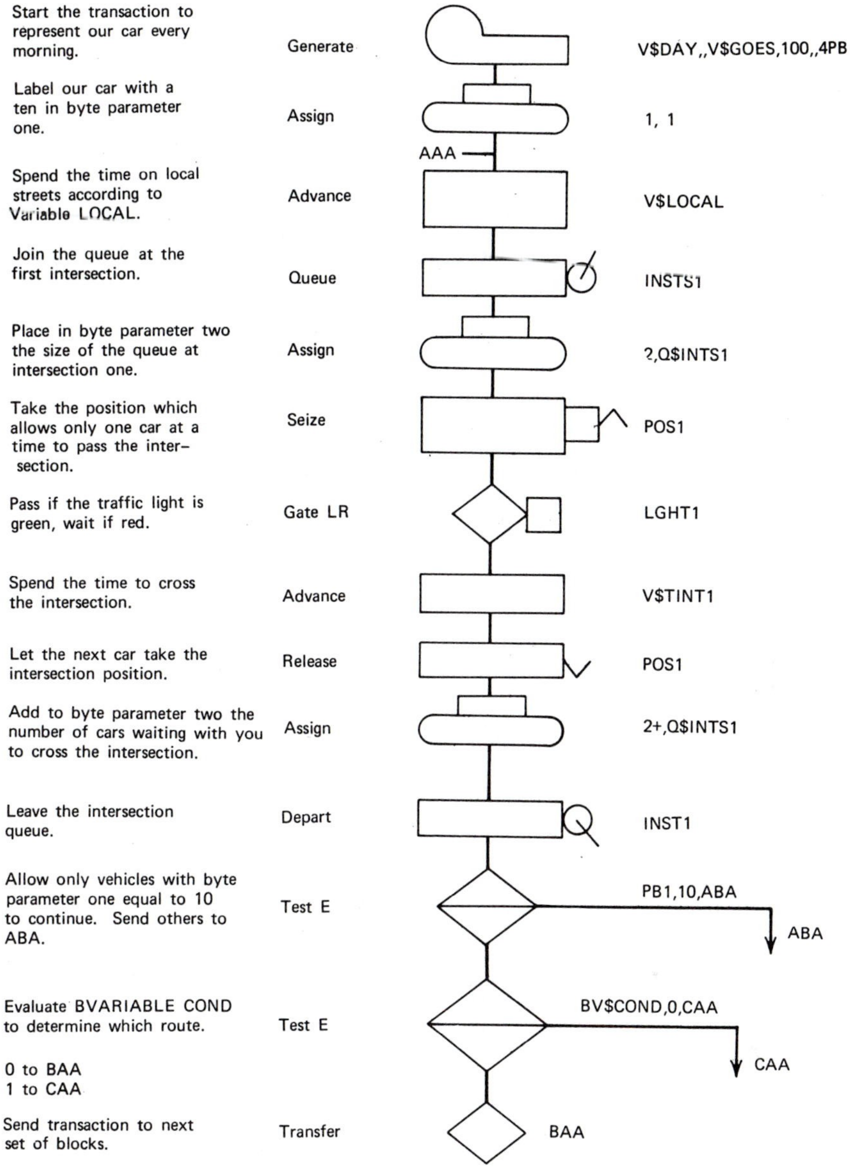

Description	Block		Operand
Start the transaction to represent our car every morning.	Generate		V$DAY,,V$GOES,100,,4PB
Label our car with a ten in byte parameter one.	Assign		1, 1
	AAA		
Spend the time on local streets according to Variable LOCAL.	Advance		V$LOCAL
Join the queue at the first intersection.	Queue		INSTS1
Place in byte parameter two the size of the queue at intersection one.	Assign		2,Q$INTS1
Take the position which allows only one car at a time to pass the intersection.	Seize		POS1
Pass if the traffic light is green, wait if red.	Gate LR		LGHT1
Spend the time to cross the intersection.	Advance		V$TINT1
Let the next car take the intersection position.	Release		POS1
Add to byte parameter two the number of cars waiting with you to cross the intersection.	Assign		2+,Q$INTS1
Leave the intersection queue.	Depart		INST1
Allow only vehicles with byte parameter one equal to 10 to continue. Send others to ABA.	Test E		PB1,10,ABA
			ABA
Evaluate BVARIABLE COND to determine which route.	Test E		BV$COND,0,CAA
0 to BAA 1 to CAA			CAA
Send transaction to next set of blocks.	Transfer		BAA

Figure 2.8 Generation of our vehicle and simulation of all traffic over local streets.

TABLE DURTN
ENTRIES IN TABLE 100

MEAN ARGUMENT 14.669

STANDARD DEVIATION 1.125

SUM OF ARGUMENTS 1467.000

UPPER LIMIT	OBSERVED FREQUENCY	PER CENT OF TOTAL	CUMULATIVE PERCENTAGE	CUMULATIVE REMAINDER	MULTIPLE OF MEAN
10	0	.00	.0	100.0	.681
11	0	.00	.0	100.0	.749
12	0	.00	.0	100.0	.817
13	0	.00	.0	100.0	.886
14	68	67.99	57.9	32.0	.954
15	14	13.99	81.9	18.0	1.022
16	1	.99	82.9	17.0	1.090
17	17	16.99	100.0	.0	1.158

REMAINING FREQUENCIES ARE ALL ZERO

Figure 2.9 Standard form of GPSS table output.

Both sequences recombine after the second intersection at block BAB. From here there is the ADVANCE block, which represents the final time expenditure of the trip. The remaining blocks, TABULATE, DURTN, DUBOR, and DURAL, are for gathering durations of trip lengths for all, original and alternate routings, respectively.

```
        TABULATE      DURTIN
        TEST E        PB3 1 BCA
        TABULATE      DUROR
        TERMINATE
    BCA TABULATE      DURAL
        TERMINATE
```

The TABULATE blocks require an additional descriptor, the TABLE statement.

Statement Name	Statement Type	Field A B C D
DURIN	TABLE	V$TRIPT,10,1,35

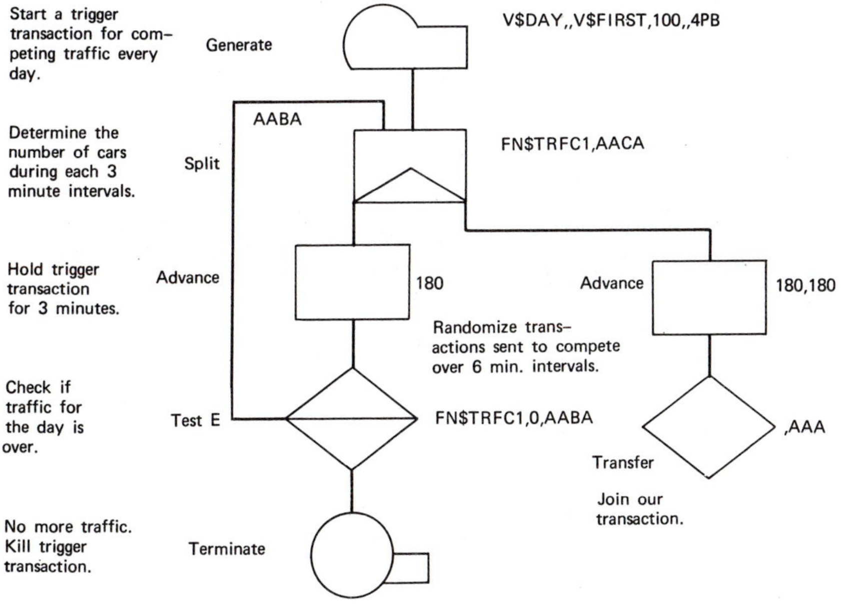

Figure 2.10 Generation of competing traffic (local segment).

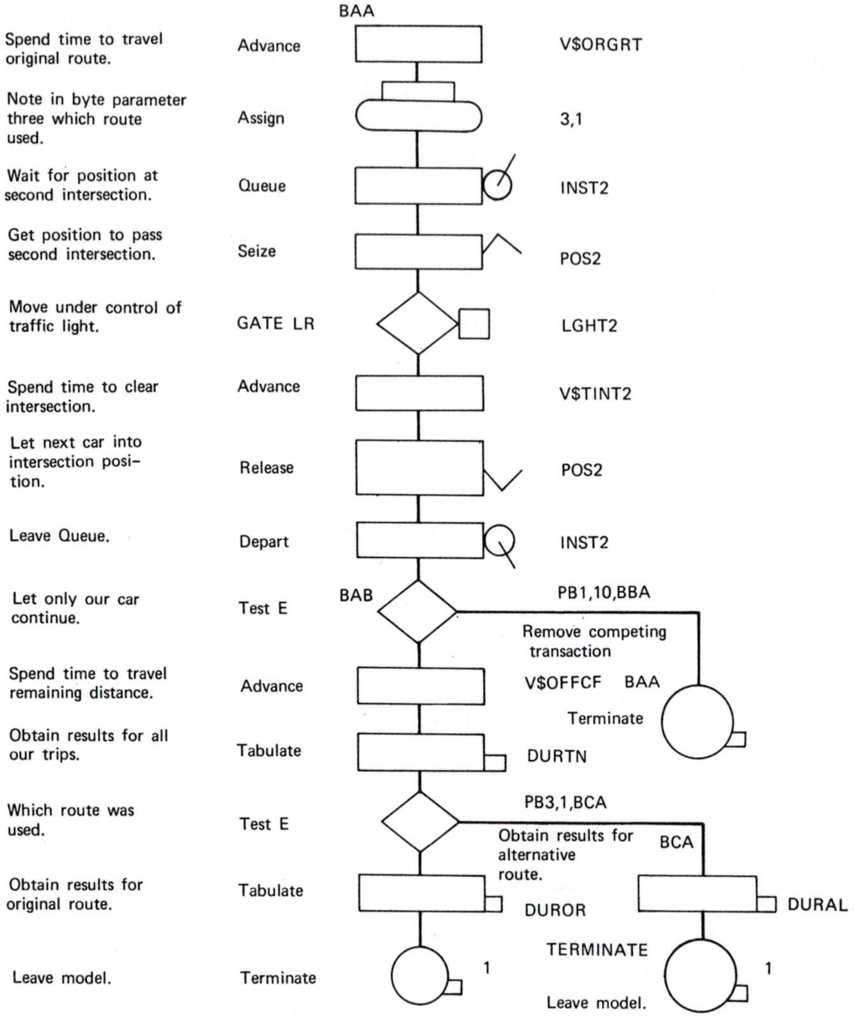

Spend time to travel original route.	Advance		V$ORGRT
Note in byte parameter three which route used.	Assign		3,1
Wait for position at second intersection.	Queue		INST2
Get position to pass second intersection.	Seize		POS2
Move under control of traffic light.	GATE LR		LGHT2
Spend time to clear intersection.	Advance		V$TINT2
Let next car into intersection position.	Release		POS2
Leave Queue.	Depart		INST2
Let only our car continue.	Test E	BAB	PB1,10,BBA
			Remove competing transaction
Spend time to travel remaining distance.	Advance		V$OFFCF BAA
			Terminate
Obtain results for all our trips.	Tabulate		DURTN
			PB3,1,BCA
Which route was used.	Test E		
			Obtain results for alternative route. BCA
Obtain results for original route.	Tabulate		DUROR DURAL
			TERMINATE
Leave model.	Terminate		1 1
			Leave model.

Figure 2.11 Original and office route segments.

The A field determines the basis for the table—in this case **V$TRIPT**. This is a variable used to compute the trip duration in hours and minutes. This makes use of one of the built-in features of GPSS, M1, the elapsed time since the transaction entered the model.

TRIPT VARIABLE   M1/60/60*100+M1@3600/60+1

The results from the model described by the TABULATE block and TABLE statement are in a standard GPSS format shown in Figure 2.9. The B field of the TABLE statement sets the upper limit at 10 minutes, the C field determines the increments—1 minute, and the D field limits the size of the table to 35 increments. Thus TABLE DURTN covers the results of the simulation in one-minute steps from 10 to 45 minutes for the trip. Since the tabulation is according to the upper limit of each frequency class, a 1 was added to V$TRIPT to place fractional values in their proper frequency class.

All elements of the model have now been covered. These are shown

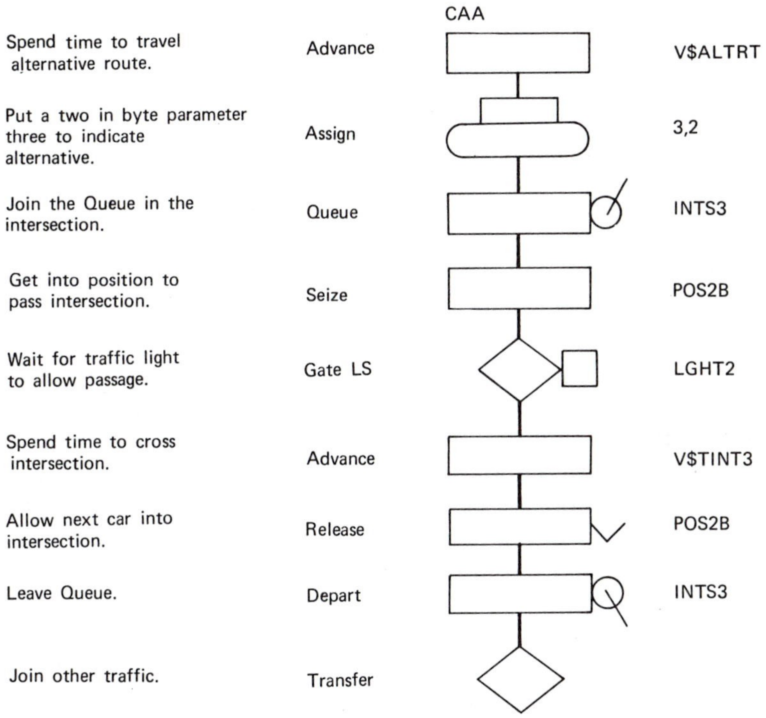

Figure 2.12 Alternate route segment.

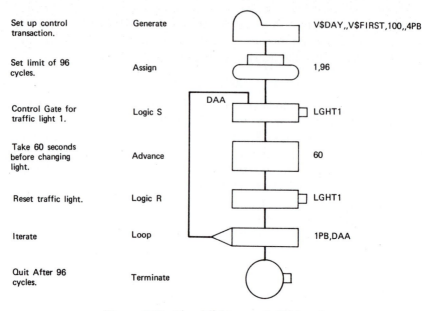

Set up control transaction.	Generate	V$DAY,,V$FIRST,100,,4PB
Set limit of 96 cycles.	Assign	1,96
Control Gate for traffic light 1.	Logic S	LGHT1
Take 60 seconds before changing light.	Advance	60
Reset traffic light.	Logic R	LGHT1
Iterate	Loop	1PB,DAA
Quit After 96 cycles.	Terminate	

Figure 2.13 Signal light control (light one).

in two forms as GPSS block diagrams and the GPSS assembled listing of the input deck. The GPSS block diagrams are separated by route segments with the local segment, page A, shown in Figures 2.8 and 2.10; the original segment, page B, in Figure 2.11; the alternate route, page C, in Figure 2.12; and the traffic-signals control, page D, in Figure 2.13.

The listing of the full-model description includes the variable statements and graphic functions which are not defined in the block diagram. The organization of the model into pages with the statement and input data before blocks is a matter for individual choice. This procedure does simplify debugging and communication with others, however, since when the blocks are assembled by the program all modifying statements have already been processed. The choice of the mnemonics A,B,C, and so on, for page names helps since it is consistent with GPSS block mnemonics AAA, BAA, CAA, and so on. Also it clearly establishes whether the routine continues within itself or branches out into another route as with the TRANSFER, BAB in the page C subroutine. Like all programs there is need for care with syntax, layout, and logic. The computer is unforgiving of mistakes whether trivial or serious.

```
*LOC    OPERATION   A,B,C,D,E,F,G              COMMENTS

        SIMULATE
*
*
*                            VARIABLES
*
* VARIABLE STATEMENTS ARE USED TO REPRESENT ALL NUMERICAL CONSTANTS AND
* EQUATIONS IN THE MODEL
*
 DAY    VARIABLE    60*60*24      NUMBER OF SECONDS IN A DAY
*
 FIRST  VARIABLE    7*60*60       EARLIEST POSSIBLE DEPARTURE TIME IS
* SEVEN A M. THIS STATEMENT CONVERTS SEVEN A M TO SECONDS
*
 GOES   VARIABLE    V$FIRST+30*60      SEVEN THIRTY A M DEPARTURE TIME IS
*                   VARIABLE FIRST  PLUS THIRTY MINUTES
*
 LOCAL  VARIABLE    1*60*60/25*FN$DENS1*XB$ABNR1   THE TIME REQUIRED TO
*                   TRAVEL THE DISTANCE ON LOCAL STREETS
*
 ORGRT  VARIABLE    4*60*60/35*FN$DENS2*FN$BNCH1*XB$ABNR2/10   THE TIME
*                   REQUIRED TO TRAVERSE THE ORIGINAL ROUTE SEGMENT
*
 ALTRT  VARIABLE    6*60*60/55*FN$DENS3*XB$ABNR3        THE TIME REQUIRED
*                   TO TRAVERSE THE ALTERNATIVE ROUTE SEGMENT
*
 OFFCE  VARIABLE    1*60*60/15*XB$ABNR4    THE TIME REQUIRED TO TRAVERSE
*                   REMAINING DISTANCE TO THE OFFICE
*
 TINT1  VARIABLE    1/Q$INTS1*2+PB2/2*4-(PB2-2)/2*4             TWO SECOND
*                   FOR NO TRAFFIC,FOUR SECONDS FOR TRAFFIC
*
*
 TINT2  VARIABLE    1/Q$INTS2*2+PB2/2*4-(PB2-2)/2*4             TWO SECOND
*                   FOR NO TRAFFIC, FOUR SECONDS FOR TRAFFIC
*
 TINT3  VARIABLE    1/Q$INTS3*2+PB2/2*4-(PB2-2)/2*4             TWO SECOND
*                   FOR NO TRAFFIC, FOUR SECONDS FOR TRAFFIC
*
 TIME   VARIABLE    C1@86400/60/60*100+C1@3600/60 INTERNAL CLOCK TO
*                   DAILY TIME IN HOURS AND MINUTES
*
 TRIPT  VARIABLE    M1/60/60*100+M1@3600/60+1 TRIP TIME - HOURS & MINUTES
*
 COND   BVARIABLE   (Q$INTS1'G'1)+(PB2'G'5)+(W$AAA'G'10)+(XB$ABNR1'G'3)
*
*
* OTHERS ARE AT THE INTERSECTION WITH OUR CAR
* THE QUEUE AT THE INTERSECTION CONTAINED MORE THAN FOUR CARS
* A TOTAL OF TEN CARS ARE ON THE LOCAL STREET AT THE SAME TIME
* ABNORMAL CONDITIONS EXCEED A FACTOR OF THREE
*          ANY ONE OF THE ABOVE CONDITIONS SELECTS ALTERNATIVE ROUTE
*
*
```

GPSS Simulation Results

The drive-to-the-office model was run to simulate the passage of 100 days. From this run the results indicated that the range of trip durations was from 14 to 17 minutes; with the mean value being 14.669 minutes. All these results were shown in Figure 2.8, describing TABLE DURTN. GPSS allows graphical presentation of these results also.

```
*LOC    OPERATION  A,B,C,D,E,F,G                    COMMENTS


*                             FUNCTIONS
*
* FACTORS IN THE SIMULATION WHICH CANNOT BE HANDLED WITH EQUATIONS ARE
* PLOTTED AND HANDLED USING GPAPHICAL RELATIONSHIPS - FUNCTIONS
*
*
 TRFC1 FUNCTION    V$TIME,C7    DENSITY OF TRAFFIC IN MORNING
0,0/645,0/700,2/715,10,/800,10/830,4/930,0
* TIME IN MORNING                          NUMBER OF CARS IN 3 MINUTES
*
 TRFC2 FUNCTION    V$TIME,C7    DENSITY OF TRAFFIC ON ORIGINAL ROUTE
0,0/645,0/700,3/715,15,/800,10/830,6/930,0
* TIME IN MORNING                          NUMBER OF CARS IN 3 MINUTES
*
 TRFC3 FUNCTION    V$TIME,C7    DENSITY OF TRAFFIC ON ALTERNATIVE ROUTE
0,0/645,0/700,4/715,20,/800,20/830,12/930,0
* TIME IN MORNING                          NUMBER OF CARS IN 3 MINUTES
*
 DENS1 FUNCTION    W$AAA,C4     SPEED ON LOCAL STREETS
0,1/25,1/250,10/5000,10
* NUMBER OF CARS ON STREET                 TRAFFIC SLOWDOWN FACTOR
*
 DENS2 FUNCTION    W$BAA,C4     SPEED ON ORIGINAL ROUTE
0,1/50,1/500,15/50000,15
*NUMBER OF CARS ON ROUTE                   TRAFFIC SLOWDOWN FACTOR
*
 DENS3 FUNCTION    W$CAA,C5     SPEED ON ALTERNATIVE ROUTE
0,1/100,1/2000,3/6000,4/20000,30
* NUMBER OF CARS ON ROUTE                  TRAFFIC SLOWDOWN FACTOR
*
 BNCH1 FUNCTION    P2,C4        SLOWDOWN CAUSED BY INTERSECTION TRAFFIC
0,10/2,10/20,20/100,20
* FIRST INTERSECTION QUEUES                BUNCHING FACTOR
*
* INITIAL INPUT DATA - ABMORMAL CONDITIONS - HALFWORD SAVEVALUES
            INITIAL   XB$ABNR1,1      NO ABNORMAL CONDITIONS
            INITIAL   XB$ABNR2,1      NO ABNORMAL CONDITIONS
            INITIAL   XB$ABNR3,1      NO ABNORMAL CONDITIONS
            INITIAL   XB$ABNR4,1      NO ABNORMAL CONDITIONS
*
*
* PAGE A DISTANCE TO FIRST INTERSECTION AND PROCEDURE TO GET THROUGH IT
*
            GENERATE  V$DAY,,V$GOES,100,,4PB        START THE TRANSACTION
*                               TO REPRESENT OUR CAR EVERY MORNING
            ASSIGN    1,10                LABEL OUR CAR WITH A TEN
   AAA      ADVANCE   V$LOCAL      TIME SPENT ON LOCAL STREETS TO REACH
*                                  FIRST INTERSECTION
            QUEUE     INTS1        JOIN QUEUE AT INTERSECTION ONE
            ASSIGN    2,Q$INTS1    FIND OUT SIZE OF QUEUE AT INTERSECTION
            SEIZE     POS1         GET INTO POSITION TO PASS INTERSECTION
            GATE  LR  LGHT1        SIGNAL LIGHT CONDITION
            ADVANCE   V$TINT1      MOVE THROUGH FIRST CAR POSITION
```

Figure 2.14 and 2.15 show two ways to present the same data. Figure 2.14 shows the popularity of each trip duration in 2 minute classes. The same data are presented in Figure 2.15 as a cumulative distribution in terms of the percentage of trips completed in less than X minutes.

Additional data are available from this model. There were queues established at each intersection. GPSS provides queue statistics. The queue statistics from this model are shown in Figure 2.16. These statis-

```
*LOC-   OPERATION  A,B,C,D,E,F,G                  COMMENTS

        RELEASE    POS1            LET NEXT CAR INTO POSITION
        ASSIGN     2+,0$INTS1      NOTE IF CAR LEFT INTERSECTION ALONE
        DEPART     INTS1           LEAVE INTERSECTION
        TEST E     PB1,10,ABA         KEEP TRACK OF OUR CAR
*                                  ONLY OUR CAR REMAINS IN THE MODEL
        TEST E     BV$COND,0,CAA   CHECK CONDITIONS TO DETERMINE ROUTE
        TRANSFER   ,BAA            CONTINUE ON NEXT SEGMENT
  ABA   TERMINATE                  COMPETING TRAFFIC NOW REMOVED
*     START THE COMPETING TRAFFIC ON LOCAL STREETS
*
        GENERATE   V$DAY,,V$FIRST,100,,4PB  OTHER TRAFFIC EVERY DAY
  AABA  SPLIT      FN$TRFC1,AACA   NUMBER OF CARS DURING EACH THREE
*                                  MINUTE INTERVAL
        ADVANCE    180             WAIT TILL NEXT INTERVAL
        TEST E     FN$TRFC1,0,AABA NO MORE TRAFFIC WAIT FOR TOMARROW
        TERMINATE
  AACA  ADVANCE    180,180         RANDOMIZE THE DISTRIBUTION OF CARS
        TRANSFER   ,AAA            SEND THE COMETING TRAFFIC TO MEET
*                                  OUR CAR ON LOCAL STREETS
*
*
* PAGE B PORTION OF DRIVE OVER ORIGINAL ROUTE SEGMENT
*
  BAA   ADVANCE    V$ORGRT         TIME TO TRAVEL ORIGINAL ROUTE
        ASSIGN     3,1             NOTE ORIGINAL ROUTE USED
        QUEUE      INTS2           JOIN QUEUE AT SECOND INTERSECTION
        ASSIGN     2,Q$INTS2       SIZE OF QUEUE
        SEIZE      POS2            GET INTO POSITION TO PASS INTERSECTION
        GATE LS    LGHT2           SIGNAL LIGHT 2 CONDITION SET
        ADVANCE    V$TINT2         MOVE CAR TO INTERSECTION
        RELEASE    POS2            LET NEXT CAR INTO POSITION
        DEPART     INTS2           LEAVE INTERSECTION
  BAB   TEST E     PB1,10,BBA      KEEP TRACK OF OUR CAR
*                                  OUR CAR CONTINUES IN MODEL
        ADVANCE    V$OFFCE         FINISH REST OF TRIP
        TABULATE   DURTN           RETAIN DATA OF EACH TRIP DURATION
        TEST E     PB3,1,BCA       SEPARATE ACCORDING TO ROUTE USED
        TABULATE   DUROR           ORIGINAL ROUTE DURATION
        TERMINATE  1               ONE MORE DAY
  BCA   TABULATE   DURAL           ALTERNATIVE ROUTE DURATION
        TERMINATE  1
  BBA   TERMINATE                  CLEAR OUT OTHER VEHICLES
*     START THE COMPETING TRAFFIC ON THE ORIGINAL ROUTE
*
        GENERATE   V$DAY,,V$FIRST,100,,4PB  OTHER TRAFFIC EVERY DAY
  BAAB  SPLIT      FN$TRFC2,BAAC   NUMBER OF CARS IN THREE MINUTES
        ADVANCE    180             WAIT TILL NEXT INTERVAL
        TEST E     FN$TRFC2,0,BAAB NO MORE TRAFFIC WAIT FOR TOMORROW
        TERMINATE
  BAAC  ADVANCE    180,180         RANDOMIZE TRAFFIC OVER NEXT SIX MINUTES
        TRANSFER   ,BAA            SEND COMPETING TRAFFIC TO MEET OUR CAR
*
*
```

tics provide a reading of the number of vehicles passing through each intersection, the average of how long each vehicle stayed in the queue, the average contents of the queue, and the maximum number of vehicles in the intersection queue at any one time. These last data are shown for all intersections in Figure 2.17. Numerous additional data are available. Some, such as the breakdown of trip duration by route taken, is summarized in TABLES DUROR and DURAL, Figures 2.18 and

```
*LOC   OPERATION  A,B,C,D,E,F,G                COMMENTS


       RELEASE    POS1         LET NEXT CAR INTO POSITION
       ASSIGN     2+,Q$INTS1   NOTE IF CAR LEFT INTERSECTION ALONE
       DEPART     INTS1        LEAVE INTERSECTION
       TEST E     PB1,10,ABA      KEEP TRACK OF OUR CAR
*                              ONLY OUR CAR REMAINS IN THE MODEL
       TEST E     BV$COND,0,CAA   CHECK CONDITIONS TO DETERMINE ROUTE
       TRANSFER   ,BAA         CONTINUE ON NEXT SEGMENT
 ABA   TERMINATE               COMPETING TRAFFIC NOW REMOVED
*     START THE COMPETING TRAFFIC ON LOCAL STREETS
*
       GENERATE   V$DAY,,V$FIRST,100,,4PB   OTHER TRAFFIC EVERY DAY
 AABA  SPLIT      FN$TRFC1,AACA    NUMBER OF CARS DURING EACH THREE
*                              MINUTE INTERVAL
       ADVANCE    180          WAIT TILL NEXT INTERVAL
       TEST E     FN$TRFC1,0,AABA  NO MORE TRAFFIC WAIT FOR TOMARROW
       TERMINATE
 AACA  ADVANCE    180,180      RANDOMIZE THE DISTRIBUTION OF CARS
       TRANSFER   ,AAA         SEND THE COMETING TRAFFIC TO MEET
*                              OUR CAR ON LOCAL STREETS
*
*
* PAGE B PORTION OF DRIVE OVER ORIGINAL ROUTE SEGMENT
*
 BAA   ADVANCE    V$ORGRT      TIME TO TRAVEL ORIGINAL ROUTE
       ASSIGN     3,1          NOTE ORIGINAL ROUTE USED
       QUEUE      INTS2        JOIN QUEUE AT SECOND INTERSECTION
       ASSIGN     2,Q$INTS2    SIZE OF QUEUE
       SEIZE      POS2         GET INTO POSITION TO PASS INTERSECTION
       GATE LS    LGHT2        SIGNAL LIGHT 2 CONDITION SET
       ADVANCE    V$TINT2      MOVE CAR TO INTERSECTION
       RELEASE    POS2         LET NEXT CAR INTO POSITION
       DEPART     INTS2        LEAVE INTERSECTION
 BAB   TEST E     PB1,10,BBA   KEEP TRACK OF OUR CAR
*                              OUR CAR CONTINUES IN MODEL
       ADVANCE    V$OFFCE      FINISH REST OF TRIP
       TABULATE   DURTN        RETAIN DATA OF EACH TRIP DURATION
       TEST E     PB3,1,BCA    SEPARATE ACCORDING TO ROUTE USED
       TABULATE   DUROR        ORIGINAL ROUTE DURATION
       TERMINATE  1            ONE MORE DAY
 BCA   TABULATE   DURAL        ALTERNATIVE ROUTE DURATION
       TERMINATE  1
 BBA   TERMINATE               CLEAR OUT OTHER VEHICLES
*     START THE COMPETING TRAFFIC ON THE ORIGINAL ROUTE
*
       GENERATE   V$DAY,,V$FIRST,100,,4PB   OTHER TRAFFIC EVERY DAY
 BAAB  SPLIT      FN$TRFC2,BAAC    NUMBER OF CARS IN THREE MINUTES
       ADVANCE    180             WAIT TILL NEXT INTERVAL
       TEST E     FN$TRFC2,0,BAAB  NO MORE TRAFFIC WAIT FOR TOMORROW
       TERMINATE
 BAAC  ADVANCE    180,180      RANDOMIZE TRAFFIC OVER NEXT SIX MINUTES
       TRANSFER   ,BAA         SEND COMPETING TRAFFIC TO MEET OUR CAR
*
*
```

2.19, respectively.   Other data describe the number of times each block was entered, where transactions are, the contents of the savevalues, and, if desired, a listing of each transaction currently in the model.

These results apply to only one set of 100 days and one set of input conditions.   The simulation could be forced to send our car on either path for all 100 days and determine whether there was a gain from choosing the route segment.   The simulation could also be run for

```
*LOC    OPERATION   A,B,C,D,E,F,G                COMMENTS


*  PAGE C PORTION OF DRIVE OVER ALTERNATIVE ROUTE SEGMENT
*
   CAA    ADVANCE     V$ALTRT      TIME TO TRAVEL ALTERNATIVE ROUTE
          ASSIGN      3,2          NOTE ALTERNATIVE ROUTE USED
          QUEUE       INTS3        JOIN QUEUE AT INTERSECTION
          ASSIGN      2,Q$INTS3    SIZE QUEUE AT INTERSECTION
          SEIZE       POS2B        GET INTO POSITION TO PASS INTERSECTION
          GATE LR     LGHT2        SIGNAL LIGHT 2 CONDITION RESET
          ADVANCE     V$TINT3      MOVE CAR TO INTERSECTION
          RELEASE     POS2B        LET NEXT CAR INTO POSITION
          DEPART      INTS3        LEAVE INTERSECTION
          TRANSFER    ,BAB         JOIN OTHER ROUTE TRAFFIC
*
*      START THE COMPETING TRAFFIC ON THE ALTERNATIVE ROUTE
*
          GENERATE    V$DAY,,V$FIRST,100,,4PB   OTHER TRAFFIC EVERY DAY
   CAAB   SPLIT       FN$TRFC3,CAAC    NUMBER OF COMPETING CARS EVERY
*                                      THREE MINUTES
          ADVANCE     180          WAIT TILL NEXT INTERVAL
          TEST E      FN$TRFC3,0,CAAB  NO MORE TRAFFIC - WAIT FOR TOMARROW
          TERMINATE
   CAAC   ADVANCE     180,180      RANDOMIZE TRAFFIC OVER NEXT SIX MINUTES
          TRANSFER    ,CAA         JOIN ALTERNATIVE ROUTE TRAFFIC
*
*  PAGE D SIGNAL LIGHT CONTROL
*
          GENERATE    V$DAY,,V$FIRST,100,,4PB   SET UP SIGNAL LIGHT ONE
          SPLIT       1,DBA        PROVIDE TRANSACTION FOR SECOND LIGHT
          ASSIGN      1,96         LIMIT LIGHTS TO NINETY SIX CYCLES
   DAA    LOGICS      LGHT1        SIGNAL LIGHT GREEN
          ADVANCE     60           LIGHT ON FOR 60 SECONDS
          LOGICR      LGHT1        SIGNAL LIGHT RED
          ADVANCE     60           LIGHT ON FOR 60 SECONDS
          LOOP        1PB,DAA      CONTINUE CYCLING
          TERMINATE
*
   DBA    ASSIGN      1,108        LIMIT LIGHTS TO 108 CYLES PER DAY
   DBAA   LOGICS      LGHT2        SIGNAL LIGHT GREEN
          ADVANCE     45           LIGHT ON FOR 45 SECONDS
          LOGICR      LGHT2        SIGNAL LIGHT RED
          ADVANCE     60           LIGHT ON FOR 60 SECONDS
          LOOP        1PB,DBAA     CONTINUE CYCLING
          TERMINATE
*
   DURTN  TABLE       V$TRIPT,10,1,35   DURATION OF ALL TRIPS
   DUROR  TABLE       V$TRIPT,10,1,35   DURATION ON ORIGINAL ROUTE
   DURAL  TABLE       V$TRIPT,10,1,35   DURATION ON ALTERNATIVE ROUTE
*
```

another 100 days. Another possibility lies in varying the abnormal conditions and sensing which route is better. None of these additional runs require any extensive changes to the model. To run for an additional 100 days another START 100 card is added. For all cars to take either original or alternate routes, the statement for the binary variable is changed to either BVARIABLE 0 or 1. Other statements could provide a different set of conditions governing the choice. Finally, to show deterioration of road conditions in the INITIAL XB$ABNR1, 1 is changed from 1 to 10 for the worst case.

Later chapters will cover other GPSS models. This same model will

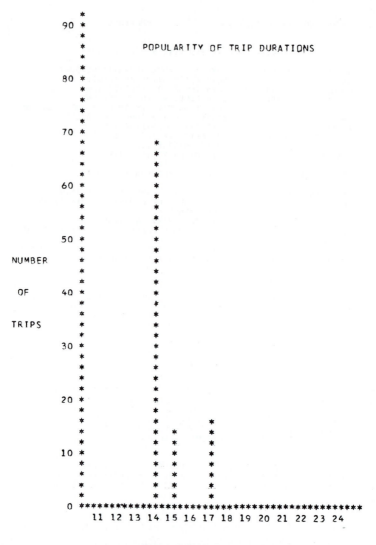

Figure 2.14   Histogram of simulation results.

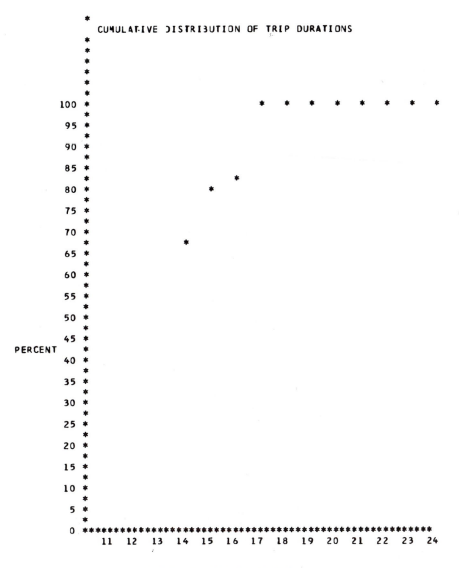

CUMULATIVE DISTRIBUTION OF TRIP DURATIONS

DRIVE DURATION IN MINUTES

Figure 2.15 Plot of simulation results.

QUEUE	MAXIMUM CONTENTS	AVERAGE CONTENTS	TOTAL ENTRIES	ZERO ENTRIES	PERCENT ZEROS	AVERAGE TIME/TRANS	$AVERAGE TIME/TRANS	TABLE NUMBER	CURRENT CONTENTS
INTS1	9	.063	27331		.0	20.058	20.058		3
INTS2	12	.102	36294	1	.0	24.283	24.284		
INTS3	15	.137	61943	1	.0	19.100	19.101		

$AVERAGE TIME/TRANS = AVERAGE TIME/TRANS EXCLUDING ZERO ENTRIES

Figure 2.16   Queue statistics—standard GPSS output.

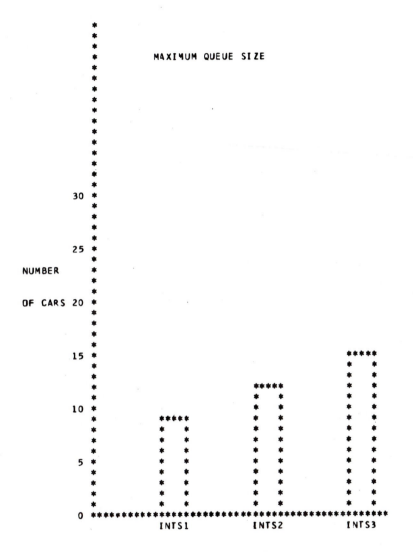

Figure 2.17 Plot of part ia queue statistics.

TABLE DUROR
ENTRIES IN TABLE 24   MEAN ARGUMENT 16.458   STANDARD DEVIATION .883   SUM OF ARGUMENTS 395.000

UPPER LIMIT	OBSERVED FREQUENCY	PER CENT OF TOTAL	CUMULATIVE PERCENTAGE	CUMULATIVE REMAINDER	MULTIPLE OF MEAN
10	0	.00	.0	100.0	.607
11	0	.00	.0	100.0	.668
12	0	.00	.0	100.0	.729
13	0	.00	.0	100.0	.789
14	0	.00	.0	100.0	.850
15	6	25.00	25.0	75.0	.911
16	1	4.16	29.1	70.8	.972
17	17	70.83	100.0	.0	1.032

REMAINING FREQUENCIES ARE ALL ZERO

Figure 2.18 Standard GPSS output for trip duration using the original route.

TABLE DURAL
ENTRIES IN TABLE 76   MEAN ARGUMENT 14.105   STANDARD DEVIATION .308   SUM OF ARGUMENTS 1072.000

UPPER LIMIT	OBSERVED FREQUENCY	PER CENT OF TOTAL	CUMULATIVE PERCENTAGE	CUMULATIVE REMAINDER	MULTIPLE OF MEAN
10	0	.00	.0	100.0	.708
11	0	.00	.0	100.0	.779
12	0	.00	.0	100.0	.850
13	0	.00	.0	100.0	.921
14	68	89.47	89.4	10.5	.992
15	8	10.52	100.0	.0	1.063

REMAINING FREQUENCIES ARE ALL ZERO

Figure 2.19 Standard GPSS output for trip duration using the alternate route.

be treated from different programming points of view in Chapters 6 and 7.

PROBLEMS

1. Make the following changes to the GPSS model:
   (a) Instead of one vehicle entering the local street each day, substitute one vehicle every 120 seconds until 100 have entered the model.
   (b) Change the interval between vehicles in (a) to anywhere from 0 to 120 seconds with equal probability.
   (c) Identify each vehicle in (a) with a sequence number and place that number in halfword parameter 1.
   (d) To characterize each vehicle and driver arbitrarily place them in one of four nominally equal classes. Store the decision in byte parameter five.
   (e) Now use the classification of (d) to vary the time spent on local streets according to the usual factors and this additional classification.
   (f) Determine the maximum number of vehicles on the local streets at any one time.
   (g) Increase the number of choices at the intersection to include another alternate route.
2. What are the effects on the statistics gathered in 1 (a) if only 100 vehicles are allowed into the model instead of gathering the statistics after 100 vehicles have finished? Does it make any difference if more vehicles can enter the model?
3. Suggest another method for introducing the competing traffic which does not preserve the scenario of pseudorandom numbers.
4. Modify the model to sequence number each competing vehicle starting with 100.
5. Change the setting of light 1 to operate from a pressure pad when a vehicle enters the queue.
6. (a) Add weather factors to the road condition indicator to that there is equal probability of a value of 1 to 10 without a reference to current weather.
   (b) Change the probability of road conditions to the following: 1 50%, 2 20%, 3 10%, 4 6%, 5 4%, 6–101%.
   (c) Since weather tends to continue in a pattern, take current weather and road conditions into account by arbitrarily continuing the current weather 50% of the time. Let the remaining 50% vary according to the rules of (a).

7. The model describes a very specific route. How could this model be revised to allow route selection from an overall array of streets and highways?

8. The system designer may require more information than is provided by the average trip duration or the distribution of trip times. Specifically he may want to know the worst case and how it occurred. How could the model be modified to provide a detailed history of the five longest trips?

9. Generalize the model to measure the travel times encountered by all traffic operating over specified routes.

BIBLIOGRAPHY

1. A Computer Simulation Model of Driver-Vehicle Performance at Intersections, K. L. Laughery, T. E. Anderson, and E. A. Kidd, *Procedures of the 22nd Association for Computer Machines Conference,* 221–231, Thompson, Washington, D.C., 1967.
2. Attitudes of Drivers Toward Alternative Highways and Their Relation to Route Choice, R. M. Michaels, *Highway Res. Rec.,* **122,** 5074 (1966).
3. Relationships Between Drivers Attitudes Toward Alternate Routes and Driver and Route Characteristics, M. Wachs, *Highway Res. Rec.,* **197,** 70–78 (1967).
4. *GPSS V User's Manual,* IBM Corporation Form No. SH20-0851-0, 1970.

3

Nomenclature of Computer Simulation—If Only Everybody Used the Same Term to Mean the Same Thing

The system designer when using simulation has to communicate with two differently oriented groups—one concerned with the system and the other with the computer. With the former the system designer must evolve a durable system definition; with the latter he must cooperate to obtain results. To help resolve difficulties in describing and analyzing complex systems, peculiar jargons have evolved for each major class of system. Whether the system is a telephone exchange, aircraft fleet, road, TV repair service, or chemical-processing plant, however, there are fundamental similarities in many of the internal functions and system criteria. These may range from specific items such as service-delay times, queue sizes, equipment utilizations, and personnel requirements to the general areas of system performance, overall cost, and dependability. The jargons for these separate systems may have almost nothing in common, but the experienced system designer will recognize the basic elements described. When a simulation language is used, many system items adopt common meanings from the simulation language. This use of accepted terminology benefits the system design team in internal communication and classification of items in the system.

The second group the system designer communicates with, programmers, holds the key to successful use of the simulation tool—the computer. This diverse group includes programmers who developed simulation languages and those who control use of a computer installation. They also have their multiplicity of jargons. This chapter will not overcome all the communication difficulties which confront the system designer; only close working relationships can do that. The object is to provide some insight into what things are called and to translate some of the jargon.

System Designer's Environment

The interesting thing about each of the various books on how to design large-scale systems is that the same goals are sought and similar techniques are used, but usually with different terminology. Generally the terminology varies according to the subject area.[1-3] There are basic rules underlying the methodology for system design. So far no one set of system design tools has won overwhelming approval.

The responsibilities of the system designer depend on the type of system being designed, but the tasks fall into several general classes:

1. Problem definition data gathering, determining interactions, and formulation of relationships.

2. System design or syntheses—trade-offs, analytical expressions, simulation, and prediction of performance.

3. System analysis—formulization of the system, after it has been designed, into a prediction of how it will behave.

4. System catalytic action—compromise the parochial tendencies of specialists on the design team in order to develop a balanced configuration for complex system.

System design teams use a wide variety of terms to mean the same thing. This will continue since the customer's terms are the ones most often used. Moreover since the system designer rarely spends his professional career designing only one type of system, the basic terms should be portable so as to carry from one system design to the next. However, similar factors can be grouped together.

Cost

The prime question usually is what will the system cost? The many parts of this question are divided into design, development, and manufacturing nonrecurring costs before the system is used: costs of ownership once the system is in use; and costs to modify the system after it is in use. The system designer has to have the particular cost framework clearly defined since rarely does he have an influence over the cost criteria to be used.

Performance

The next question is what will the system do? Unlike cost each system has its own measure of performance. Some examples are output, dependability, accuracy, effectiveness, and service. These general terms do not exist by themselves. They must be combined with costs to form the basis for the trade-off analyses which guide the system design process.

Detailed System Characteristics

After these two basic areas, there are many fundamental details which the system designer must evaluate. Some examples are utilization, delays, reliability, replacement time, repair time, spares, skill, resources, ' and time. These terms can be broken down into more detailed items uniquely defining each system.

System Generalities

There are a large number of basic rules governing the design of complex systems. Some of these are flexibility, generality, adaptability, growth, and simplicity. While the specific characteristics of a complex system design may seem to contradict these rules, they are the basis for the long-term evolution of systems handling a multitude of factors.

Modeling Terms

There are many simulation languages. The terms used are different. Frequently the same terms mean different things. Each language has developed its own terminology. This makes it harder for the uninitiated to learn to use simulation, and almost as hard to learn to use another language.[4-10] The object of this section is to provide the reader with a translation of some of the major terms used in simulation languages.

The number discussed is limited to the most common. The reader will be able to make better use of this information after he has selected a particular language to learn and use.

Entity

An entity is any thing or being which does the work or is worked on. An entity can be a machine tool, a machine tool operator, or the metal being worked on by both the operator and the machine tool. The drive to the office has as its set of entities our car, competing traffic, roads, the limited space before each intersection, and the traffic control signals. The set of entities can be further broken down into those which are permanent features of the model such as the road and temporary entities which are passing through the simulation. The machine tool is always in the factory, but the operator is there only during certain hours. Both types of entities must be part of the structure and organization of a model and the distinction between the two can be quite artificial. Since developers of languages are so fond of jargon, it might be noted that the derivation of the word *entity* is from the word for *thing* which probably would have been as good a term.

Attribute

The word *thing* is not sufficiently descriptive; therefore labels, parameters, and attributes are needed. The machine tool is the one over by the door, an attribute. The operator is skilled or qualified to use this machine tool. This alloy has a particular cutting-speed setting. These are the attributes, labels, or parameters which enable the simulation to distinguish one thing from all the other things.

The labels used in a simulation are not restricted to physical characteristics or properties. Frequently it is desirable to include labels for the convenience of the model, such as the size of the queue at each intersection when the car enters. There should be flexibility in using attributes to help simulate the real world. There is also the need to keep track of each attribute, especially those used for internal convenience in the model.

Set

It is helpful when the thing called machine tool is identified as being different from the machine operator. This can be achieved by grouping things into collections with common properties. The concept of entity grouping, the set, is helpful since it organizes and structures the problem by relating items through their common characteristic. The number of cars on each route section of the drive to the office is a set. Each car joins a set as it starts over the route and it leaves that set when it reaches the intersection.

Set identification using English is preferred over numerical label, since this use of symbolic notation simplifies keeping track of lists of labels and makes for a more readable, communicable model.

Event

Within the framework of discrete event simulation languages an event has a number of interrelated meanings, all based on the fact that the "time" during simulation is not related to continuous real-world time, but is forced to jump discontinuously from event to event. These events may be sequential or simultaneous in time and independent or conditionally related. Our car left home every morning to join other cars already on the road. At the intersections cars cross one after the other. The traffic lights everywhere could either change at the same instant or not change unless activated by a car. The next significant event need not be predicted, nor is it usually of concern to the designer. It could be another car reaching an intersection or the condition of a changing signal light. The simulation language sorts these events, keeps track of their order, automatically handles the sequencing for the designer, and so controls the representation of the real world.

The events in a simulation are those particular times when something happens or should have happened. The evaluation during an event may result in a branch to a new sequence of further actions, the domino effect; a single action; or the conclusion that at this moment in time there can be no action and, therefore, jump time to the next event. The progression from an event is not predetermined. Some events must follow others in a direct manner; others depend on the conditions that develop at each event. The automatic event-sequencing ability of simulation languages reduces the task of the system designer to a manageable one of establishing general guidelines, while leaving the computer program to do the detailed follow up.

State

A simulation does not go forever. It has some initial conditions and a series of situations to go through to a "conclusion." During the simulation there are "occasions" when the condition or state of either portions or the entire simulation is of interest. At one of these moments a snapshot of the simulation may serve to relate various items which would not be possible with purely dynamic or historical data. Some of these moments could be especially introduced into the model as hourly, daily, or weekly system snapshots.

Sometimes the need arises for working the simulation backward. Wait until a particular state—the longest queue at the intersection, for example—and then start over from some point. Only this time record how the situation developed. The records would show only the factors contributing to this situation, not the entire simulation.

The process of getting from an initial state, where nothing has yet come into the simulation, to the normal state requires some preliminary running. Sometimes once this state has been reached, the complete status of the system can be recorded and then used as a reference point for starting additional simulations. Cars have to be introduced into the model to reach the normal level before our car enters the simulation. The normal level may be a complete record of the state of the simulation after stabilizing or the steady-state condition after sufficient time has passed to complete initialization. When long-term factors are being considered, it may take considerable simulated time to reach a steady-state condition. Frequently, however, the normal state is a state of change, a set of different transient states tied together.

Terminating the simulation should be the result of reaching a particular final state. Sometimes this cannot be predicted in advance; therefore it is necessary to preserve the state of the simulation and possibly restart after reviewing the results. This implies that the state of the simulation can be saved and continued as if there were no interruption. When all the reasons for terminating the simulation are known in advance, they can be introduced so that the simulation will automatically end after 100 days, one success, or one thousand trials, or whichever condition arises first.

Rules

The simulation represents a sequence of actions according to some measures of control and constraint. In representing the nondeterministic complex system, these chiefly logical elements form the fundamental rules governing the simulation. There are, however, several different ways to express rules. In the classical situation rules can be expressed as mathematical equations and simulation is usually not required. But, in combinatorial and logical situations, both decision trees and GO, NO-GO rules are much more readily implemented using simulation.

The variety of rules is endless. The requirements are to handle any possible sequence of conditions. The original route is always preferred except when the following quantified rules indicate selection of the alternate route.

- The traffic has been *unusually* heavy.
- The queue at the intersection *seems* long.
- Road conditions are *abnormally* poor.

In addition the element of choice may be omitted entirely when our car approaches a green traffic light. There are numerous possibilities, but the simulation processes only those warranted by the particular instance. Each decision is reduced to an individual application of the rules, rather than a statistical probability. The logic for the sequence of individual actions, each one based on the previous result, forms the decision tree. From this structure the rules governing model areas develop and are filled out by analytical expressions.

Exogenous and Endogenous

Exogenous is a term borrowed from the biological sciences meaning generated from without. Common words such as external, outside, or independent could possibly have been used, but today this borrowed term is established as part of the simulation jargon. The daily departure time from home is an example of an input over which the model has no control—an exogenous event.

Since there are items being generated outside the simulation and brought in, there must also be those developed within the simulation. Endogenous events are those caused by the internal workings of the simulation. These terms may be great for impressing the uninitiated, but the distinction frequently is meaningless as one exogenous event triggers a number of endogenous events, which are in turn coupled to their instigator. The treatment of both types of events by simulation languages is either identical or similar. The distinction may, however, be helpful in the problem definition phase.

Monte Carlo

Many of the items of a simulation are neither known in advance nor can they be determined analytically. A suitable means is required to select today's departure time and it probably should be different from yesterday's. The method used to choose a single nondeterministic element from the range of all possible values consists of two steps:

1. A number is obtained from a random number source.
2. It is inserted as the independent variable in a relationship to obtain a resultant value.

When these numbers are generated by a computer program and are always in the same sequence, they are termed pseudorandom. This use of random numbers in simulation is called by some a Monte Carlo technique. This usage is different from the original concept of the term Monte Carlo by von Neumann, where instead of obtaining a solution for all conditions, a limited number of paths through the problem were solved.[11] These paths were selected using random numbers.

Report Generator

Simulation requires communicating the results to a human. The form of the computer output is determined by a report generator. This may range from eye-appealing graphs or charts to a heavy tabular listing of everything that went on during the simulation. The output may have been selective under the system designer's control. A pictorial presentation of the distribution of driving times to get to the office might be required for one user, while another requires a number of items including driving times, routes selected, intersection delays, and weather conditions.

Computer Terms

This brief review of some of the more common terms in simulation is intended to provide a feeling of confidence when reading an instruction manual. Unfortunately there is no standardization or even choice of common English for simulation terminology. It is hoped the future will see more terms from the basic language and less from specialized jargon.

Higher-order simulation languages have the advantage of enabling the system designer to both develop the model and run it without considerable outside assistance. While some knowledge of computers is helpful, however, detailed knowledge of the rules of the computer installation and its operating system is not required. Nevertheless an understanding of some widely used computer terms may help.[12,13]

Operating System

Access to the computer is controlled by a housekeeping system which may be called the executive, monitor, supervisor, or operating system. This is a program to control the sequence of tasks performed and to check numerous items. Among these are the questions is the program being run still within the estimated running time, do the magnetic tapes called for have proper labels, are work and file areas available for the program and numerous record-keeping tasks? Beyond these minimal tasks, the operating system can provide access to the simulation language, FORTRAN, COBOL, and particular models and files.

The operating system is the monitor of the computer system. Depending on the particular system it can range from controlling one job at a time to overseeing a system juggling several jobs during one interval. Knowledge of the particular system where your simulations will be run is necessary. Fortunately a set of procedures is normally available to guide the beginner in using the computer system.

The basic addition to the model is a set of control cards inserted ahead of the simulation deck or tape to provide instructions for the operating system. Typical information provided is job name, location of input data, purpose of each storage unit, source of payment for computer time, record keeping, and use of a particular simulation language (Figure 3.1). After the model there are a few control cards instructing the operating system that this job is finished.[14]

The model description used by the system designer is not suitable for the computer. Therefore it is necessary for the card deck, the source program, to be converted by the computer into a form suitable for its internal processing. Any model developed in FORTRAN must be converted from its source form into the object program. A special program, a compiler, is designed to convert the few statements of FORTRAN source code into the hundreds of individual machine instructions required to implement the program. Models using SIMSCRIPT I are converted into equivalent FORTRAN expressions and then compiled. SIMSCRIPT II does not use the intermediate step of FORTRAN; rather it uses the SIMSCRIPT language itself as the vehicle of compilation.[15,16]

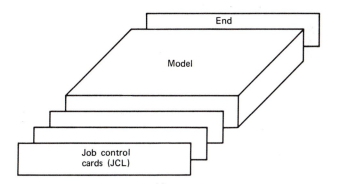

Figure 3.1 Card deck-model and its control cards.

GPSS is not set up to be compiled. Instead the GPSS program accepts the block structure and data of a particular model. These items are interpreted by the general GPSS language into a specific description or set of machine instructions. Basically these are structured lists of the different elements of the model.

The designer is usually not required to attend to trivial details such as assigning a number to a particular block since this is a very reasonable computer task. The assigning of specific block and statement numbers is performed by an assembly program. In a further step the names of the entities may be given a form of English name and the computer then converts these symbolic names to the particular number assigned to each entity. The assembler serves to convert English forms of machine language code to computer code on a statement-for-statement basis.

List Processing

The various items in a simulation have to be related according to their real-world relationship. Unfortunately this is a difficult task which involves control over the next item to be processed, that is, the next event to occur, and adherence to priority relationships as a minimum. The method used to process these items is to establish an ordered list for each particular function. Then it is convenient, quick, and efficient to seek the next item for processing from the first place on the list. Likewise, new items may be entered into the list at the end, beginning, or relative to items already on the list.

The ability to handle specific items within the simulation according to real-world rules depends greatly on the internal list processing capability of the simulation language. The uses for list processing extend throughout the simulation. Some examples are the following:

1. *Keep track of all future events that are specifically scheduled.* Each new item to be scheduled is inserted into its proper chronological sequence in future events list.

2. *Control all items which could occur at this current instant of time.* One item being processed may trigger a whole set of different conditions, causing the sequencing of items processed at this clock instant to start over again from the beginning. In addition, the structure should process items with a higher priority before those with less priority.

3. *Control over potential future events which cannot be scheduled.* There are numerous items which are inhibited from becoming active through a set of restricting conditions. This may require a separate

list for each restriction, but this only implies the need to process a hierarchy of lists. If the event, in fact, is to occur then it will no longer be on any of the restrictive lists.

4. *Ordering of any other factors in the simulation.* Inventory conditions of people or materials may be considered a list, each addition being inserted into its relative location.

Simulation languages make use of list processing to handle their internal structure and make the possibilities of sorting, merging, and extracting particular items or events convenient under both automatic and manual generated structures.

Storage

The computers used for simulation have several storage devices from the list of possible devices, magnetic cores, tapes, drums, disks, and solid-state registers. These vary in speed and size, with the fastest being the smallest, and the largest the slowest. Therefore a degree of compromise is required regarding what will reside in the fastest memory at any one time. The fastest memory available to the modeler, usually the core storage, is used to hold the parts of the program and those data that are being worked on. When different data are needed, or different parts of the program required, these original items are partially or wholly removed from core storage and new items substituted from direct access devices, such as magnetic drums or disks, or sequential access devices, such as magnetic tapes. This overlay procedure keeps the fastest part of the computer in use, but requires time for transfer of information from slow storage to core. On tapes the information must be sequentially searched and retrieved; while drums and disks are capable of being directly or randomly accessed. This latter capability provides time advantages over tapes especially for simulation purposes. Since the next item may not be in logical sequential order relative to the previous item. The forms of memory have varying degrees of reliability, with the tapes being the least reliable part of the computer system.

Computer

The computer itself is far from the last item, although frequently the one over which the system designer has the least choice. The limitations of the computer are primarily in size of core storage, total available storage, and speed of operation. These factors affect the modeler principally by forcing the partitioning of the model to fit the size and speed of the available computer. In addition simulation languages are available on some computer systems, but not others. The result from the system designer's viewpoint is that he looks for a convenient computer which uses the language he prefers and had adequate storage to accommodate his model. Unfortunately models grow bigger and bigger, while the computer core storage tends to remain fixed. This has led to a series of compromises using the overlay of core storage, with a consequent increase in the elapsed running time.

A further computer complication lies in partitioning core storage to permit multiple use of the computer. The modeler may no longer have the entire core storage of a large computer available to him. Instead there are several jobs simultaneously sharing the computer core storage. This leaves the modeler with part of the core storage, a partition, for his model to use. Frequently these partitions are very small in relation to a modeler's requirements. GPSS V, for example, requires at least 50,000 bytes of core storage to run a minimum model. Typical moderate-sized models require twice that amount. But there are numerous FORTRAN and COBOL jobs which require at the most 50,000 bytes. This poses a problem for the modeler if priority is given to jobs requiring small amounts of core relative to those requiring several partitions to be combined. Then the situation may arise where models are postponed until other jobs are finished. As a result these models may never run during the prime shift. The characteristics of particular computer installations, therefore, have an effect on the choice of simulation language, partitioning of the model, and selection of a back-up facility.

Graphics

The facility choices open to the modeler also extend to the degree of man-machine interaction. In the classical "closed shop" a deck is submitted for running together with its estimated running time. The deck becomes part of a "batch" which will, in due time, be run, and the results printed out. Any errors will not become apparent until the end of the process requiring a resubmittal of the deck. Because of the many uncertainties associated with large-scale simulations and their long running times, it is desirable to be able to follow the progress of the simulation while it is still in the computer. This has become feasible through the use of terminals and graphic devices. These can be updated to show the current status of parts of the simulation and in this way provide information to the modeler. He can then make the decision to continue or terminate the run. Improved man-machine coupling is one key element in developing and gaining acceptance for large-scale simulation.

PROBLEMS

1. Indicate how costs influence system design. Suggest several figures of merit based on cost factors that may be used to evaluate alternatives.
2. Compare an automobile service station, food supermarket, and retail dry-cleaning store and plant for measures of performance. Indicate how graphs may be used to compare alternatives.
3. For the three systems in **2**:
   (a) Define the permanent and temporary entities.
   (b) Describe the attributes for each entity.
   (c) Identify those entities which should be grouped together into sets.
   (d) List the significant events characteristic of each system.
   (e) Explain what system conditions govern the selection of time increments.
   (f) List the factors which could determine the length of the simulation.
   (g) List separately independent and interrelated system factors.
4. Given an unordered shopping list, define and flow chart three different sets of rules to control which aisle the shopper may choose next.
5. Queues frequently are the important measure of system performance. Define three different systems where queues are important. Then define the specific data to be obtained from the system simulation of the queues and their properties. How do these data relate to the overall systems?
6. Compare sequential and random-access data-storage methods from the viewpoint of their use in the simulation of complex systems with large amounts of data.

BIBLIOGRAPHY

1. Simulation Applied to a Court System, J. G. Taylor, J. A. Navarro, and R. H. Cohen, *IEEE Trans. Sys. Sci. Cyber.*, **SSC-4**, 4:376–379 (1968).

2. A Description of Steamship Cargo Operation, R. J. Parente and D. F. Boyd, *Simulation*, **11**, 4:195–202 (1968).

3. Simulation: Uses, R. L. Sisson, in *Progress in Operations Research Vol. III*, J. S. Aronofsky (ed.), Wiley, New York, 1969.

4. *Computer Simulation Techniques*, T. H. Naylor, D. Balintfy, D. S. Burdick, and K. Chu, Wiley, New York, 1966.

5. *Essentials of Simulation*, J. H. Mize and J. G. Cox, Prentice-Hall, Englewood Cliffs, N.J., 1968.

6. *Computer Modeling and Simulation*, F. F. Martin, Wiley, New York, 1968.

7. Simulation, Emulation, and Translation, R. H. Hill, *Simulation*, **10**, 2:81–84 (1968).

8. *Computer Simulation Models*, J. Smith, Hafner, New York, 1968.

9. *System Simulation*, G. Gordon, Prentice-Hall, Englewood Cliffs, N.J., 1969.

10. *Design and Use of Computer Simulation Models*, J. R. Enshoff and R. L. Sisson, Macmillan, New York, 1970.

11. Various Techniques Used in Connection with Random Digits—Monte Carlo Method, J. von Neumann, *Nat. Bur. Stand. Appl. Math. Ser.*, **12**, 36–38 (1951).

12. *A Dictionary of Computers*, A. Chandor, J. Graham, and R. Williamson Penguin, Baltimore, Md., 1970.

13. *Computer Organization*, I. Flores, Prentice-Hall, Englewood Cliffs, N.J., 1969.

14. *System/360 Job Control Language*, G. D. Brown, Wiley, New York, 1970.

15. *A Comparative Study of Programming Languages*, B. Higman, American Elsevier, New York, 1967.

16. *Programming Languages: History and Fundamentals*, J. E. Sammet, Prentice-Hall, Englewood Cliffs, N.J., 1969.

4

Pseudorandom Numbers—What They Are; Why Use Them; and How to Get Them

The system designer meets the problem of designing in an area of uncertainty by basing decisions on insight gained from both probabilistic relationships and specific interactions of individual cases. In using simulation there must be a convenient method to select individuals from the mass. Simulation languages aid the system designer in the process by providing pseudorandom numbers. Pseudorandom numbers are numbers in sequence which meet certain criteria for randomness, but always start with a particular number and continue in a repeatable sequence.

An example of pseudorandom number use is a procedure to distribute competing traffic. A transaction representing several competing vehicles enters the simulation every three minutes. This initial transaction duplicates itself into a batch of traffic. Each vehicle enters the simulation after it is separated from the initial transaction by anywhere from 0 to 6 minutes. The determination of each vehicle's delay depends on the draw of a random number. The following equation accomplishes this objective for a random number ranging between 0 and 999 and provides from 0 to 360 seconds with equal probability:

$$6*60*RN1/999 \quad (4-1)$$

Random numbers could be obtained from a sack of numbered beads as in bingo; from a table, such as RAND Corporation's "One Million Random Digits"; from a random electronic noise source whose output is quantized periodically; or from a computer program. The comparison of alternate system designs can be based either on statistical confidence or relative performance under controlled conditions. Since the former can be quite an inordinate consumer of computer time, the latter technique is preferred where possible. Therefore the scenario can be the same if the sequence of random numbers is always generated in the same order pseudorandom numbers. The advantages of pseudorandom numbers are as follows:

- Enable the study of one parameter at a time while holding others constant.
- Provide the easiest and cheapest method for digital computer programs.

Before discussing how pseudorandom numbers may be generated, it is helpful to consider both why these numbers are required and the criteria that may be applied to evaluating a series of such numbers.

Generation of Pseudodata

If simulations had to be conducted only after *all* the required data had been collected, there would probably be little point to simulating, since by then the solutions would be apparent. This does not imply that simulation is based on an absence of facts, only that significant facts may be hard to obtain, unobtainable, obtainable later, or known only in general probabilistic relationships. Since in the real-world analysis and decisions are usually made before *all* the facts are available, simulations are based on knowledge expressed as general or historical relationships. The pseudorandom number provides the means to separate an individual case from data that includes all cases.

An example of the use of pseudodata is a method to establish the random separation between competing vehicles as they enter the simulation. The selection of the randomizing delay is obtained from using either Equation 4-1 or a graphical relationship. The random number can be the same. When the continuous cumulative probability distribution shown in Figure 4.1 is interrogated by one random number—for example 600—a delay of 216 seconds is obtained. The same result obtained by using Equation 4-1. Other input data including traffic density, spacing between vehicles, and even the setting of traffic signals could also be established using random numbers.

Modifying Factors

In a simulation there are numerous instances when the duration of an event is known to be within a range, but the individual values have to be determined. For example, how long will a car take to cross an intersection after stopping for the traffic light, or how long to retrieve

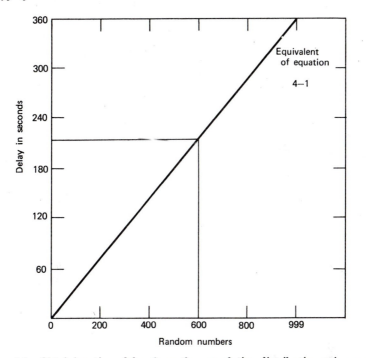

Figure 4.1 Obtaining time delay from the cumulative distribution using random numbers.

a record from a direct access device or a disk unit. These conditions are bounded and require modification within a range. The time for the first car to start and move past an intersection after the light has changed to green is usually independent of other factors such as accidents and breakdowns. Therefore a pseudorandom number could be used to modify the transit time. If it were equally probable for all cars to require 3 to 9 seconds; then the individual value could be obtained from the following equation:

$$\text{TIME} = 3 + 6\text{*RN1}/999 \quad (4\text{-}2)$$

Where RN1 is a pseudorandom number from 0 to 999, if the choice were not equally probable, a weighted curve could be used to select the individual time. For either case, as shown in Figure 4.2, the one chance in a thousand selection of the number 500 would give the same value—6 seconds. In the weighted case, however, numbers below 100 would result in considerably less delay, while numbers greater than 900 would have much longer delays. Notice, however, that both distribu-

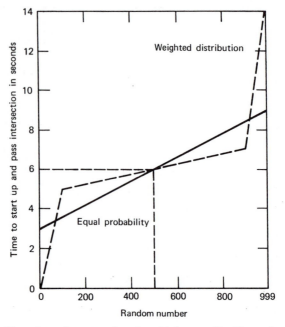

Figure 4.2 Use of random number to obtain specific times from cumulative distributions.

tions have the same average value, while the worst cases are quite different.

A variation of this might be to determine the physical length of the queue waiting for the light to change. The number of cars in the queue are known from the simulation. The contribution of each car to the physical length is determined from a relationship as shown in Figure 4.3, the relative popularity of each car length to the nearest foot. These data would indicate half the cars are 17 feet long. Again a pseudorandom number selects individual values for each car in the queue. These procedures serve to develop data which would otherwise be tedious and difficult to introduce into the model. It is quite practical, when setting up the input data, to select each car length with a simulation program rather than individual decisions.

Choosing a Path

The widespread use of decision logic in simulation requires a method of selecting one path over another. When the selection is based on deterministic factors, the equations are evaluated and the choice made. In nondeterministic situations decision logic based on equations cannot be used and the choice is based on a probabilistic relationship with a pseudorandom number selecting the individual case. For example the decision rules used on the drive to the office indicated selection of the alternate route when there was heavy traffic, abnormal weather, or a long queue at the light. Instead one out of every four cars, on the average, could have selected the alternate route. In this case those vehicles drawing a number between 0–249 would be routed to the alternate, the remaining would stay on the original route. And, furthermore, both sets of criteria could be used simultaneously.

There are times when lack of data and the data form force the system designer to assume independence of choice and use pseudorandom numbers. Later the system designer can go back over this area with better insight and change the decision rules. The process is an evolutio͞ ͞ry one.

Multiple Pseudorandom Number Sources

A comparison between two alternate systems requires the system designer to control the number of variables. Pseudorandom numbers

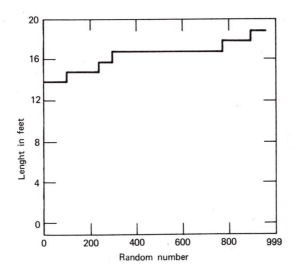

Figure 4.3 Cumulative popularity of cars by their overall lengths.

rather than random numbers help to achieve this aim by providing the same numerical sequence. There are some complications, however. When there is only one source of pseudorandom numbers, the numbers used are always in the same order. A change in decision rules from one system to another could change the results based on different uses for the sequence of pseudorandom numbers. Figures 4.2 and 4.3 both use pseudorandom numbers. Each competing vehicle to enter the model must be fully described. If the same number of vehicles enter the model under each system, the number sequence remains the same. If the model were stopped for the day after the car reached the office, however, this would change the number of vehicles entering the model and therefore the number sequence. In one case there are 10 competing cars generated at random intervals, each requiring a random number. Under different traffic conditions, however, 15 cars might be generated. The decision rules will be different now for each car since its conditions are now defined by pseudorandom numbers elsewhere in the sequence.

The problem of control over pseudorandom numbers can be handled in a variety of ways. One technique, the scenario approach, is to prepare the sequence of departure times for the entire period and record the data. Later, when the model is run, read each day's data into the model as needed. The disadvantage this approach has is the extra effort to set up a second model, record the input data, and introduce the data into the simulation. Another approach is to include in the simulation language provision for more than one pseudorandom number source. Then, if there are decisions which can have an effect on later system behavior, these can be isolated to the same pseudorandom number source. Other sources would be used for additional numbers. Possibly one pseudorandom number source would be used for each day's traffic. Under these circumstances, however, it would be necessary to ensure that each day's sequence of pseudorandom numbers is different.

Criteria for Pseudorandom Numbers

The criteria for the selection of pseudorandom number sources have in the past been highly academic. Therefore needs peculiar to simulation, as contrasted with poker hands, roulette runs, and arbitrary statistical criteria, have not yet received adequate consideration.[1-4] Moreover the capability of the computer to cheaply generate quantities of numbers has too neatly tied in with using the computer to statistically analyze the numbers. This relative ease and convenience led to studies of pseudorandum number generators over huge quantities of numbers, even their full cycle. Unfortunately these are not the criteria of the system designer. Instead he requires rather simple criteria:

- Frequency of number occurrence.
- Freedom from bias.
- Independence.
- Documentation for limited sets of numbers.

Therefore, in spite of previous efforts, where there is extreme concern over the influence of the pseudorandom numbers on results, individual studies must be made of *that* particular sequence of numbers.

Frequency

One basic requirement for the pseudorandom number source is to show equal occurrence for each number. Consider a pseudorandom number source producing three digit numbers from 0 to 999; some of the requirements are as follows:

- Each number from 0 to 999 occurs.
- Numbers are selected in accordance with the laws of probability. The initial series of 1000 numbers should consist of about 780 different numbers.
- The popularity should conform to an expected distribution when large series of numbers are selected.

Figure 4.4 shows, for a particular pseudorandom number source GPSS V, the varying popularity of each number when the series consist of 3000 numbers. The four instances shown are for different starting numbers, seeds, 1, 37, 39, and 741. Figure 4.5 contrasts 3000 and 10,000 numbers from seed 1. There are few times that a simulation uses 10,000 numbers.

Bias

For the same pseudorandom number source producing three digit numbers from 0–999, it is desirable that the average value of the numbers, as more and more are generated, approach 500 whether 1000 numbers or 100,000 numbers have been generated. Unfortunately this does not seem to be one of the conventional criteria for the evaluation of pseudorandom number sources. The tests performed show that as the series of numbers selected increases, starting from the initial portion of the full cycle used for simulation, a bias occurs and continues. A 50/50 split, divided on the basis of the series of pseudorandom numbers, produces instead a continued bias which gets relatively worse as the series of numbers increases. Figure 4.6*a–d* illustrates the data from over 50,000 numbers for the same GPSS V seeds. So far this problem does not seem to have aroused the interest of those who have developed pseudorandom number generators. The solution at this time requires the analyst to study the pseudorandom number generator available for that quantity of numbers. Then, when the quality does not meet the requirements, other seeds can be tried.

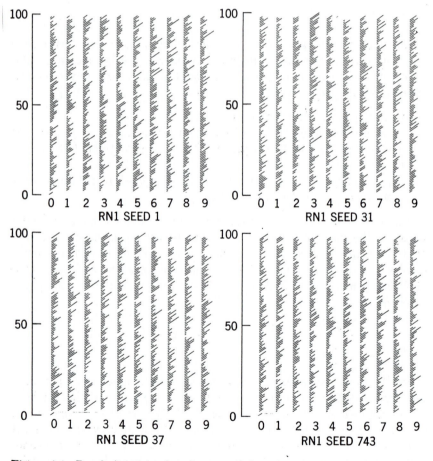

Figure 4.4 Popularity of numbers from a pseudorandom number generator. Data obtained from the GPSS/360 random number generator.

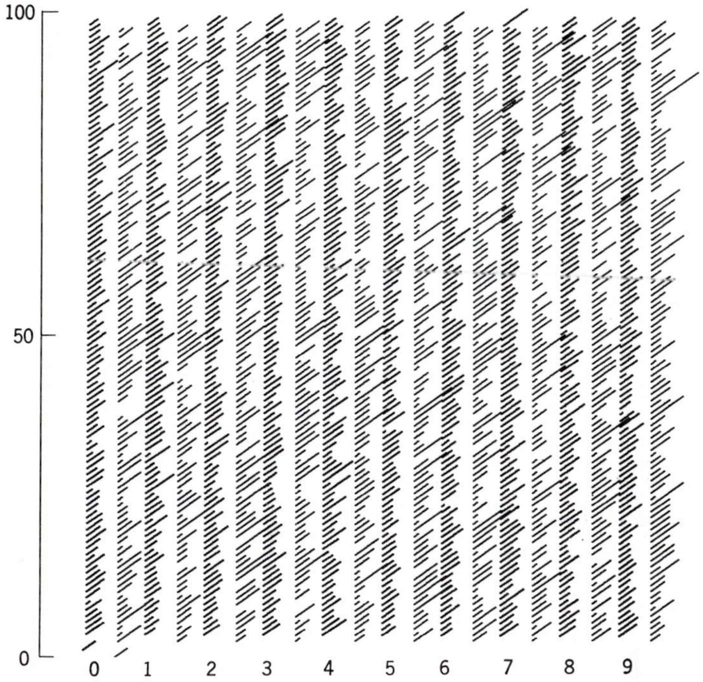

Figure 4.5 Popularity of pseudorandom numbers for 10,000 and 3,000 trials.

Independence

Another criteria of randomness is that there must be independence between successive numbers in the series. This can be tested by comparing the second, third, fourth number, and so on, with the first number of the series to ascertain if the frequency and bias tests are preserved; then continuing the comparison with the second number in the series; and eventually using each number in the series as a basis for comparison. Since this test is used in the development and checking of pseudorandom number sources, the independence requirement has been fairly well met.

Sources of Pseudorandom Numbers

There have been many methods used to obtain series of both random and pseudorandom numbers. These efforts are well documented in the literature and are of historical interest only. Briefly the more popular

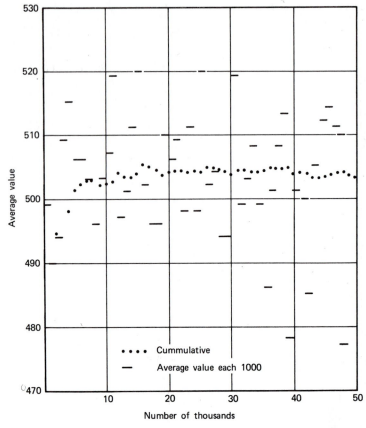

RMULT 1

Figure 4.6a

techniques relied on are, the following: mechanically removing numbered balls from a container (similar to the technique of a lottery, except for returning each ball immediately after its removal); using a list of numbers (clearing-house statements, for example) and picking only the last digit from every tenth listing; and the approach used by RAND Corporation in developing their series of "One Million Random Digits," in which the noise produced by the electronic emission was passed through an electronic quantizer and recorded.[5] All of these techniques generated random rather than pseudorandom digits, but this could be overcome by using the listing of the numbers in sequence. Each of these techniques involved considerable effort for the development and checking of the numbers and obtaining printed listing.

The digital computer changed the techniques and expense of obtaining random numbers. The early digital computer approach was to record previously verified lists of numbers on magnetic tape and read them as the program required. Since magnetic tape units were a source of trouble and it was costly to record long lists of numbers on tape, the logical development was to generate the numbers as needed. This approach has turned out to be the most practical source of obtaining pseudorandom numbers.

There are many different methods used to generate uniform pseudorandom numbers[6-35] and normal and exponential variables directly.[36,37] It is not within the scope of this volume to cover these techniques. One of the widely used techniques, the multiplicative con-

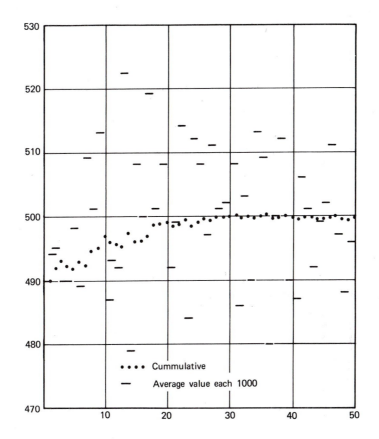

RMULT 31

Figure 4.6*b*

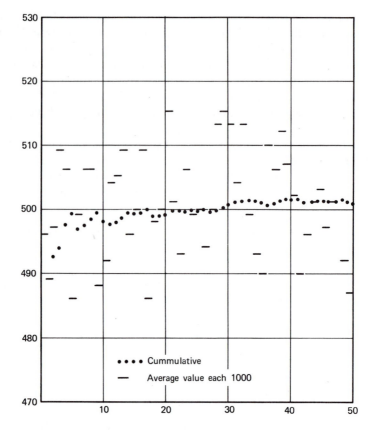

RMULT 37

Figure 4.6c

gruential method, produces a fully deterministic sequence of numbers by starting with an initial value, a seed, which has been previously loaded into the computer and multiplying it by a constant. When this is done in the computer, the product fills two registers. For a pseudo-random number of less than the size of one register, the higher-order register contents are discarded and only the lower-order register contents converted into the number. This number is also retained to act as the next initial value to start the process over to obtain the next number. The process is repeated over and over, but eventually it will repeat the series of numbers. The cycle, until it repeats, is dependent on the size of the internal computer registers and the selection of the internal multiplier. Cycles of billions of numbers are readily obtained. This method produces pseudorandom numbers quickly, cheaply, and in sufficient number for almost any simulation. Many sources of pseudorandom numbers may be obtained by duplicating the constant and using a different seed. The only unfavorable characteristic for satisfactory pseudorandom number generation is the dependence on which initial value is used. It is especially important to ensure that the first few numbers behave in a suitable manner since these are the values used for model debugging. Strange wild-goose chases through the model can occur during debugging when the first few values are particularly skewed.

The pseudorandom number generator used in GPSS implemented for the IBM Systems 360 and 370 performs in a slightly different fashion. It has eight pseudorandom number generators, specified as RN1 through

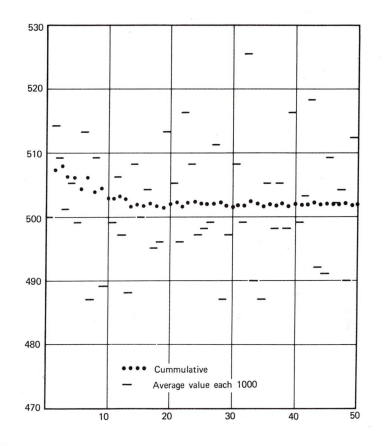

RMULT 743

Figure 4.6d

RN8. Each will produce either the same sequence of numbers or may be altered to provide up to eight unique sequences of pseudorandom numbers. The GPSS program maintains 8×1 arrays:

1. The *base number* array contains 8 words, each different. The first of these words is called the *seed*.

2. The *multiplier* array contains 8 words, one for each random number generator. Each multiplier is initially one, unless an alternate initial multiplier is supplied by the analyst.

3. The *index* array contains 8 words, one for each random number generator; each is initially zero.

When a random number is called for, the following procedure is used:

1. The appropriate word in the index array points to one of the 8 numbers in the base number array. Since the index array words are initially zero, the first base number used will be the seed. All 8 random number generators use this common seed.

2. The appropriate number in the multiplier array is multiplied by the base number chosen in step 1.

3. The low-order 32 bits of this product are then sometimes subjected to another transformation. If the highest-order bit of these 32 bits is 1, the two's complement of these 32 bits replace the original 32 low-order bits of the 64 bit product.

4. The low-order 31 bits of this possibly partially complemented product are stored in the appropriate word of the multiplier array, to be used the next time a random number is called for.

5. Three bits of the high-order 16 bits of the product after step 3 are stored in the appropriate word of the index array, for future use. This number (0–7) points to one of eight words of the base number to be used the next time a random number is called for.

6. (a) If the random number required is a fraction, the middle 32 bits of the product after step 3 are divided by 10^6, and the remainder becomes the six-digit fractional random number. (b) If the random number required is an integer, the middle 32 bits of the product after step 3 are divided by 10^3, and the remainder becomes the three-digit random number.

Unlike the multiplicative congruential method, the GPSS/360 Uniform Random Number Generator (URNG) does not use the previously generated random number as the new multiplier, but rather uses a separate, although overlapping, segment of the 64-bit product. In contrast no indexing is used in a power-residue method of generating random numbers. The second "randomizing factor," a feature of GPSS/360, does not make the generator appreciably slower. One difference it does make is that this approach does not lend itself to analytic methods of analysis. In fact even the period of the generator for different multipliers is not known.

A number of multipliers were tested,[38] including the default multiplier, 1. No pattern in the location of satisfactory multipliers was found. The following eight multipliers appear satisfactory:

31	6352	1366595713
37	751860533	1626487679
743	1088707833	

Of these eight initial multipliers only **31**, **37**, and **743** can be used in the version of GPSS for the 360. The multiplier 6352 violates the condition that the multipliers must be odd, and the remaining four are over five decimal digits in length.

Tables 4.1–4.3 summarize the number of times a multiplier's two se-

Table 4.1. Summary of Testing of the Multiplier 31

Size of Batch	Number of Trials	Length of Tested Sequence	Failures Expected per Test	Tests of Integers Sequence 1 2 3 4 5 6 7 8 9 10	Tests of Decimal Fraction Sequence 1 2 3 4 5 6 7 8 9 10
250	1	250	0.05		
1,000	1	1,000	0.05		
2,000	1	2,000	0.05		
5,000	1	5,000	0.05	1	
10,000	1	10,000	0.05		
15,000	1	15,000	0.05		1
25,000	1	25,000	0.05		
35,000	1	35,000	0.05		1
50,000	1	50,000	0.05	1	
250	20	5,000	1	1 1 3 2 1 2 1 2 1 1	3 1 1 1 1 1 1 1 1
1,000	50	50,000	2.5	3 4 4 2 4 5 4 4 3 2	4 5 3 2 4 4 5 3 3 5
2,000	25	50,000	1.25	2 1 1 2 1	3 2 1 2 2 2 2 2 2

The entries under each test indicate the number of failures of that test under each procedure. The auto correlation tests (Tests #5–10) are not independent of each other.

Test 1—One Dimensional Uniformity	Test 6—Auto correlation, Lag 1
2—Uniformity of Pairs	7—Auto correlation, Lag 2
3—Mean	8—Auto correlation, Lag 3
4—Runs	9—Auto correlation, Lag 4
5—Auto correlation, Lag 0	10—Auto correlation, Lag 5

Table 4.2. *Summary of Testing of the Multiplier 37*

Size of Batch	Number of Trials	Length of Tested Sequence	Failures Expected per Test	I1	I2	I3	I4	I5	I6	I7	I8	I9	I10	D1	D2	D3	D4	D5	D6	D7	D8	D9	D10
				Tests of Integers Sequence										Tests of Decimal Fraction Sequence									
250	1	250	0.05																				
1,000	1	1,000	0.05																				
2,000	1	2,000	0.05																				
5,000	1	5,000	0.05																				
10,000	1	10,000	0.05																				
15,000	1	15,000	0.05																				
25,000	1	25,000	0.05																				
35,000	1	35,000	0.05																				
50,000	1	50,000	0.05																				
250	20	5,000	1		2	1	1	2	1	2	2	2	3 1				1	1	1	1	1	1	1
1,000	50	50,000	2.5	2	2		6		1	1				1	2		1	1		2	3	1	1
2,000	25	50,000	1.25	1	3	1		1	2		1	1		3	3	3		3	3	2	2	3	3

quences (decimal fraction and integer) "failed" various tests. A failure was considered to be when a statistic was outside the appropriate 95% confidence limits. Five percent of the statistics calculated should be failures. None of the above multipliers show unusual or systematic deviations from the 5% figure. The tests were designed to statistically investigate the properties of the uniform random numbers in accordance with a user's expectations.

Table 4.3. *Summary of Testing of the Multiplier 743*

Size of Batch	Number of Trials	Length of Tested Sequence	Failures Expected per Test	I1	I2	I3	I4	I5	I6	I7	I8	I9	I10	D1	D2	D3	D4	D5	D6	D7	D8	D9	D10
				Tests of Integers Sequence										Tests of Decimal Fraction Sequence									
250	1	250	0.05																				
1,000	1	1,000	0.05																				
2,000	1	2,000	0.05																				
5,000	1	5,000	0.05											1									
10,000	1	10,000	0.05	1																			
15,000	1	15,000	0.05																				
25,000	1	25,000	0.05																		1		
35,000	1	35,000	0.05																				
50,000	1	50,000	0.05																				
250	20	5,000	1	2	1									2	1	2	1	1	2	2	2	2	2
1,000	50	50,000	2.5	4	2	4	6	1	2	1	3	1	1	2		1	2	2	2	1	3	3	2
2,000	25	50,000	1.25	2	1	1	2							1	1	1	1	1	1	1	2	1	1

- Individual number uniformity over the range using a χ^2 test with the number of intervals selected to provide the 95% level of confidence.
- Uniformity of pairs of consecutive numbers using a χ^2 test based on each number being located within only 10 equal intervals with both numbers of the pair being used to determine the location of the product in a 10×10 array.
- Comparison between batch mean and the expected distribution according to $\sigma\mu/\sqrt{n}$ where $\sigma\mu$ is the standard deviation of the uniform distribution and n is the quantity of numbers used.
- Consecutive runs of numbers above and below the mean evaluated by a χ^2 test based on the total number of expected runs $(n + 1)/2$ where n is the quantity of numbers and the expected run length above the mean length j is $(n - j + 3)/2j + 2$.
- A series of auto correlation tests for lags between numbers from 0 to 5 when the number sequence is x_i, $i = 1, \ldots n$ and for lag to the auto correlation is defined as

$$A_h = \frac{1}{n - h} \sum_{i=1}^{n-h} x_i x_{i+n}$$

There can be no assurances that the GPSS multipliers will not fail other statistical tests. For some purposes other multipliers may be better suited. Table 4.4 summarizes the behavior of the sequence generated

Table 4.4.   *Summary of Testing of the Multiplier 1*

Size of Batch	Number of Trials	Length of Tested Sequence	Failures Expected per Test	Tests of Integers Sequence										Tests of Decimal Fraction Sequence									
				1	2	3	4	5	6	7	8	9	10	1	2	3	4	5	6	7	8	9	10
250	1	250	0.05																				
1,000	1	1,000	0.05																				
2,000	1	2,000	0.05												1				1				
5,000	1	5,000	0.05																				
10,000	1	10,000	0.05														1						
15,000	1	15,000	0.05														1						
25,000	1	25,000	0.05	1			1		1														
35,000	1	35,000	0.05	1			1	1		1	1	1		1									
50,000	1	50,000	0.05	1	1		1																
250	20	5,000	1	1	2		1	2	2	1	1	1		2		1	1		1		1	1	
1,000	50	50,000	2.5		4	6	3	5	4	5	5	4	5	2	4	3	4	2	1	1	2	2	4
2,000	25	50,000	1.25	1		4	1	2	2	2	2	2	3	3		2	2	1	2	2	3	3	2

THESE ARE THE FIRST 200 RANDOM NUMBERS FROM RN1 WITH MULTIPLIER 1

MATRIX HALFWORD SAVEVALUERNDM1

	COL. 1	2	3	4	5	6	7	8	9	10
ROW 1	573	675	337	177	871	160	719	667	340	420
2	431	36	449	202	693	580	218	428	125	96
3	340	318	343	961	610	512	339	485	507	777
4	795	820	407	384	277	784	943	16	140	859
5	.954	239	958	743	250	705	90	404	399	515
6	491	818	715	843	372	754	755	637	772	386
7	103	628	971	340	142	715	205	607	550	681
8	862	950	19	536	838	51	543	577	101	343
9	890	351	18	245	508	571	125	945	918	643
10	872	630	627	374	802	770	364	562	238	927
11	512	655	419	829	299	811	740	267	148	152
12	950	328	207	753	727	33	249	61	264	397
13	28	60	326	223	95	740	579	398	991	610
14	663	588	396	80	577	428	336	495	729	467
15	126	32	686	466	354	568	462	705	277	155
16	163	648	895	610	494	321	415	817	492	932
17	615	447	482	914	558	62	193	933	802	593
18	954	311	194	169	757	662	500	958	139	884
19	188	382	446	266	130	915	572	803	813	178
20	635	660	831	618	252	974	392	540	400	884

THESE ARE THE FIRST 200 RANDOM NUMBERS FROM RN2 WITH MULTIPLIER 31

MATRIX HALFWORD SAVEVALUERNDM2

	COL. 1	2	3	4	5	6	7	8	9	10
ROW 1	778	926	103	97	407	312	507	294	829	837
2	838	105	605	261	530	783	606	838	115	780
3	606	286	434	850	212	603	686	39	38	703
4	731	54	3	219	329	871	211	802	749	120
5	446	495	857	300	954	221	857	830	51	945
6	97	39	962	100	815	191	708	328	902	137
7	861	229	626	846	327	29	230	19	843	616
8	871	364	430	904	506	890	198	535	57	661
9	929	127	909	845	854	532	74	604	304	564
10	258	122	751	304	671	551	679	719	161	75
11	641	713	636	490	17	831	657	410	66	140
12	169	565	396	535	98	111	496	759	352	959
13	618	486	41	836	407	173	134	54	116	184
14	822	383	521	203	185	991	389	77	151	774
15	868	407	132	752	638	649	18	293	914	449
16	920	338	966	335	283	712	502	501	854	793
17	622	19	743	7	144	344	385	197	693	874
18	927	607	743	908	551	918	793	675	142	244
19	416	986	954	850	402	565	316	306	691	57
20	451	290	1	505	601	596	490	76	916	35

THESE ARE THE FIRST 200 RANDOM NUMBERS FROM RN3 WITH MULTIPLIER 37

MATRIX HALFWORD SAVEVALUERNDM3

	COL. 1	2	3	4	5	6	7	8	9	10
ROW 1	219	567	502	436	17	537	585	967	926	60
2	387	84	288	289	216	682	669	557	447	194
3	671	107	916	314	749	806	184	115	589	491
4	822	76	609	607	375	581	939	884	506	831
5	193	138	633	861	874	189	501	847	650	983
6	798	430	347	489	738	815	227	423	190	463
7	380	823	489	477	784	213	391	940	674	937
8	658	127	456	378	94	62	185	788	805	761
9	118	166	797	348	593	336	340	322	440	629
10	682	15	275	564	538	70	313	288	913	41
11	207	914	162	354	834	843	180	598	264	982
12	566	271	966	827	332	283	745	598	65	318
13	81	631	64	533	472	365	220	218	98	244
14	742	180	555	970	435	823	143	374	282	907
15	145	637	812	593	980	796	751	883	49	203
16	73	223	741	56	121	31	949	102	284	111
17	286	979	940	905	699	454	69	319	924	295
18	183	512	342	861	597	640	912	447	507	161
19	628	878	53	264	211	902	737	439	958	857
20	451	526	328	933	244	388	408	477	323	497

THESE ARE THE FIRST 200 RANDOM NUMBERS FROM RN4 WITH MULTIPLIER 743

MATRIX HALFWORD SAVEVALUERNDM4

	COL. 1	2	3	4	5	6	7	8	9	10
ROW 1	863	952	683	155	10	45	633	667	116	925
2	540	308	687	897	65	209	312	253	295	552
3	278	41	684	655	578	993	809	296	16	752
4	156	713	279	516	59	313	760	851	410	241
5	60	696	134	937	862	718	224	352	688	824
6	511	497	562	610	374	516	508	684	952	827
7	972	539	163	955	904	805	964	719	17	380
8	822	643	591	313	612	149	539	250	230	99
9	632	689	919	194	519	417	190	312	227	882
10	933	889	940	932	832	967	485	86	448	662
11	553	801	13	554	149	468	524	474	383	627
12	516	532	530	557	61	88	474	221	157	971
13	268	393	86	143	937	743	26	443	957	750
14	279	305	566	265	760	507	402	572	131	232
15	100	736	384	200	574	0	494	331	62	963
16	345	170	269	574	767	744	121	537	467	496
17	235	315	595	563	827	290	199	231	550	650
18	330	186	624	429	856	90	0	212	443	179
19	416	482	579	453	888	494	618	837	353	663
20	661	926	131	273	959	542	471	848	289	195

by the default multiplier, 1. Because of its repeated failure on the χ^2 test of the decimal random numbers for uniformity and its systematic failures on the mean test and various auto correlation tests when more than 15,000 numbers in the integer sequence were considered. Cautious use of it should be made.

Simulation is based on use of pseudorandom numbers. Therefore it is necessary to both use and be warned of associated risks when using any particular random number generator.

PROBLEMS

1. The following list of areas for pseudodata generation may be separated into continuous or discrete distributions. Select and discuss which distribution to use in a simulation.

   - Distances between cities.
   - Probability of failure for circuits, equipments, or businesses.
   - TV signal attenuation.
   - Trip duration.
   - Task-work times.
   - Height of men, women, population, buildings.
   - Length of vehicles.
   - Telephone-call holding times.
   - Duration of interval between telephone calls during busy hours, over the entire day.

2. Suggest three methods of providing a fixed scenario to control travel times for some of the vehicles in a simulation.

3. Set up repair time duration based on strategies employed when the repair man knows what caused the fault and how to fix it in contrast to the occasions when he is uncertain.

4. For a 365-day year, show how to use numbers from a uniform random-number generator with a range of 0–999 to select a series of specific days for a particular calendar year.

5. Compare the first 200 numbers provided by the GPSS/360 random-number generator, using multipliers 1, 31, 37, and 743.

BIBLIOGRAPHY

1. Generation and Testing of Pseudo-Random Numbers, O. Taussky and J. Todd, in *Symposium on Monte Carlo Methods,* H. A. Meyer (ed.), Wiley, New York, 1956.

2. *The Art of Simulation,* K. D. Tocker, Van Nostrand, Princeton, N.J., 1963.
3. *Monte Carlo Methods,* J. M. Hammersly and D. C. Handscomb, Methuen, London, 1964.
4. A History of Distribution Sampling Prior to the Era of the Computer and Its Relevance to Simulation, D. Teichroew, *J. Am. Statis. Assoc.,* **60,** 309:27–49 (1965).
5. *A Million Random Digits with 100,000 Normal Deviates,* RAND Corporation, Free Press, Glencoe, Ill., 1955.
6. Random Number Generation, J. Moshman, in *Mathematical Methods for Digital Computers,* Vol. 2, A. Ralston and H. S. Wilf (eds.), Wiley, New York, 1967.
7. On Sequences of Pseudo-Random Numbers of Maximal Length, J. Certaine, *J. Assoc. Comput. Mach.,* **5,** 4:353–356 (1958).
8. A Periodic Property of Pseudo-Random Sequences, E. Bofinger and V. I. Bofinger, *J. Assoc. Comput. Mach.,* **5,** 3 (1958).
9. Random Number Generation and Testing, IBM Form C20-8011, 1959.
10. Serial Correlation in the Generation of Pseudo-Random Numbers, R. R. Coveyou, *J. Assoc. Comput. Mach.,* **7,** 1:72–74 (1960).
11. A New Pseudo-Random Number Generator, A. Rotenberg, *ibid.,* 75–77.
12. An A Priori Determination of Serial Correlation in Computer Generated Random Numbers, M. Greenberger, *Math. Comput.,* **15,** 76:383–389 (1961); *Corrigenda,* **16,** 126–406 (1962).
13. Bias in Pseudo-Random Numbers, P. Peach, *J. Am. Statis. Assoc.,* **56,** 295:610–618 (1961).
14. Mixed Congruential Random Number Generator for Decimal Machines, J. L. Allard, A. D. Dobell, and T. E. Hull, *J. Assoc. Comput. Mach.,* **10,** 2:131–141 (1963).
15. Generating Discrete Random Variables in a Computer, G. Marsaglia, *Commun. Am. Assoc. Comput. Mach.,* **6,** 1:37–38 (1963).
16. Mixed Congruential Random Number Generator for Binary Machines, T. E. Hull and A. R. Dobell, *J. Assoc. Comput. Mach.,* **11,** 1:31–40 (1940).
17. Calculations with Pseudo-Random Numbers, F. Stockmal, *ibid.,* 41–52.
18. Uniform Random Number Generator, M. D. MacLaren and G. Marsaglia, *J. Assoc. Comput. Mach.,* **12,** 1:83–89 (1965).
19. Method in Randomness, M. Greenberger, *Commun. Am. Assoc. Comput. Mach.,* **8,** 3:177–179 (1965).
20. A New Uniform Pseudorandom Number Generator, D. W. Hutchinson, *Commun. Am. Assoc. Comput. Mach.,* **9,** 6:432–433 (1966).
21. Multiplicative Congruential Pseudo-Random Number Generator, D. Y. Downlan and F. D. K. Roberts, *Comput. J.,* **10,** 1:74–77 (1967).
22. Testing a Random Number Generator, S. Gorenstein, *Commun. Am. Assoc. Comput. Mach.,* **10,** 2:111–118 (1967).
23. Fourier Analysis of Uniform Random Number Generator, R. R. Coveyou and R. D. MacPherson, *J. Assoc. Comput. Mach.,* **14,** 1:100–119 (1967).
24. A Uniform Random Number Generator Based on the Combination of Two Congruential Generators, W. J. Westlake, *J. Assoc. Comput. Mach.,* **14,** 2:337–340 (1967).
25. Some New Results in Pseudo-Random Number Generation, A. Van Gelder, *J. Assoc. Comput. Mach.,* **14,** 4:785–792 (1967).

26. Additive Congruential Pseudo-Random Number Generator, J. C. P. Miller and M. J. Prentice, *Comput. J.*, **11**, 3:341–346 (1968).

27. A Comparison of the Correlational Behavior of Random Number Generator for the IBM 360, J. R. B. Whittlesey, *Commun. Am. Assoc. Comput. Mach.*, **11**, 9:641–644 (1968).

28. One-Line Random Number Generators and Their Use in Combination, G. Marsaglia and T. A. Bray, *Commun. Am. Assoc. Comput. Mach.*, **11**, 11:757–759 (1968).

29. More on Fortran Random Number Generators, L. R. Grosenbaugh, *Commun. Am. Assoc. Comput. Mach.*, **12**, 11:639 (1969).

30. Coding the Lehmer Pseudo-Random Number Generator, W. H. Payne, J. R. Rubung, and T. P. Bogyo, *Commun. Am. Assoc. Comput. Mach.*, **12**, 2:85–86 (1969).

31. Extremely Portable Random Number Generator, J. B. Kruskal, *Commun. Am. Assoc. Comput. Mach.*, **12**, 1:93–94 (1969).

32. A Pseudo-Random Number Generator for the System/360, P. W. Lewis, A. S. Goodman, and J. M. Miller, *IBM Syst. J.*, **8**, 2:136–146 (1969).

33. The Behavior of Pseudo-Random Sequences Generated on Computers by the Multiplicative Conquential Method, V. D. Barnett, *Math. Comput.*, **16**, 77:63–69 (1962).

34. Random Numbers Generated by Linear Recurrence Modulo Two, R. C. Tausworth, *Math. Comput.*, **19**, 90:201–209 (1965).

35. Expressing a Random Variable in Terms of Uniform Random Variables, G. Marsaglia, *Ann. Math. Statis.*, **32**, 31:894–898 (1961).

36. A Fast Procedure for Generating Normal Random Variable, G. Marsaglia, M. D. MacLaren, and T. A. Bray, *Commun. Am. Assoc. Comput. Mach.*, **7**, 1:4–9 (1964).

37. A Fast Procedure for Generating Exponential Random Variable, M. D. MacLaren, G. Marsaglia, and T. A. Bray, *Commun. Am. Assoc. Comput. Mach.*, **7**, 5:298–300 (1964).

38. The GPSS/360 Random Number Generator, H. Felder, 2nd Conference on Application of Simulation, New York, 1968.

5

Discrete-Event Simulation Languages—What Is Wrong with FORTRAN

The purpose of higher-order programming languages is to simplify the user's task by shifting to the computer some of the routine elements of programming. The computer becomes a tool, and, it is hoped, a helpful assistant. Higher-order simulation languages provide the key elements of structure, data base, and convenience. Today there are higher-order, discrete event simulation languages available. Compared to FORTRAN these languages provide a considerable improvement in making the computer a tool for system design. This is understandable since FORTRAN was specifically developed for the solution of equations. The complex problems associated with system design cannot be solved by a set of equations. Alternate techniques, such as combinatorial mathematics, list processing, and logic models, are required. These are the basis for the structure of all simulation languages. As a result simulation provides needed design insight, a means of expressing problem definition, and an evaluation of potential behavior of possible systems under a variety of constraints.

While FORTRAN, PL/I, and ALGOL are not suitable in their basic form for simulation, many people have used their structure to develop simulation languages. Some examples of languages structured in FOR-TRAN are GASP, CSL, and SIMSCRIPT I.[1-4] PL/I has been used for HAWSIM and SPL.[5,6] ALGOL has been used to develop SIMULA, ESP, and SOL.[7-9]

Continuous simulation languages—MIDAS, PACTOLUS, DSL-90, and CSMP—have been developed for solving continuous simulation problems using digital instead of analog computers. DYNAMO represents a different equation technique for using the digital computer to simulate business activity. It assumes that business activity can be represented by a continuous simulation of the feedback process. Therefore it has a severe restriction when used to simulate interrelated business activity.

While an average value may be obtained as a result, historical worst cases and measures of risk cannot be obtained. SIMSCRIPT II has its own compiler written in SIMSCRIPT II. GPSS is the prime example of an interpretive discrete event language. It is the only widely used simulation language implemented in assembly languages, FORTRAN, and ALGOL.

User's Requirements

It is necessary to review the conditions under which system design teams operate before developing criteria to compare simulation languages.[10-14] The user of simulation rarely has developed the language. Usually he is part of a team of specialists working simultaneously on the system design. Those with system responsibility should develop the computer representation of the problem; they then have the opportunity to perform the system design trade-off analyses. They may not have determined the conclusions, but they have set up a range of choices for the decision maker.

Over a period of time, the system designer will have been a member of numerous teams and worked on a variety of different systems. As a result simulation should be part of his background and training. For the system designer the advantage gained from experience will be his ability to use concepts from one area when engaged in the design and development of trade-off models for other systems, portable concepts with a portable tool.

Problems stay with the system designer in various forms for a long period of time. He has time to grow proficient in the use of simulation. It is unreasonable to expect this sophisticated technique to be available after a trivial effort and indoctrination. The system designer should have the opportunity to learn to use a language during the initial stages, before the problem becomes too detailed, use the same language to modify and detail the problem representation, and extend the application to broader areas while developing increasing skill and proficiency.

Simulation Language Criteria

Having the Computer Do What the System Designer Wants

The discrete event simulation language must enable the system designer to represent a complex system conveniently and comfortably. There are requirements for solving equations, preserving interrelationships, representing decision logic and physical characteristics of systems. The burden of accommodation must lie with the simulation language. It has to become a convenient and realistic tool. In particular the language must not *demand* simplifying assumptions where complexity is present.

System trade-off analysis should be useful over a wide range of problem areas without restricting the structure to a single approach. Parallel occurrences provide interactions that can become extremely complex. Consider the difficulty of representing the interactions of numbers of individual cars, arrays of traffic signals, and a wide variety of abnormal conditions, all affecting system behavior at the same instant of time. This is the particular area in which the system designer must have a choice of simulation methods to evaluate and represent the latticework of possible existing and interacting conditions.

Debugging Aids

It is naive to assume that simulation language will be useful without some difficulties derived from the characteristics of the system being analyzed and the ambiguity in problem definition. In turn, these factors are affected by the structure and compromises within the simulation language. Since all these factors are combined in the designers operating environment, the following items require special attention:

1. *Realistic results.* Represent the problem to avoid difficulties of language interpretation. Since there may be difficulties of system interpretation, these should emerge from a quick comparison of simulation results with those that were anticipated. When the result is other than expected, additional diagnostic aids should be available to obtain intermediate results in order to investigate particular details.

2. *Aids for insight.* Examine the reasons for the worst case. Historical recording of the system behavior could be overwhelming, but the sequential historical record of one worst case could provide system insight and aid in debugging.

3. *Debugging the model.* Separate bugs introduced by the system designer from those of the language.

4. *Computer environment.* The criteria for debugging aids depend on the capability and characteristics of the computer installation. When the system designer waits 24 hours before getting the results, he must try to make sure that the model does not have a bug in it before it is run. On the other hand where the computer is immediately available,

as in a time-shared environment, the computer can assist in finding bugs.

Flexibility

Once a simulation language has been understood and applied to a variety of system design problems, there is a strong tendency to use prior solutions whether or not they apply to new problems. There is a desire to go ahead with a rework of a somewhat applicable prior simulation, or even a classic analytic solution that might be useful. Sometimes these approaches are correct, but the system designer should not be forced to use solutions that are available from other work. He should have the freedom to be supported, quickly and responsively, for each new system with system design tools. This desirable condition places the burden on the simulation language of being flexible, expandable, and adaptable to the structure of the new system and the environment of the system designer.

Flexibility is also a hedge in the face of uncertainty. When it is not clear how the problem is structured, the simulation language should represent a broad case, which later can be detailed, restructured, differently constrained, and made specific to newer rules and greater knowledge.

One key limitation in simulation languages pertains to the development of very large models. It is reasonable to start with a small or coarse model, but what happens later on? Is the original framework still useful? Can the model grow? Will it still be within the storage and speed capability of the computer? As simulation becomes more widely used, there is a trend toward bigger and bigger models. These cover greater areas, have more details, and require larger and faster computers. Will models which start out with part of the system be able to grow to encompass the scope of the eventual system?

In the simulation of a large system there are numerous options which, though existing, may rarely be entered. Does the simulation language demand a consistent structure to be anticipated which will apply throughout the model, or is it possible to have a "don't care" structure which allows these potential relationships to be ignored until they are actually used? If the problem has to be more fully structured before the computer can be used, there is much more effort required. If, on the other hand, the language has the "don't care" characteristic, there is significant benefit in getting computer results with minimum effort and time. The overall gain lies in obtaining the coarse overview results first and particular details as they arise in the evolution of a particular system simulation.

Data Banks

The system designer frequently has a severe lack of data. This forces him to consider the ranges of potential data values and generate considerable pseudodata, as has been discussed previously. There are other occasions, however, when the system designer has access to large amounts of valid historical data. In these cases it would be desirable for the simulation languages to readily accept data from these data banks without requiring separate data manipulation efforts. Furthermore, where data are common to a number of models, these data should be readily available from a data library.

It is reasonable to expect that state or city traffic departments would have records concerning traffic flow, density, distance traveled, and intervals between cars for both intersections and routes for different times of the day. It is desirable for the simulation language to accommodate data directly.

Sometimes raw data are not directly useful. The simulation language should include techniques which can be used to process the data—regression analysis, statistical analysis, smoothing, and processing. When only a limited part of the available data is required, the simulation language should enable the system designer to select, retrieve, or modify a particular item or class of items rather than process all the data to obtain the desired information. Overall the system designer does not want to be restricted by the constraints of the present form of data, the lack of data, or too much data.

Graphic Aids

A system design results from the coordinated efforts of a number of individuals who comprise the design team. No one individual is able to know all the problem details. Use of simulation has to be considered in terms of how it contributes to the total system design. There are less useful gains from this sophisticated and capable tool when only one member of the team is knowledgeable. The simulation language must provide a medium for exchange of information and the grounds for a common understanding of the system by team members. The requirement for dissemination of system design concepts implies a change in the importance of graphic display techniques. Up to now graphic display of simulation results has been considered secondary. With the current availability of sophisticated display equipment at reduced costs, it is now practical to use the display console to support two distinct efforts:

1. *System designer.* During model development he needs the advantage of seeing how the model is progressing. This permits computer runs to be interrupted or terminated when errors are indicated. The human element contributes a sophisticated pattern-recognition capability; then, using the dynamic representation, the designer can gauge simulated system performance. This is a strong advantage. It gives the designer a sense of participating, understanding, and being able to visualize various trade-offs.

2. *System design team.* Communication must be established among system designers and other team members. Here interaction is necessary to determine what should be done and how. The interaction with the computer and display system serves to provide quick and ready assistance to understand the implications of potential modifications and alternatives.

The graphic output from simulation languages should not be limited to displaying results. The spectrum of information associated with simulation should be available—input data, in process status data, and accumulated results. Once these data are accessible the design team can cooperatively use simulation.

Transferability

A model can be generic, with form and structure that other design teams can also use. This concept can be extended to enable others to use the model and results for analysis in allied areas. These additional users may belong to the operating level rather than the design team. Predictive models for use by the decision maker are an example of simulation being employed by the ultimate user rather than the model developer.

Another factor of system design environment which affects the simulation language is the dynamic change in the complement of the system design team. As models have different functions during the early stages of system design as opposed to during eventual system implementation, so the use of simulation languages should provide continuity for the user team through variations of the model and historical records of expected performance at each stage of system design.

Computer Time

Up to now computer time generally has been considered a very valuable and expensive commodity. When the value gained from computer usage is compared to total system costs, however, the savings are significant. Experience has shown that the principal cost of the simulation is not computer time, but manpower required to develop, assemble, and obtain the problem definition and system design. This situation has resulted from the rapid decrease in computer time costs, in spite of the great increase in the total amount of computation required to perform a complex simulation.

Distribution of Available Computers

The attributes of the computer simulation languages have so far been discussed only in terms of desirable language characteristics. One more fundamental area must be considered. For techniques as complex and difficult as simulation languages to develop, the users must be numerous—attacking a variety of problems and possessing a wide range of skills. Unless the users of a simulation language exceed a critical number, that language will not be debugged properly and uncertainties regarding the meaning and behavior of that language will remain. The implication of this factor is that efforts must be concentrated on getting adequate instruction manuals, training, debugging, and potential applications about a few languages instead of proliferating languages. One language will not do everything—certainly not initially—but if the basic structure of the language permits, more features can be added to round it out.

As a concomitant to the above point the language should be available on a wide variety of computers. If the language is available from only one manufacturer, on a particular type of computer, circulation is limited. Schools, government, and industry all desire a broadly useful tool.

In the long run discrete event simulation languages will improve through a process of evolution. No language starts out to be the answer to everything for everybody. In time, however, with development, implementation, and experience a few well-qualified languages will accommodate system designers requirements.

Review of Simulation Languages

The previous general comments about simulation have to be correlated to specific languages to provide guidance in selecting the system design tool. The general comments may be reduced to four basic characteristics: short-term results, ability of the system to represent the real world, long-term results, and effort required. The languages selected for comparison range from assembly language (that is, no higher-order language at all) through the general purpose higher-order languages—FORTRAN, ALGOL, and PL/I—to widely used discrete event simulation languages—GPSS III through GPSS V, SIMSCRIPT and SIMSCRIPT II.

Assembly Languages

Historically the first simulations were implemented with assembly languages. Reports in the literature indicate that there is still considerable effort expended using assembly language for simulation. Therefore it is of more than academic interest to see how this approach serves the system designer.

1. *Short-term results.* The individual using assembly language must be an accomplished programmer.

- Structure of the simulation must be developed to keep track of internal events and statistics.
- Detailed programming is required to use available data banks.
- Slight opportunity exists to minimize effort by using subroutine libraries.
- Methods for developing pseudodata must be individually developed.
- Flexibility to accommodate the usual changes in problem definition result in considerable rework.
- Debugging is entirely dependent on the previous experience of the programmer. There are no aids or structures to ensure convergence of the debugging process.
- Graphics use is almost automatically ruled out, since the effort adds to what is already considerable effort.

2. *Ability of simulation to represent the real world.* The only possible advantages may be, if the programmer is very familiar with Boolean methods, for expressing complex logical interactions and decision trees has only a small computer available.

- Mathematical capability would depend on the freedom to interface the assembly language with higher-order programming languages.
- List processing would be a severe additional burden. This would

practically eliminate the ability to retain historical worst cases and rank results.

- Maximum model size is open; the allocation of computer storage is entirely within the programmer's design. Large models requiring complex core overlays can be developed and must be tailored to the specific hierarchy of available storage devices.
- Hybrid computer systems can be used, since the interface considerations are being programmed in assembly language, allowing any computer system to be used; thus allowing the tying together of discrete event and continuous simulation.

3. *Long-term results.* In this area there are only disadvantages.

- Documentation is completely up to the individual. Since the level of detail is so much greater, there is slight likelihood of anything approaching adequacy. In most cases, the original programmer will not be able to follow the program six months or a year later.
- Computers with other assembly languages would not be usable.

4. *Effort required.* Considerable, since there are no convenience features available. In addition, assigning several programmers to the task increases communication difficulties rapidly and reduces the rate of model development per programmer.

Procedure-Oriented Languages

FORTRAN is one of the most widely used languages for expressing mathematical relationships. It, along with ALGOL, has been used to provide a structure for special simulation languages. FORTRAN, ALGOL, and PL/I will be considered as a group for simulation.

1. *Short-term results.* The system designer must have a good programming background in the language before he can use it for simulation.

- Structure is not available when using these languages alone. Each subroutine must have its development tailored to the purpose of the simulation. Modularity permitting broad aggregation of subroutines requires consistent control of information transfer.
- Statistics, both input and internal, have to be processed, providing considerable additional programming burden.
- Data banks may be in a usable form.
- Pseudodata represents considerable effort for organization, tie-in with subroutines, and format.

- Flexibility depends on the size and scope of the simulation; small models can be changed and modified easily. The complex systems cause considerable difficulty especially when subject to a rapid sequence of changes.
- Debugging aids are limited. When parts of old models are reused there may be debugging aids which carry over. In general there is the process of finding one bug and eliminating it to expose another one. Considerable effort must be expended in developing the debugging structure.
- Graphic presentation of input data and results requires considerable additional effort. This is in contradiction to having the results quickly available and in a form for others to evaluate.

2. *Ability of simulation to represent the real world.* Almost any real-world condition could be represented; the effort goes up greatly with complexity in a nonlinear relationship.

- Logical situations can be represented depending on the version of the language used. Not all language versions have equivalent Boolean capabilities.
- Mathematical capability is excellent. There are numerous special-purpose techniques for data smoothing, linear programming, and other forms of data manipulation which are both available and accessible for the simulation.
- List processing is weak in FORTRAN but it is part of other languages; and for simulation purposes this requires investigation, depending on the degree of list processing each simulation requires, as well as, the basic minimum required to structure the model.
- Maximum model size is under the control of the programmer. He can make trade-offs between storage hierarchy and speed. This requires increasing efforts as the limits of core storage are approached.
- Hybrid systems may be tied together when the interface accommodates higher-order languages. This is a special condition which is unique to each type of installation. Otherwise, part of the tie must be made in assembly language.

3. *Long-term results.* In this area there is the advantage of language generality.

- Documentation is under the control of the individual. There are aids in the form of cross-reference files, but the individual must still set up his comments and do so thoroughly.
- System designers other than the original model developer now have a chance at following the logic and detail of the simulation, but

with considerable effort. There is no inherent way of making the model transferable to someone else.

Computers of many different manufacturers can use the same higher-order language program. Usually there is some requirement for rework, but in the overall size of effort this could be considered trivial. If PL/I were the language used, there is a greater inherent capability for simulation. However, there has been relatively little effort to use PL/I for simulation. The overall usefulness of PL/I would depend on the availability of basic simulation subroutines, graphics, data manipulation and housekeeping routines.

4. *Effort required.* Less than for assembly language simulations, but considerable. Several programmers can work on the simulation in parallel if the conventions governing the transfer of data between subroutines are well planned in advance. Thorough and detailed advance planning is needed for any complex system program. The problem must be clearly defined and conventions established *first*. Often this is an impossible set of conditions for a complex, poorly defined problem.

General-Purpose Simulation System GPSS III–V

GPSS in its various versions is the most widely used and one of the oldest simulation languages. Several versions have different specific characteristics and may be considered as subsets of GPSS V. The comments apply to all versions of GPSS except where specifically limited to the most advanced version GPSS V. It should be noted that these comments do not apply to GPSS II.

1. *Short-term results.* GPSS is designed to get useful results quickly.

- Structure for the simulation is strongly evident. The language is highly structured, with a limited number of generalized concepts. The convenience features are built-in, providing format, organization, diagnostics and direct relationships to specific model elements.
- Statistics from the internal workings of the model are automatically maintained and presented both during the simulation and at its conclusion.
- Data banks were awkward to use until the GPSS versions, possessing in array capability in the form of matrix savevalues. This form of data bank can be set up externally to GPSS but requires programming help. Large data banks can be accommodated using disk storage.

- Pseudodata can be readily introduced with the use of up to eight internal pseudorandom number generators. Graphical relationships can be used with the pseudorandom numbers to structure data using the GPSS function.
- Flexibility was one of the prime objectives of the language development. This generality makes it relatively easy to change logic, data, and select results.
- Debugging aids are many, covering the phases of simulation from model construction and assembly through execution errors. The language has an assembler and symbolic notation capability, reducing the possibility and time required to correct human error. Routine bugs are found quickly. Logical errors, however, sometimes are compounded by the need to understand the internal workings of GPSS to explain results where simultaneous events are improperly ordered and sequenced.
- Graphic presentation is part of the language. Some input data and internal workings may be shown in graphic form in addition to the results.

2. *Ability of simulation to represent the real world.* A limitation of GPSS III and 360 was in the size of simulation that can be used on available computers. It required the active program, data, model, and activity all to be in core storage at the same time. However, the latest versions of the language such as GPSS/360-NORDEN, GPSS V and XDS permit parts of the model and its data to be stored on random access devices.[15] Therefore, the desired level of simulation detail can be achieved. Efficient use of core storage is achieved through the structuring of data in byte, halfword and fullword arrays.

- Logical situations can be well represented through Boolean equations.
- Mathematical capability is adequate for problems that do not require complex equations such as those requiring double precision arithmetic. When complex mathematical needs arise, they must be handled outside of GPSS through the HELP block. When there is a large need for this form of mathematical assistance, the HELP block can be set up to call FORTRAN or other programming languages routines.
- List processing is readily handled with both the Set and Chain concepts. These permit complex data to be processed and handled on First In, First Out or Last In, First Out (FIFO and LIFO) basis, or by any arbitrary ranking. Any item anywhere in the system may be treated and modified.
- Maximum size of the simulation is primarily tradeoff between com-

puter available core storage and elapsed execution time. When really large-scale models are constructed in GPSS, it is necessary to first consider how the model will fit into core, what will overflow onto auxiliary devices, and the effect on simulation running time.

- Hybrid system can only be accommodated through the HELP block or the use of special ties between the systems. This leaves open the basic question of whether GPSS or the hybrid system has control of the system. Since there is no guarantee that either will finish its tasks in an allotted time slot, the results could be catastrophic. On the other hand, if the size of the computing system is large enough to be truly multiprogrammed, this problem might easily be solved.

3. *Long-term results.* Here there are benefits from the highly structured GPSS language. The intent of the model maker is fairly obvious even when comments are not complete.

- Documentation capability is very good. Comment statements can be part of every item. The internal support through symbolics and assemblies are useful.
- System designers other than the initial designer can follow the GPSS block diagram, both when it is being developed, and later on when trying to revive an old model.
- Computers of the IBM 360 series with 65,000 bytes of core and up can handle varying size GPSS models. Various other manufacturers have versions of GPSS available: UNIVAC, GPSS II; Honeywell, GPSK III; RCA, FLOWSIMULATOR, General Electric, GSIM; Xerox Data System, GPSS. Other manufacturers have yet to announce their position on this language.

4. *Effort required.* GPSS offers the most working model for the effort expended.

SIMSCRIPT

Next to GPSS, SIMSCRIPT is the most widely used simulation language. Developed by the RAND Corporation it has been in use since 1962.

1. *Short-term results.* SIMSCRIPT is designed for use with relatively large models. The designer should have programming competence in FORTRAN. Extensive and detailed problem definition are required before a SIMSCRIPT model is developed.

- Structure for the specific simulation is largely left to needs of the system designer. The basic structure is left to the programmer. The language does provide means with which to build the model relationships, events, sets and functions.
- Statistics during and after the running of the model are introduced by the system designer. He has access to almost anything but has to specify what statistics are to be gathered—maximum, minimum, or mean, and when.
- Data banks are formed into arrays of data since the programmer must provide the data organization.
- Pseudodata can be introduced, since pseudorandom number sources are available. Graphic relationships can be used to structure input data.
- Flexibility is geared to major changes in the model, when different subroutines can be used; for trivial changes, different input data can be used. There is accessibility to change logic or to obtain different outputs.
- Debugging aids are limited, since the model structure is developed by the system designer rather than be a prescribed structure. This lack of structure can make debugging a time-consuming and difficult process. Greater use of English statements does help in the debugging. Routine bugs are exposed by the compiler. Special diagnostic aids are available when required.
- Graphic presentation can be added by the system designer; there is a flexible report generator; input and output must be programmed for either display or presentation of relationships.

2. *Ability of simulation to represent the real world.* Even though it provides capability for storage overlay and dynamic storage allocation, SIMSCRIPT has a limitation on the size of model which can be run on available computers. Since greater detail means considerably more effort, there is less tendency to over-detail the problem in SIMSCRIPT.

- Logical situations can be well represented with those versions of SIMSCRIPT having Boolean capability. The excellent set capability of the language enables complex situations to be represented.
- Mathematical capability is wide open, since other programming languages can be called by the model as needed.
- List processing handles complex data with the capability for FIFO, LIFO, and ordered processing.
- Maximum size of the simulation is a function of computer core storage. Overlays may be used and can slow down the running

time. Skilled programming competence is especially required for fitting large models into overlay structures, since location of data is critical.

- Hybrid systems can be accommodated with the problem of which system has and can preserve system control. This is an area that is quite unexplored. Specific hardware would change the practicality greatly.

3. *Long-term results.* SIMSCRIPT has the problems of a rather freely structured modeling language.

- Documentation capability is very much up to the individual. Since actions are not obvious without documentation, there must be considerable effort to document the model if anyone is going to use it at a later date.
- Users other than the initial programmer can follow SIMSCRIPT models that are carefully structured and documented. Complex situations, where many things are going on simultaneously, can lead to confusion, even when special care is taken.
- Computers of many manufacturers have SIMSCRIPT and SIMSCRIPT I.5 compilers.

4. *Effort required.* SIMSCRIPT requires considerably more effort to develop a medium to large sized simulation than does GPSS.

SIMSCRIPT II

SIMSCRIPT II is a new programming system rather than an improved version of SIMSCRIPT. Developed by the RAND Corporation it has been available since 1969.

1. *Short-term results.* SIMSCRIPT II was designed as a higher-order programming system instead of being "merely" a simulation language. As such it may be used for applications other than simulation as well as a broad range of simulation applications. To use the language the designer must develop a programming competence in SIMSCRIPT II. This is closer to an understanding of FORTRAN rather than GPSS. Like SIMSCRIPT there is no inherent simulation structure. Therefore an extensive and detailed problem definition must be developed before programming the model.

- Structure for the specific simulation is developed by the system designer. Model relationships are set up through the entity-attribute-set relationships.

- Statistics obtained either during or after the model execution are programmed by the system designer. The language allows access to anything and at any time, but the specific structure has to be added by the user.
- Data banks are formed through the organization of data to define the entities and their attributes.
- Pseudodata may be readily introduced, since there are both pseudo-random number source and a library of common functions. In addition graphic relationships can be used.
- Flexibility is tied in with the basic subroutine structure of the programming approach. Individual subroutines can be compiled separately and added or substituted into the model. The same model may be run with different input data. Small changes can be made at the assembly language level.
- Debugging aids require experience and extensive preplanning. The model structure is established by the programmer, therefore, he must also establish the debugging approach. Once planned, events may be followed to debug the model. The routine bugs exposed by the compiler are of the less sophisticated type, syntax and structure.
- Graphic presentation is hampered by the need to also develop a complete graphic program. The location of data for presentation must be specified for each application. The flexible report generator makes it easy to display data after the model has been run.

2. *Ability of the simulation to represent the real-world.* The size of model that can be contained in core at one time is maximized through the use of ragged tables and assembly language code. These require clear understanding of the problem before programming is started. This can result in significantly greater effort, since the effort must be distributed over a number of people to program a large system.

- Logical situations can be well represented through the Boolean capability. Complex situations can be structured.
- Mathematical capability will vary with the experience of the particular installation, because SIMSCRIPT II is a separate programming system. Subroutines in SIMSCRIPT II will have to be developed. Interfaces to other higher-order languages also have to be developed.
- List processing capabilities are strong, owing to the structured data storage system with ability to select the desired data. FIFO and LIFO and ordered data structures are easily established.
- Maximum size of the simulation depends on the available core stor-

age. There are provisions for overlay of routines to make larger models possible. Utilization of core storage is good, since only items under the control of the programmer are stored. For sparse data tables, core storage is required for ragged tables rather than all possible locations. The amount of data packing is under the programmer's control.

- Hybrid systems can be accommodated when either the continuous simulation is in SIMSCRIPT II or links are established to existing languages.

3. *Long-term results.* SIMSCRIPT II is the closest approach yet to a readable English in a simulation language. However, the structure of the model is determined by the programmer. For complex problems the readability does not convey the relationships among different factors.

- Documentation at the subroute level is excellent. SIMSCRIPT II syntax has been developed for both the compilation and execution of the logic of the model and to provide a readable English. The statements are understandable directly to someone with a minimum exposure to the language. Explanatory comments may be added as needed and with great freedom as to location and quantity.
- Users other than the initial programmer should be able to follow model detail with relative ease. However, simultaneous events are still extremely difficult to document. Changing the logic of an existing model is difficult for those changes which require restructuring of data and system attributes. The lack of a cross-reference table makes it difficult to discover all factors affected by a change.
- Compilers for SIMSCRIPT II will become more available as time progresses. The language is available for the IBM System 360, and RCA Spectra 70 series.

4. *Effort required for SIMSCRIPT II.* This requires less effort to produce a model than the earlier ~~versions of SIMSCRIPT~~ and SIMSCRIPT I.5. Extremely large models of a highly replicated nature could be quite easy to build. In general when compared to GPSS SIMSCRIPT II models of trivial size require the same effort as in GPSS, those of medium size require more effort, and large ones require considerably more effort to develop. In all cases SIMSCRIPT II requires a well-defined problem statement before the model can be programmed. The structure to aid in debugging the model is the responsibility of the programmer. For complex models an additional debugging structure must be included with the problem definition and set of system constraints.

Alternate Approaches

There is another possible method to develop specific simulations. This approach consists of using a highly structured language such as GPSS to define the problem, develop a set of flow charts to cover the problem describe the logic, and establish the mathematical relationships. After these tasks are completed the actual model is programmed in another language to reduce execution time and increase the size of model that can be in core memory at one time. SIMSCRIPT, FORTRAN, or assembly languages could be used.

The advantages and disadvantages of this approach probably would require a model for the trade-off analysis. The main advantage is faster execution of production runs. The disadvantage for a complex model is the additional cost and time to program the problem a second time, the need for an individual with additional programming skills, and potentially an inflexible model.

An alternative to making the model twice is to determine particular areas of the model that run slowly or require excessive core memory and substitute special subroutines for these. The basic model would be in GPSS, and HELP routines would be used for the subroutines. The disadvantage of this approach is the extensive special programming.

The general question of running time for a simulation is unclear. When assembly language is used there is the potential for rapid running times. This is valid for simple models. For complex models—the more normal case—there are no clear guidelines. Once core storage becomes a limited resource, trade-offs are necessary between how much of the simulation is core resident and how often data are transferred to and from core. Under these complex conditions there is no clear guide to which approach will require the least running time. In fact the representation of the same function using different approaches in the same language has widely varying running times. The consolation is that, except when simulating computers, running time does not become the major cost consideration. Labor costs are, and will continue to be, the major cost element.

Other Languages

There is no shortage of simulation languages; it seems to be fashionable to begin developing a new language. The use of these languages, however, generally introduces greater difficulty and fewer benefits.

There are activity-oriented languages which. could be used when the problem area is well understood and not changing rapidly. Models for evaluation of time-shared computer systems could be one such area. MILITRAN represented an attempt to develop a conversational language for war gaming, but it is far from a complete simulation language. SIMULA I has many of the properties of SIMSCRIPT, is based on ALGOL, and has little in the way of convenience aids, report generators, graphics, man-machine interaction, or the rest of the items that turn a concept into a useful tool. SIMULA 67 is a more useful and general version of the language. GASP, SOL, OPS3, and WASP all have individual groups declaring their merits, but little general usage. They are based on general purpose languages. Familiarity with FORTRAN, for example, would simplify using GASP. Unfortunately there is the major weakness that these languages cannot provide the user with the sophisticated tool he usually requires and can get from GPSS V, SIMSCRIPT I.5, and SIMSCRIPT II.

In summary there is a specific need for improvement in the existing languages. Weaknesses are in the man-machine interface, model size, and running times. Advanced concepts are needed to interface these languages in hybrid systems.[16] But these languages have been around for over 5 years, have been modified to the extent that they no longer resemble the original, and are still evolving which indicates many problems have been successfully solved. Standardization of simulation languages is premature, but models written in one language should run on various computers. A set of criteria for languages should evolve as more extensive experience is gathered together and evaluated. Finally general specifications for GPSS and SIMSCRIPT should be established to emphasize the goal of machine independence.

PROBLEMS

1. Compare the queue structures with regard to internal discipline, priority, single and multiple server service, and interaction among queues for the following:
   Supermarket checkout.
   Limited access highway toll station.
   Cafeteria.
   Movie theater.
   Automobiles waiting to make left turns.
   Crossing runways.
   Automobiles leaving a parking area after a ball game.

2. Flow chart the critical element of discrete event simulation: the selection of the next instant of time from the viewpoint of either a transaction or an event.

3. Add to the flow chart the logic to resolve conflicts among items scheduled to happen at the same time both with and without a priority structure.

4. Could the model of the drive to the office be salvaged to provide a design aid for the evaluation of a multilevel parking garage?

5. State those applications in which the simulation requires several runs to converge to an answer. Consider the usefulness of regression analysis, learning curves, and pattern recognition techniques.

6. Flow chart a simulation for an airport serviced by two crossing runways. Include the following:

   (a) Aircraft departure schedule—bunch aircraft to depart shortly after the hour and half-hour.

   (b) Aircraft ground servicing—during the busy hours of the day maximum time for an aircraft on the ground is 1.5 hours with 0.75 hours as the average. There is a minimum time interval between flights, however, of 0.33 hours.

   (c) Runway scheduling—both runways are used when wind conditions permit; otherwise one runway is used for both takeoff and landing.

   (d) Aircraft landing—under Visual Flight Rules (VFR) the interval between aircraft ranges from 60 to 75 seconds. Landings under Instrument Flight Rules (IFR) require 75 to 110 seconds. Weather conditions determine VFR and IFR.

   (e) Crossing runway scheduling—when both runways are in use one is used only for landings and the other for takeoffs. At least 20 seconds must pass from the time the landing aircraft has touched down and passed the runway intersection point and the time the aircraft taking off will reach the intersection. Once the aircraft taking off has passed the intersection, 40 seconds are required until the landing aircraft can touch down.

7. Indicate through flow charts how the system simulation in No. 6 could be used to determine:

   (a) Hourly runway acceptance rates, hourly and daily utilization statistics.

   (b) System performance under both VFR and IFR conditions

   (c) Influence of the passage of a storm front over the airport and the evaluation of alternate strategies to return the airport to operation.

(d) Evaluation of grouping of aircraft into landing and take-off queues during single-runway operation.

(e) Accept historical weather data for both wind and visibility conditions in order to determine runway-usage rules

8. List a series of debugging aids in two groups:

(a) Those which should be inherent in the simulation language.

(b) Specific items added to aid in following the characteristics of this specific problem as defined in No. 6.

BIBLIOGRAPHY

1. Control and Simulation Language, J. N. Buxton and J. G. Laski, *Comput. J.*, **5**, 3:194–199 (1962).
2. Extended Control and Simulation Language, A. T. Clementson, *Comput. J.*, **9**, 3:215–220 (1966).
3. *SIMSCRIPT: A Simulation Programming Language*, H. M. Markowitz, H. W. Karr, and B. Hausner, Prentice Hall, Englewood Cliffs, N.J., 1963.
4. *Simulation with GASP II: A FORTRAN Based Simulation Language*, A. A. B. Pritsker and P. J. Kiviat, Prentice-Hall, Englewood Cliffs, N.J., 1969.
5. HAWSIM Simulation of Social Services, B. A. Brenner and W. F. Eicker, *3rd Conf. Appl. Simul.*, Los Angeles, 396–401, 1969.
6. On A Simulation Language Completely Defined on the Programming Language PL/I, L. Petrone, in *Simulation Programming Languages*, J. N. Buxton (ed.), North-Holland, Amsterdam, 1968.
7. SIMULA—An ALGOL-Based Simulation Language, O. J. Dahl and K. Nygaard, *Commun. Am. Assoc. Comput. Mach.*, **9**, 671–678 (1968).
8. The Elliot Simulation Package (ESP), J. W. J. Williams, *Comput. J.*, **6**, 4:328–331 (1964).
9. SOL—A Symbolic Language for General Purpose System Simulation, D. C. Knuth and J. L. McNeley, *IEEE Trans. Electron. Comput.*, **EC 13**, 4:401–414 (1964).
10. Simulation Languages, J. McNeley, *Simulation*, **9**, 2:95–98 (1967).
11. A Language for Modeling and Simulating Dynamic Systems, R. J. Parente and H. S. Krasnow, *Commun. Am. Assoc. Comput. Mach.*, **10**, 9:559–567 (1967).
12. A Comparison Between SIMULA and FORTRAN, J. Palme, *BIT* (*Nordisk Tipskrift for Informationbehandling*), **8**, 3:201–209 (1968).
13. On Simulating Networks of Parallel Process in Which Simultaneous Events May Occur, D. L. Parnas, *Commun. Am. Assoc. Comput. Mach.*, **12**, 9:519–531 (1969).
14. Simulation: Languages, K. D. Tocher, in *Progress in Operations Research*, Vol. III, J. S. Aronofsky (ed.), Wiley, New York, 1969.
15. GPSS/360-Norden: An Improved System Analysis Tool, W. A. Walde, D. Eig, and S. R. Hunter, *IEEE Trans. Sys. Sci. Cyber.*, **SSC-4**, 4:442–445 (1968).
16. Combined Discrete Event Systems Simulation, D. A. Faberland, *Simulation*, **14**, 2:61–72 (1970).

6

Several Looks At One Model—
How to Structure the Model
for Use

The drive-to-the-office example in Chapter 2 presented only one implementation to establish basic concepts. This chapter will take the same illustration and show how it can be modified to be more general, explore particular aspects of the problem, and provide greater utility to the system designer. At the same time GPSS will be more extensively used so that its capabilities for much more complex problems becomes obvious. Finally the purpose is to prepare for the illustrative examples of simulation applications.

The GPSS implementation of the drive-to-the-office problem in Chapter 2 was specific. Exactly one set of characteristics was represented. Flexibility to represent alternatives, however, is needed. Therefore the first major modification of the model is to develop a more generalized version. Initially the model was a one-for-one map of the route. Additional route segments require adding an extensive number of blocks. These blocks would resemble the set of blocks or subroutine representing each segment and intersection. Another set of blocks would be needed for each additional segment. Obviously as the overall route approaches the complexity of the real world, the number of blocks required becomes unwieldy. This presents a difficulty because the number of route segments may easily exceed fifty. Under the original approach a large number of simular block subroutines and their associated cards would have to be organized, punched, and kept track of. Eighteen blocks are required to define the route segment. This approach would eventually require 900 blocks, obviously, an awkward procedure. The computer should be used to reduce the effort and provide organization for the data describing each route.

Development of Generalized Model

Two aspects of GPSS help the system designer handle repetitive elements in the model in a generalized manner—indirect addressing and array structures. For example the time spent to traverse each route segment is represented by the same block type performing the same function:

```
ADVANCE V$LOCAL      Local segment
ADVANCE V$ORGRT      Original route segment
ADVANCE V$ALTRT      Alternate route segment
ADVANCE V$OFFCE      Office segment
```

The VARIABLE statements provide the unique characteristics for each route segment:

```
LOCAL VARIABLE      1*60*60/25*FN$DENS1*XB$ABNR1
ORGRT VARIABLE      4*69*60/35*FN$DENS2*FN$BNCH1*
                    XB$ABNR2/10
ALTRT VARIABLE      6*60*60/55*FN$DENS3*XB$ABNR3
OFFCE VARIABLE      1*60*60/15*XB$ABNR4
```

Each of the statements is of the form

DISTANCE/SPEED*TRAFFIC DENSITY FACTOR*BUNCHING FACTOR*
ABNORMAL CONDITIONS FACTOR,

with some of the factors not present in each statement. If a single subroutine were generalized to represent any route segment, it would have to include data representing any segment. While there may not be full data for each route segment, the subroutine must be able to handle any segment. This could be implemented through a new VARIABLE statement of the form

$(DISTANCE)_n/(SPEED)_n*(TRAFFIC\ DENSITY)_n*(BUNCHING)_n*$
$(ABNORMAL\ CONDITIONS)_n$

Where input data exist to describe each of the n routes, the specific n route would be obtained through indirect addressing indicated by n. The convenient method to introduce these data into the model is by using an array. In GPSS this array is a matrix savevalue. A single MATRIX called ROUTE could be structured to provide some of the data needed for the VARIABLE statement as well as related data describing each route segment and intersection.

MATRIX ROUTE

Column	1	2	3	4 Time Traffic Light Green	5 Time Traffic Light Red	6 Road Conditions
Row	Distance	Speed	Bunching			
1 LOCAL	1	25	9	60	60	1
2 ORGRT	4	35	5	45	60	1
3 ALTRT	6	55	9	60	45	1
4 OFFCE	1	15	9	—	—	1

To use this array it is necessary to address each item by referencing its row and column. MB\$ROUTE (2,1) contains the value four which represents the original distance. It is possible to retain the previous form of route segment VARIABLE statements and insert the data from the MATRIX. For example the VARIABLE statement for the original route segment becomes:

```
ORGRT VARIABLE MB$ROUTE(2,1)*6*60*60/MB$ROUTE(2,2)*
FN$DENS2*FN$BNCHG*MB$ROUTE(2,6)/10
```

where MB\$ROUTE(2,3) is the location in the array defined by row 2 and column 3 and contains the number of the bunching function, 5.

Now if the two specific terms FN\$DENS2 and FN\$BNCHG for class of traffic density, and vulnerability to bunching effects, were generalized, one expression could apply to any route segment. To generalize the entire expression its current route segment, as represented by the vehicle transaction, must specify current location. Byte parameter 4 updated by ASSIGN blocks would be used to insert the numerical equivalent of LOCAL, ORGRT, ALTRT, or OFFCE—1, 2, 3, or 4—in the vehicle transaction. Substituting the value for PB4 and generalizing the VARIABLE now assumes the form:

```
ROUTE VARIABLE   MB$ROUTE(PB4,1)*60*60/MB$ROUTE(PB4,2)*
                 FN*PB4*FN$BNCHG*MB$ROUTE(PB4,6)/10,
```

where FN*PB4 refers to the appropriate one of four traffic slowdown functions: LOCAL, ORGRT, ALTRT, and OFFCE.

The generalized expression also requires the FN\$BNCHG statement to be modified according to the characteristics of the particular route segment. This may be accomplished by assigning an additional byte parameter 5, the value 5, the number of the BNCHG FUNCTION— which has the original data—or the value of the number of a new function BNCHO, which always has the value of unity. The latter effectively removes the bunching factor for those segments which do not require it. As a result, the ROUTE VARIABLE becomes:

```
ROUTE VARIABLE   MB$ROUTE(PB4,1)*60*60/MB$ROUTE(PB4,2)*
                 FN*PB5*V$EXTRA
EXTRA VARIABLE   FN*PB4*MB$ROUTE(PB4,6)
```

Two variable statements were required since one statement is limited to 72 characters and the full statement exceeds that limit. The final result is a generalized expression which could refer to any route segment. Moreover the route segment data are separated from the model structure. Now it is possible to assemble a model, change the matrix savevalue data, and execute a series of different cases. The data required are introduced by a series of INITIAL statements.

```
INITIAL    MB$ROUTE(LOCAL,1),1
```

This statement places in Column 1 of the LOCAL row the distance to be covered, 1 mile. If the model were extended to cover additional route segments, these could be identified by name and inserted as additional items into the ROUTE array. For convenience all data describing a segment are organized in one row of the MATRIX SAVEVALUE. This includes the number of seconds the traffic signals are green or red. In this case the data are not used by the transactions representing control of the various traffic signals. A single organized data structure has been made to cover a variety of needs throughout the model. As a result it is quite easy to follow the model, learn what the data represent, where they are used, and, as a nontrivial benefit, enable more members of the system design team to follow and modify the simulation.

In addition to organizing the data structure into a generalized form it is also necessary to develop a similar generalized approach to the block subroutines. It is not the goal to represent all conditions through one set of blocks. Rather those sequences of blocks which perform similar functions can be converted into a single sequence in the model. Then, by adding to the transaction data the identification of the current route segment, the transaction may address the different elements of the subroutine and specify which segment is being selected. This method of indirect addressing is implemented by inserting in different parts of the model ASSIGN blocks which place into byte parameter 4 the value associated with the symbolic name for that route segment, 1, 2, 3, and 4, respectively.

```
ASSIGN      4,Local,PB
ASSIGN      4,ORGRT,PB
ASSIGN      4,ALTRT,PB
ASSIGN      4,OFFCE,PB
```

During the movement of our vehicle through the model its transaction will begin with byte parameter 4 having a value 1, LOCAL, then continue on to have either 2 or 3, ORGRT or ALTRT, and finally 4, OFFCE. At each stage of its passage through the model it will go through the same subroutines. On each passage, however, that same subroutine will represent a different route segment.

```
AAA    ENTER       PB4
       ASSIGN      5,MB$ROUTE(PB4,3)
       ADVANCE     V$ROUTE
       LEAVE       PB4
       QUEUE       PB4
       ASSIGN      2,Q*PB4,PB
       SEIZE       PB4
       TEST NE     PB4,2,ABB
       GATE LS     PB4
ABC    ADVANCE     V$INTSN
       RELEASE     PB4
       ASSIGN      2+,Q*PB4
       DEPART      PB4
       TEST E      PB1,10,ABA
       TEST E      PB4,1,BAB
       TEST E      BV$COND,0,CAA
       TRANSFER    ,BAA
ABB    GATE LR     PB4
       TRANSFER    ,ABC
```

Block ENTER PB4 will place the transaction in a storage depending on the value of byte parameter 4 which can represent any one of the four possible segments. The remaining blocks in the sequence are all similarly addressed by the same value in byte parameter 4. The TEST E PB4,1,BAB is used to separate the transaction after it has gone over either the original or alternate route and now must continue on to the final segment. Its unique data are introduced at BAB ASSIGN 4,OFFCE.

The one other expression, to generalize, governs the choice of route, COND BVARIABLE. There are two approaches to make this general.

- Establish a series of predetermined most useful routes.
- Provide within the model the rules for route selection at each intersection.

Either method has advantages and disadvantages. The choice depends on the problem. The establishment of predetermined routes through the model requires advance planning and runs the risk that potentially good, but not obvious, routes may not be considered. The inclusion of the selection rules within the model extends the amount of detail and complexity required to construct the model. Since the latter approach was used before it is retained.

```
COND BVARIABLE  (Q*PB4'6')+(PB2'6'4)+(S*PB4'6'10)+
                (MB$ROUTE (PP4,6)'6'3)
```

The general form of the model is more complex in one major area. Previously the number of vehicles in the route segment were readily determined from the number of transactions currently in each ADVANCE block, W$AAA. This can no longer be used. Now one ADVANCE block contains transactions representing all route segments. This problem may be overcome by any of the following approaches, each based on the specific route being identified by the value of byte parameter 4.

- Associate each transaction as it entered the model with a set representing that segment. This requires using the JOIN*PB4 and REMOVE*PB4 blocks to have the transaction join a group and then be removed from it before joining another group.
- Establish a storage for each route segment. Then the transaction will ENTER*PB4 and LEAVE*BP4, the storage. The current contents of the storage provides the indication of the number of vehicles on the route.
- Provide in addition to the queues for intersections a set of route segment queues for those vehicles currently in each segment through additional QUEUE and DEPART blocks. The number of vehicles would be the number in the queue at that time.
- Utilize halfword savevalues to keep a record of the number of active vehicles on a segment through incrementing SAVEVALUE *PB4,1,H by one when entering, and decrementing SAVEVALUE *PB4,1,H again by one when leaving the segment.

```
*LOC    OPERATION  A,B,C,D,E,F,G                    COMMENTS


        SIMULATE                                              1/30/71
*                       SYMBOLIC EQUIVALENTS
*
 LOCAL  EQU        1,S,F,Q,L,Z      STORAGE,FACILITY,QUEUE,LOGIC,FUNCTION
 ORGRT  EQU        2,S,F,Q,L,Z,T    ALL ABOVE AND TABLE
 ALTRT  EQU        3,S,F,Q,L,Z,T
 OFFCE  EQU        4,Z              FUNCTION
 ROUTE  EQU        1,V,Y,T          VARIABLE,HALFWORD MATRIX,TABLE
 BNCHG  EQU        5,V,Z            VARIABLE, FUNCTION
 TRLCL  EQU        6,Z,T            FUNCTION, TABLE
 TRORT  EQU        7,Z,T            FUNCTION, TABLE
 TRALT  EQU        8,Z,T            FUNCTION, TABLE
 BNCHO  EQU        9,Z              FUNCTION
*
*                          VARIABLES
*
* VARIABLE STATEMENTS ARE USED TO REPRESENT ALL NUMERICAL CONSTANTS AND
* EQUATIONS IN THE MODEL
*
 DAY    VARIABLE   60*60*24     NUMBER OF SECONDS IN A DAY
*
 FIRST  VARIABLE   7*60*60      EARLIEST POSSIBLE DEPARTURE TIME IS
* SEVEN A M. THIS STATEMENT CONVERTS SEVEN A M TO SECONDS
*
 GOES   VARIABLE   V$FIRST+30*60      SEVEN THIRTY A M DEPARTURE TIME IS
*                                     VARIABLE FIRST  PLUS THIRTY MINUTES
*
 ROUTE  VARIABLE   (MB$ROUTE(PB4,1))*60*60/(MB$ROUTE(PB4,2))*V$EXTRA/10
 EXTRA  VARIABLE   FN*PB4*(MB$ROUTE(PB4,6))*FN*PB4  THE TIME REQUIRED TO
*                                     TRAVERSE ANY ROUTE SEGMENT
*
 INTSN  VARIABLE   1/Q*PB4*2+PB2/2*4-(PB2-2)/2*4     TWO SECONDS FOR NO
*                              TRAFFIC AND FOUR SECONDS FOR TRAFFIC
*
 TIME   VARIABLE   C1@86400/60/60*100+C1@3600/60     INTERNAL  CLOCK TO
*                              DAILY TIME IN HOURS AND MINUTES
*
 TIMET  VARIABLE   C1@86400/60/60*100+C1@3600/60+BV3*40 INTERNAL CLOCK
*                              TO DAILY TIME IN HOURS AND MINUTES
 MINIT  VARIABLE   C1@3600/60
 TRIPT  VARIABLE   M1/60/60*100+M1@3600/60+1    TRIP TIME-HOURS &MINUTES
*
 COND   BVARIABLE  (Q*PB4'G'1)+(PB2'G'5)+(S*PB4'G'10)+BV$COND1
 COND1  BVARIABLE  MB$ROUTE(PB4,6)'G'3
*      OTHERS ARE AT THE INTERSECTION WITH OUR CAR
*      THE QUEUE AT THE INTERSECTION CONTAINED MORE THAN FOUR CARS
*      TOTAL OF TEN CARS ARE ON THE LOCAL STREET AT THE SAME TIME
*      ABNORMAL CONDITON FACTOR EXCEEDS THREE
*          ANY ONE OF THE ABOVE CONDITIONS SELECTS ALTERNATIVE ROUTE
*
 2      BVARIABLE  N$FAB'L'3
 SIGNL  BVARIABLE  Q$SIGNL'L'3
 3      BVARIABLE  V$MINIT'G'55
```

- Add an additional column to the route matrix if only values up to 127 vehicles per segment are expected MSAVEVALUE MB*ROUTE(PB4,7)+,1,XB or provide a new halfword matrix MSAVEVALUE MH$ROUTE(PB4,1)+,1,XH which are incremented and decremented as was the savevalue.

These methods are approximately equal in usefulness for this application. The advantage to using either the *storage* or *queue* approach lies also in obtaining their normal output statistics—maximum number at any one time and deviation of average stay in the route segment—that would not be obtained with the other approaches. The matrix savevalue sets up the data in an ordered manner, but there are no results available at the end of the run—only intermediate results.

The traffic signal control has also been converted into a single sequence of blocks for all intersections. This sequence obtains a control transaction from a SPLIT BV$SIGNL,DAA,,4PB block in the computing traffic sequence. Since the competing traffic is delayed before entering the model, the BVARIABLE SIGNL has the value of one when there are no control transactions; traffic signals are operational when needed.

```
*LOC    OPERATION   A,B,C,D,E,F,G                 COMMENTS


 *                          FUNCTIONS
 *
 * FACTORS IN THE SIMULATION WHICH CANNOT BE HANDLED WITH EQUATIONS ARE
 * PLOTTED AND HANDLED USING GRAPHICAL RELATIONSHIPS - FUNCTIONS
 *
  TRLCL FUNCTION   V$TIME,C7      DENSITY OF TRAFFIC IN MORNING
 0,0/645,0/700,2/715,10,/800,10/830,4/930,0
 * TIME IN MORNING                         NUMBER OF CARS IN 3 MINUTES
 *
  TRORT FUNCTION   V$TIME,C7      DENSITY OF TRAFFIC CN ORIGINAL ROUTE
 0,0/645,0/700,3/715,15,/800,10/830,6/930,0
 * TIME IN MORNING                         NUMBER OF CARS IN 3 MINUTES
 *
  TRALT FUNCTION   V$TIME,C7      DENSITY OF TRAFFIC ON ALTERNATIVE ROUTE
 0,0/645,0/700,4/715,20,/800,20/830,12/930,0
 * TIME IN MORNING                         NUMBER OF CARS IN 3 MINUTES
 *
  LOCAL FUNCTION   S*PB4,C4       SPEED CN LOCAL SEGMENT
 0,1/25,1/250,10/5000,10
 * NUMBER OF CARS ON STREET                TRAFFIC SLOWDOWN FACTOR
 *
  ORGRT FUNCTION   S*PB4,C4       SPEED ON ORIGINAL ROUTE
 0,1/50,1/500,15/50000,15
 *NUMBER OF CARS ON ROUTE                  TRAFFIC SLOWDOWN FACTOR
 *
  ALTRT FUNCTION   S*PB4,C5       SPEED CN ALTERNATIVE ROUTE
 0,1/100,1/2000,3/6000,4/20000,30
 * NUMBER OF CARS ON ROUTE                 TRAFFIC SLOWDOWN FACTOR
 *
  OFFCE FUNCTION   PB4,C2         SPEED CN OFFICE SEGMENT
 0,1/4,1
 * NUMBER OF CARS ON ROUTE                 TRAFFIC SLOWDOWN FACTOR
 *
  BNCHG FUNCTION   PB2,C4         SLOWDOWN CAUSED BY INTERSECTION TRAFFIC
 0,10/2,10/10,20/100,20
 *         INTERSECTION QUEUES             BUNCHING FACTOR
 *
  BNCHO FUNCTION   PB4,C2         IGNORE BUNCHING
 0,10/4,10
```

```
*LOC    OPERATION  A,B,C,D,E,F,G                    COMMENTS

*  INITIAL INPUT DATA - BYTE SAVEVALUE ROUTE MATRIX
*          COLUMN  1    DISTANCE
*          COLUMN  2    AVERAGE SPEED
*          COLUMN  3    BUNCHING FACTOR
*          COLUMN  4    DURATION LIGHTS GREEN
*          COLUMN  5    DURATION LIGHTS RED
*          COLUMN  6    ABNORMAL ROAD CONDITIONS
*          ROW     1    LOCAL SEGMENT
*          ROW     2    ORIGINAL ROUTE SEGMENT
*          ROW     3    ALTERNATIVE ROUTE SEGMENT
*          ROW     4    OFFICE SEGMENT
*
   ROUTE MATRIX       MB,4,6
           INITIAL    MB$ROUTE(1,1),1/MB$ROUTE(2,1),4
           INITIAL    MB$ROUTE(3,1),6/MB$ROUTE(4,1),1
           INITIAL    MB$ROUTE(1,2),25/MB$ROUTE(2,2),35
           INITIAL    MB$ROUTE(3,2),55/MB$ROUTE(4,2),15
           INITIAL    MB$ROUTE(LOCAL,3),BNCHO
           INITIAL    MB$ROUTE(ORGRT,3),BNCHG
           INITIAL    MB$ROUTE(ALTRT,3),BNCHO
           INITIAL    MB$ROUTE(OFFCE,3),BNCHO
           INITIAL    MB$ROUTE(1,4),60
           INITIAL    MB$ROUTE(2-3,4),45
           INITIAL    MB$ROUTE(1-3,5),60
           INITIAL    MB$ROUTE(1-4,6),1
*
* PAGE A DISTANCE TO FIRST INTERSECTION AND PROCEDURE TO GET THROUGH IT
*
           GENERATE   V$DAY,,V$GOES,100,,6PB        START THE TRANSACTION
           ASSIGN     1,10                 LABEL OUR CAR WITH A TEN
           ASSIGN     4,LOCAL              LOCAL SEGMENT
           QUEUE      ACTVE                KEEP A COUNT OF ALL VEHICLES IN MODEL
  AAA      ENTER      PB4                  HOW MANY ON EACH SEGMENT
           ASSIGN     5,MB$ROUTE(PB4,3)    OBTAIN NUMBER OF BUNCHING FUNCTION
           ADVANCE    V$ROUTE              TIME SPENT ON ROUTE SEGMENT
           LEAVE      PB4                  NO LONGER IN SEGMENT
           QUEUE      PB4                  JOIN QUEUE AT INTERSECTION
           ASSIGN     2,Q*PB4              RETAIN SIZE OF QUEUE AT INTERSECTION
           SEIZE      PB4                  GET INTO POSITION TO PASS INTERSECTION
           TEST NE    PB4,2,ABB            CHECK PHASE OF TRAFFIC LIGHT
           GATE LR    PB4                  SIGNAL LIGHT CONDITION
  ABC      ADVANCE    V$INTSN              MOVE THROUGH FIRST CAR POSITION
           RELEASE    PB4                  LET NEXT CAR INTO POSITION
           ASSIGN     2+,Q*PB4             NOTE IF CAR LEFT INTERSECTION ALONE
           DEPART     PB4                  LEAVE INTERSECTION
           TEST E     PB1,10,ABA           ONLY OUR CAR REMAINS IN MODEL
           TEST E     PB4,1,BAB            READY FOR CHOICE OF ROUTE
           TEST E     BV$COND,0,CAA        CHECK CONDITIONS TO DETERMINE ROUTE
           TRANSFER   ,BAA                 CONTINUE ON NEXT SEGMENT
  ABB      GATE LS    PB4                  OPPOSITE PHASE LIGHT
           TRANSFER   ,ABC                 CONTINUE ON TO INTERSECTION
  ABA      DEPART     ACTVE                NO LONGER OF CONCERN IN MODEL
  ABZ      TERMINATE
```

Once there are control transactions in the subroutine, it has the value zero, indicating no more copies of the original transaction are needed. The ADVANCE blocks, representing the duration of green and red light phases, obtain their data from Columns 4 and 5, respectively, of the ROUTE MATRIX. When these changes are implemented the model still has the identical statistics as in the original version since the same sequence of pseudorandom numbers has been used.

The version of the model using indirect addressing and data arrays is usually preferred over the original version, because there are fewer blocks, a debugged subroutine is used over and over, larger systems can more easily be represented, and the descriptive data are more readily and simply structured. But most significantly there is a separation between the logic represented by the blocks and the data represented by items in matrices, functions, and variable statements. Now the same logic can easily be used to compare sets of alternatives. The disadvantage lies in slightly longer running times. The listing of the generalized version starts on page 142 and should be compared with the original version starting on page 67.

```
*LOC    OPERATION  A,B,C,D,E,F,G                COMMENTS

 *      START THE COMPETING TRAFFIC ON LOCAL STREETS
 *
        GENERATE   V$DAY,,V$FIRST,100,,6PB    OTHER TRAFFIC EVERY DAY
        LOGICS     HALT              ELIMINATE RUNNING OF MODEL ONCE ARRIVED
        SPLIT      2,AABA,4,4PB      TRAFFIC FOR ALL SEGMENTS
 AABA   SPLIT      BV$SIGNL,DAA,,4PB  CONTROL TRANSACTION FOR EACH LIGHT
        INDEX      4,5               ADD FIVE TO PARAMETER FOUR-PLACE IN ONE
        SPLIT      BV2,FAA,,6PB      FIRST TIME THROUGH GET COPIES
 AADA   SPLIT      FN*PB1,AACA,,6PB  NUMBER OF CARS DURING EACH THREE
 *                                   MINUTE INTERVAL
        ADVANCE    180               WAIT TILL NEXT INTERVAL
        TEST E     FN*PB1,0,AADA     NO MORE TRAFFIC  - WAIT FOR TOMARROW
        TERMINATE
 AACA   ADVANCE    180,180           RANDOMIZE THE DISTRIBUTION OF CARS
        GATE LS    HALT,ABZ          SEND TRANSACTIONS TO TERMINATE
        QUEUE      ACTVE             NOTE NUMBER OF VEHICLES IN MODEL
        TRANSFER   ,AAA              SEND THE COMPETING TRAFFIC TO MEET
 *                                   OUR CAR ON LOCAL STREETS
 *
 *
 * PAGE B PORTION OF DRIVE OVER ORIGINAL ROUTE SEGMENT
 *
 BAA    ASSIGN     4,ORGRT           WILL USE THE ORIGINAL ROUTE
        ASSIGN     3,2               NOTE THAT ORIGINAL ROUTE WAS USED
        TRANSFER   ,AAA              REJOIN MODEL MAINSTREAM
 BAB    ASSIGN     4,OFFCE           OUR CAR HAS REACHED THE OFFICE SEGMENT
        DEPART     ACTVE
        LOGICR     HALT              OUR CAR FINISHED HALT ACTIVITY
        ADVANCE    V$ROUTE           TIME FOR LAST SEGMENT
        TABULATE   ROUTE             RETAIN DATA OF EACH TRIP DURATION
        TABULATE   PB3               TRIP DURATION DATA BY ROUTE USED
        TERMINATE  1                 ONE MORE DAY
 *
 * PAGE C PORTION OF DRIVE OVER ALTERNATIVE ROUTE SEGMENT
 *
 CAA    ASSIGN     4,ALTRT           WILL USE ALTERNATIVE ROUTE
        ASSIGN     3,3               NOTE THAT ALTERNATIVE ROUTE WAS USED
        TRANSFER   ,AAA              REJOIN MODEL MAINSTREAM
```

```
*LOC    OPERATION  A,B,C,D,E,F,G                 COMMENTS

* PAGE D SIGNAL LIGHT CONTROL
 DAA    QUEUE      SIGNL
 DAB    LOGICS     PB4               SIGNAL LIGHT GREEN
        ADVANCE    MB$ROUTE(PB4,4)      TIME LIGHT IS GREEN
        LOGICR     PB4            SIGNAL LIGHT RED
        ADVANCE    MB$ROUTE(PB4,5)      TIME LIGHT IS RED
        GATE LR    HALT,DAB
        TEST E     Q$ACTVE,0,DAB     EVERYBODY HAS FINISHED
        DEPART     SIGNL
*
        TERMINATE
 FAA    SPLIT      FN*PB1,FACA,,6PB VEHICLES IN THREE MINUTES
 FAB    ADVANCE    180               WAIT TILL NEXT INTERVAL
        TEST E     FN*PB1,0,FAA   NO MORE TRAFFIC - QUIT
        TERMINATE
 FACA   TABULATE   PB1               GET PICTURE CF TRAFFIC
        TERMINATE
*
*
*       MANIPULATIONS NEEDED TO MOVE MATRIX VALUES INTO REPORT GENERATOR
*
        GENERATE   ,,,1,,4PB         ONE TRANSACTION
        ASSIGN     3,24              NUMBER CF HALFWORD SAVEVALUES
        ASSIGN     1,6               NUMBER OF CCLUMNS
 EAA    ASSIGN     2,4               NUMBER OF ROWS
 EAB    SAVEVALUE  PB3,MB$ROUTE(PB2,PB1),H    MOVE FROM MATRIX
        ASSIGN     3-,1
        LOOP       2,EAB             FILL ONE CCLUMN
        LOOP       1PB,EAA           FILL NEXT COLUMN
        TERMINATE
*
*       THE FOLLOWING CARD SET THE ROUTE CAPACITY TC 200 VEHICLES
*       MAXIMUM FOR EACH ROUTE
        STORAGE    S1-S4,200
*
*       THE FOLLOWING STATEMENTS ARE USED TO DEFINE THE STATISTICS
*       GATHERING TABLES THAT HAVE BEEN ADDED TO THE STANDARD STATISTICAL
*       OUTPUT
*
 ROUTE  TABLE      V$TRIPT,10,1,35   DURATION OF TRIPS
 ORGRT  TABLE      V$TRIPT,10,1,35   DURATION OF TRIPS-ORIGINAL ROUTE
 ALTRT  TABLE      V$TRIPT,10,1,35   DURATION CF TRIPS-ALTERNATIVE ROUTE
 TRLCL  TABLE      V$TIMET,700,5,50
 TRORT  TABLE      V$TIMET,700,5,50
 TRALT  TABLE      V$TIMET,700,5,50
*
        START      100
```

Improving Running Time

Once a model is operating and production runs are required, the question is raised regarding how long it takes to execute the simulation. Usually the model production runs are not extensive enough to warrant major rework of the model, but frequently there are some modifications that do offer marked improvement for little effort.

For example in order to preserve the scenario of successive sequences of vehicles, a fixed number of vehicles are generated each day and then subjected to some randomizing effects before entering the mainstream of the model. Once the trip is completed for the day the remainder of the activity is superfluous. Thus either the generation of all competing traffic can be halted or, if the scenario is to be preserved, the competing transactions traffic can be removed immediately after obtaining random numbers. Three blocks are needed to implement the second course. Immediately after the generation of competing traffic a LOGICS HALT block is used to permit the daily flow of traffic. A GATE LR HALT,AAA sends the transactions to the route segment when they leave the ADVANCE block. This continues until the trip is completed, which is when the car reaches the last or OFFCE segment of the trip and a LOGICR HALT block. The use of these three blocks eliminates almost half the competing vehicle transactions from the model after their random number has been assigned. This simple change obviously has a significant effect on the running time of the model.

A similar approach can be used to halt the operation of the traffic signals. This case is implemented by introducing an overall queue of all vehicles in the system, QUEUE ACTVE. The queue is entered wherever vehicles enter the model, each vehicle entering only once. The traffic signals are continued until there are no more vehicles in the queue. A further refinement is to continue the traffic signals with a check against the fact that the trip has not been completed, GATE LR HALT,DAA. Only when the trip is completed and some traffic is still active is the more complex TEST E Q$ACTVE,O,DAA block used to finally halt the traffic signal activity. This modification also requires the SPLIT block to be converted to produce control transactions every day, SPLIT 1,DAA,,4PB. In this case also only slight effort is required to reduce the unneeded activity within the model and still preserve the needed controlled scenario.

Obtaining Better Insight

The use of simulation depends on the ease and speed with which modifications can be made to the structure of the model to bring out additional data. Originally the requirement was to determine how long the drive would take under a specific set of rules. That result has been obtained. The structure of the model allows ready change to vary the parameters of the problem. For example to change the distance and average speed of a route segment merely requires the following two INITIAL statements:

```
INITIAL MB$ROUTE(LOCAL,1),2
INITIAL MB$ROUTE(LOCAL,2),20
```

This changed the local segment to two miles and the average speed to 20 mph.

Controlling Choices '

Sometimes it is desirable to remove some of the elements of choice from the model. One obvious restriction would be for our vehicle to select the original or alternate route for all 100 trips. This requires changing only the COND BVARIABLE statement to either

```
COND BVARIABLE    0    or
COND BVARIABLE    1
```

The first case continues along the original route; the second only selects the alternative.

Historical Records

Frequently there is a requirement to understand more clearly what caused the results. Instead of needing the mean value for the trip duration, sometimes greater insight is obtained from the detailed record of the slowest trip. Out of the 100 trips, which were the five longest? To find these data requires slight model modification. Just before the transaction leaves the model, several blocks provide the file processing capability. A SPLIT 1,BHA,4PB,2PH block supplies a second transaction with additional halfword parameters. The copy transaction goes to an ASSIGN 1,M1,PH block, which inserts in halfword parameter 1 the number of clock units it has taken this transaction to reach the end of the trip. Then the transaction is inserted in the historical file according to the duration of the trip—shortest trips first—at the LINK HISTY,PH1 block. The original transaction left the SPLIT block before the copy, but it was detained at the PRIORITY 0,BUFFER block until the higher priority copy transaction was linked to the file. The first item to be tested is whether or not there are five transactions on file: TEST G CH$HISTY,5,BAZ. The first six transactions are filed, then there is an evaluation of which transaction should stay on the file through the UNLINK HISTY,BAZ,1 block. The UNLINK block removes the first transaction (the shortest trip duration) from the file and TERMINATES it at BAZ. At the end of the simulation the

HISTY chain will contain the five longest trip durations with their associated data in the various parameters: trip duration, route used, and queuing conditions encountered at the intersection where the choice was made between the original and alternate routes. When the trip took place is indicated from the time the transaction passed through its last ADVANCE block. If instead it were desired to analyze the five shortest trips, the only change needed would be to change the UN-LINK block to UNLINK HISTY,BAZ,1,BACK. The BACK command would remove from the list those transactions with the longest trip duration.

The list processing capability of the LINK/UNLINK combination could be extended to select only trips when weather conditions were poor or any of the characteristics of particular trips, provided these data are stored in the transaction through the use of additional parameters. Note the ease with which the list processing may be added to an existing model.

Extension of the Model

The model is designed to indicate external influences on the duration of the trip. These influences could be due to abnormal traffic, weather conditions, traffic accidents, and road construction. The first of these, abnormal amounts of competing traffic, would be handled differently from the other factors. There are a variety of different modifications possible. The following are used to stimulate consideration of different ways to change the model.

The introduction of abnormal traffic is based on modification of the SPLIT FN*PB1,AACA block. The GPSS assembler has numbered the traffic generating functions TRLCL, TRORT, and TRALT 6,7, and 8, respectively. Now instead of addressing the generating functions directly an intermediate series of functions could be inserted to select either the normal or abnormal amounts of traffic. Specifically the changes introduce different functions as 6,7,8. A key, the value of the contents of byte savevalue location 6,7, or 8, would be used to address with 1 the normal TRLCL, TRORT, and TRALT functions or, with 2, a new set of abnormal functions: TRLCA, TRORA, and TRALA. The function would appear as follows (Figure 6.1):

The remaining requirement is an initial card to set the byte savevalues to 1. A way is needed to change one or more byte savevalues to 2 to institute abnormal traffic and back again to resume normal traffic.

Weather conditions affect all routes. This would be implemented by

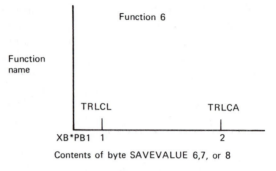

Figure 6.1

using a separate subroutine to modify the road condition factors, either according to random variables or according to historical data.

```
            GENERATE      V$DAY,,1,100,,4PH
            ASSIGN        4,4
    GAA     MSAVEVALUE    MB$ROUTE(PB4,6)+,FN$WTHR,B
            LOOP          4PB,GAA
            ASSIGN        4,4
    GAB     MSAVEVALUE    MB$ROUTE(PB4,6)-,FN$WTHR,B
            ADVANCE       V$DAY
            LOOP          4PB,GAB
            TERMINATE
```

Where the weather function represents the daily weather conditions, with a scale of values ranging from unity to 10. This allows the road condition factor to be the composite of weather and other factors. The LOOP block is used to include the weather factor in all route segments.

A change in road conditions due to road construction is a long-term situation, but not a permanent one. The subroutine to slow down traffic on the original route might be as follows:

```
            GENERATE      ,,V$NDAYS,1,,2PB
            ASSIGN        1,ORGRT
            MSAVEVALUE    MB$ROUTE(PB1,6)+,3,B
            ADVANCE       V$XDAYS
            MSAVEVALUE    MB$ROUTE(PB1,6)-,3,B
            TERMINATE
```

This applies where VARIABLES NDAYS and XDAYS control when the road construction starts and how long it lasts, respectively.

Traffic accidents could be treated in an analogous manner to the road construction. There, occurrence could be based on random selection along or tied together with current weather conditions. The traffic accident could also be made to modify the amount of traffic by diverting the competing traffic from one route to another.

At this point it is obvious that the model may be modified in one or numerous directions at the same time. The basic facts are that the effort to modify the model is small and changes can be implemented quickly unless new input data must be obtained. Before continuing to more useful examples, it is well to stop for a moment and consider how the results from this example can be presented.

Presentation of Results

GPSS has a standard output providing a count for each time a transaction enters a block, a set of statistics for each facility, storage, queue and chain, and the contents of savevalues at the end of the simulation. These are excellent outputs when debugging a model. They are full and complete summaries of the results. It is just that these are *not* the form of results to show anyone not intimately concerned with the model. Therefore additional effort is required to provide results in a more palatable form. The report generator is the tool for presentation of results. Instead of presenting the standard GPSS output a special report was set up to illustrate the presentation of both input data and output results.

The first item of output is not a result at all; rather, it is a summary of the input data.

INPUT DATA FOR EACH ROUTE SEGMENT
AND ITS INTERSECTION

Route Segment	Distance Miles	Average Speed	Bunching Factor	Seconds Green	Seconds Red	Road Conditions
LOCAL	1	25	9	60	60	1
ORIGINAL	4	35	5	45	60	1
ALTERNATIVE	6	55	9	60	45	1
OFFICE	1	15	9			1

The standard GPSS format for tabular presentation of results is illustrated below by the duration of all trips, TABLE ROUTE.

TABLE ROUTE
ENTRIES IN TABLE 100

MEAN ARGUMENT 14.669

STANDARD DEVIATION 1.125

SUM OF ARGUMENTS 1467.000

NON-WEIGHTED

UPPER LIMIT	OBSERVED FREQUENCY	PER CENT OF TOTAL	CUMULATIVE PERCENTAGE	CUMULATIVE REMAINDER	MULTIPLE OF MEAN	DEVIATION FROM MEAN
10	0	.00	.0	100.0	.681	-4.151
11	0	.00	.0	100.0	.749	-3.262
12	0	.00	.0	100.0	.817	-2.373
13	0	.00	.0	100.0	.886	-1.484
14	68	67.99	67.9	32.0	.954	-.595
15	14	13.99	81.9	18.0	1.022	.293
16	1	.99	82.9	17.0	1.090	1.182
17	17	16.99	100.0	.0	1.158	2.071

REMAINING FREQUENCIES ARE ALL ZERO

Figure 6.2 The prime result is the duration of the trip, which has been tabulated in the Route Table.

152

Since these data are usually better understood in graphical form GPSS also has the capability to present the results as a graph. This graph is produced on the high-speed printer which limits its resolution and continuity. To obtain the graph a separate set of statements following a REPORT control card are needed. These 11 statements identify the subject matter for the graph, locate its origin and set the X and Y scales. The five Statements provide the optional labeling and the last 2 items control the ending of the routine and setting up the next page.

```
    GRAPH        TF,ROUTE
    ORIGIN       48,12
    X            ,2,2,11,,14
    Y            0,5,9,5
22  STATEMENT    5,28,POPULARITY OF TRIP DURATIONS
1   STATEMENT    25,6 NUMBER
2   STATEMENT    28,2,OF
1   STATEMENT    31,5,TRIPS
22  STATEMENT    52,26,DRIVE DURATION IN MINUTES
    ENDGRAPH
    EJECT
```

Other results are tabulated in the form of a partial summary of results as shown in the following table. The coding to develop this presentation requires a statement for each line. As a result only one set of outputs are available from the GPSS report generator. Where this is an unsatisfactory situation HELP blocks have been used to provide many different output formats. A more fundamental approach was the substitution of more flexible report generators for the standard one in versions of GPSS.[5]

PARTIAL OUTPUT DATA FOR EACH ROUTE SEGMENT AND ITS INTERSECTION

Route Segment Intersection	Number of Vehicles Simulated	Seconds Vehicle in Segment	Maximum Vehicles Anytime	Seconds Intersection Delay	Maximum Intersection Queue
Local	313	144	15	23	11
Original	449	406	40	35	12
Alternative	624	389	51	40	13

One other item of input data which could be of interest is the level of competing traffic, shown in graph (Figure 6.4) for the local route segment.

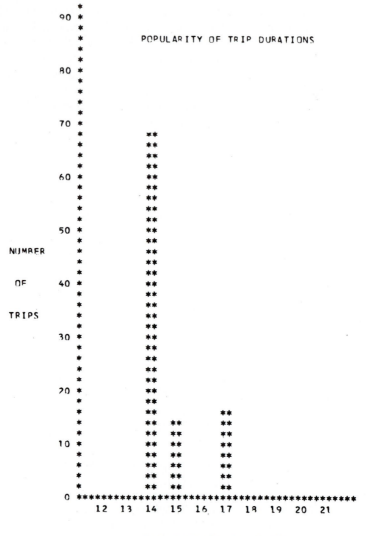

Figure 6.3 Popularity of Trip Durations

This graph, unlike the previous ones, required a model change to obtain its data. All the competing traffic transactions for the first day were directed to pass through an additional tabulate block.

The need to present lucid results has effects on the manner in which the model is constructed and organized and on some of its special sub-

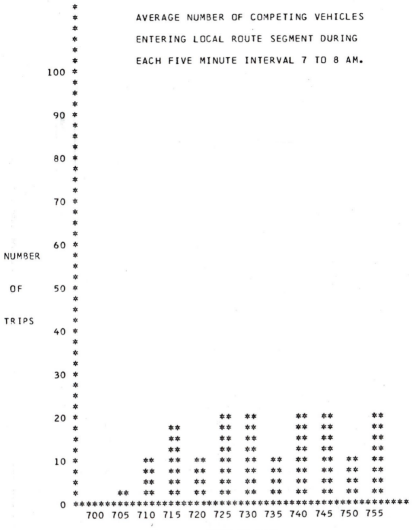

Figure 6.4 Average number of competing vehicles entering local route segment during each 5-minute interval 7 to 8 a.m.

routines. The level of effort required by the output is small but not inconsequential—and very necessary. The serious effort must be directed toward anticipating the use of the model. A model of very limited generality may be quickly implemented but it is also likely that it will become obsolete quickly. Good modeling practice suggests the need to anticipate, be general, choose flexible approaches, iterate toward better problem definition, and provide meaningful results in a clear format.

PROBLEMS

1. What is the extent to which generality is useful in setting up a problem? Discuss the advantages and disadvantages of providing for:
   (a) Any street pattern.
   (b) Radically different patterns of traffic flow.
   (c) Control of alternate route selection.

2. There is a choice of whether to predetermine the routing of each vehicle or to incorporate the rules for choice in the model. Is the method dependent on the number of combinations available to get from the origin to the destination? Are there clear benefits from one or the other approach?

3. The scenario for the introduction of competing traffic into the model is determined from the sequence of pseudorandom numbers. Consider and evaluate at least two other methods of performing this function.

4. A model of this type could be used to vary the timing sequence of the traffic signals. Indicate the preferred information feedback paths. What modifications to the model are required to incorporate these additional features?

5. For models with large numbers of transactions, running time can become a significant factor. How can the running time be reduced? What is lost by using these techniques?

6. Show how the model could be used to determine which route is least susceptible to disturbance from an accident or abnormal weather factors?

BIBLIOGRAPHY

1. A Simulation Study of a Multi-Channel Queueing System in the Hospital Environment, R. H. Holland, *Proc. 3rd Conf. Appl. Simul.*, 279–289, Los Angeles, 1969.

2. A General Simulation Model for Information Systems: A Report on a Modelling Concept, A. L. Buchanan and R. B. Waina, *ibid.*, 418–425.
3. A General Simulation Model of an On-Line Telephone Directory Assistance Information Retrieval System, W. A. Hall, *ibid.*, 434–441.
4. Simulation of an Aircraft Container System, D. M. Grant, *ibid.*, 470–479.
5. A Complete Interactive Simulation Envrionment GPSS/360-Norden, J. Reitman, D. Ingerman, J. Katzke, J. Shapiro, K. Simon, and B. Smith, *Proc. 4th Conf. Appl. Simul.*, 260–270, New York, 1970.

7

Another World View—Driving
Simulation Revisited

SIMSCRIPT II

SIMSCRIPT II is a discrete event simulation language that is a complete programming package. While aspects of the language may seem similar to FORTRAN, it has the simulation routines needed for compiling code for a broad variety of problem areas. SIMSCRIPT II is more readily understood by those who have a programming background than by those with an application orientation. SIMSCRIPT II is the second version of a SIMSCRIPT language.[1,2] The compiler does not depend on any other programming system. It is based on SIMSCRIPT II and is, therefore, very different from the original FORTRAN based SIMSCRIPT and its later modification SIMSCRIPT I.5. Its documentation has been designed for general readability to help bridge the gap between the programmer and analyst.

A language with the versatility and scope of SIMSCRIPT II can be looked at from many viewpoints—that of programmers, analysts, language developers, and so on. The language developers for instance have chosen to present the language in five somewhat arbitrary levels. A first level establishes introductory programming concepts. The next two provide a language of roughly FORTRAN capability along with a report generator. Level four provides for extensive list processing. Finally a simulation orientation is reached in level five. For simulation application it is this highest level which must be learned. The previous four levels are used to support the simulation. All aspects of the language must be learned to perform level five simulation. Thus the SIMSCRIPT II compiler, in its entirety, provides an equivalent of FORTRAN, some additional features of PL/I, a report generator, a library of widely used functions, and basic list-processing capabilities, queue structures, and timing orientation of a simulation language. While these elements are provided, there is no inherent internal structure. SIMSCRIPT II does provide time advance, event processing, statistical variates, and limited provision for accumulation of statistics. The programmer supplies frameworks both to represent the application and assist debugging. Convenience features are limited to free format for statements, 11 statistical variates, time—in day, hour, and minute—and a calendar covering the years 1900 to 1999. Outside of syntax errors the SIMSCRIPT II provisions for debugging are limited to event tracing capabilities. Beyond that debugging assists must be built with the formulation of the basic model structure. In the IBM System 360 implementation hexadecimal notation is used to locate routines and errors within routines.

The SIMSCRIPT approach is based on an entity-oriented data base which describes the current state of the system and an event-oriented data base which either now or in the future will disturb the system state. The static system definition is based on a structure of named entities having attributes which may belong to sets. A car which belongs to us may be in a queue. Dynamic system activity progresses on the basis of event routines that contain logic and change the state of the system. When the light changes the first car in the queue will be removed. SIMSCRIPT II permits the programmer to develop a tight internal data base through relating entities, with their attributes, together in sets.

Rather than treat SIMSCRIPT II as an abstract language the specific model we have been considering will again be simulated. This allows learning to use all levels of the language in an application derived order. Like any language it has specific rules which must be observed—and in proper sequence.

A complex problem cannot be programmed in a smooth, ordered sequence. First, there is the need to set up the broad model outline and then develop the parts. The same outline developed previously for the drive to the office will again be implemented, this time using SIMSCRIPT II instead of GPSS. While the model may be developed in any order, a SIMSCRIPT II program must follow a fixed sequence and unlike GPSS requires a full appreciation for the entire problem before beginning. The model is structured with the following gross parts:

1. *Preamble, which may include:*

- Permanent and temporary entity data bases described through a series of basic model declarations.
- Set structure, storage allocation, packing and equivalence.
- Event notices and event control data base with their priorities. Incidents which can change the system state.

- Analytic factors covering mathematical relationships and analysis procedures.
- Definitions which apply throughout the model.
- Additional items added to aid in debugging.

2. Main routine which has broad control over simulation execution:

- Load and initialize model with starting data.
- Provide for transfer of control to the timing routine after initialization.
- Capability to reset the computer system for the next model or the next run of this model.

3. Initialization routine to establish the particular model structure:

- Process input statements according to the organization of the preamble to bring data needed for the model structure.
- Establish initial conditions.
- Schedule initial events.

4. Routines controlling the simulation:

- Either internal or external events enter routines to execute that portion of the simulation.
- The routine, in turn, may schedule another event before returning to the event list.

5. Monitoring, debugging, and analysis aids:

- Structure for transfer of data between routines.
- Periodic event timing.
- Collect output statistics according to the structure provided.
- Perform mathematical analysis.
- Aids inserted for data gathering during the debugging.

The flow chart in Figure 7.1 shows the terminology and structure used for the SIMSCRIPT II implementation of the drive to the office. This brief description overview should be augmented by the detail available in the SIMSCRIPT II programming language description, reference manual, and description of the IBM 360 implementation.[3]

The SIMSCRIPT II syntax, while based on keywords, was developed to permit freedom when stating the model. There are numerous flexible conventions which allow the model to read like English since the keywords have synonyms which may be used interchangeably. These are used in sentences to automatically provide documentation, as well as

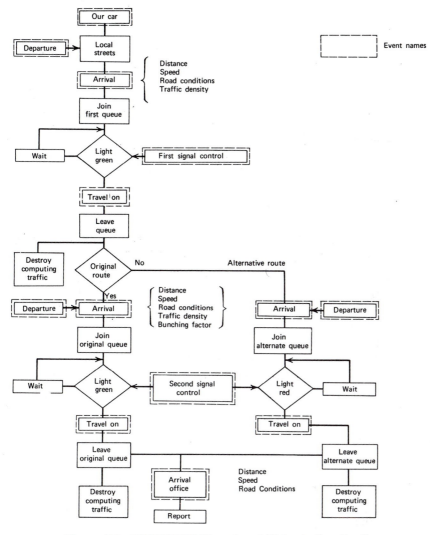

Figure 7.1 SIMSCRIPT II version of "drive to the office."

be relatively free of arbitrary form requirements. A statement can begin anywhere on the card and all 80 columns may be used. Blanks can be used to improve readability and periods may be liberally sprinkled. There is a restriction on a statement word, not the statement itself, to no more than 80 characters. However, internally only the first five to seven characters are used. The compilation order is the same as the physical sequence of statements in the deck.

```
· ''SIMULATION OF A DRIVE TO THE OFFICE
  ''A SIMSCRIPT II EXAMPLE
```

Comments follow the double apostrophe marks and continue either to the end of the card or to the next set of double apostrophe marks on the right.

The order to be followed in presenting the drive to the office conforms to the SIMSCRIPT II compiler requirement of starting with the preamble, then the main routine followed by the individual simulation subprograms, and finally the additional routines for report and special processing. These elements are processed by the compiler before input data are read in. Since the model requires extensive consideration of simultaneous events, the order of presentation is artificial. Starting with the preamble has the benefit of first establishing the ground rules.

Preamble

The declarative section of the model, the preamble, provides a mandatory heading for all aspects of the model. It represents the overall simulation structure and as such it is not the first item in model development. It is the first area which must be debugged, since the compiler halts at a faulty preamble. The preamble consists of the following: permanent and temporary entities, sets, event notices, event control, and variable, debugging, analysis, and miscellaneous declarations.

Before listing the mandatory items it is desirable to set the background condition for whether the variables are to be treated as real or integers. In this case integer.

```
NORMALLY, MODE IS INTEGER...
```

Permanent Entities

This statement declares that the attributes of a collection of entities are stored in arrays which require storage allocations. Permanent entities cannot be destroyed individually; all attribute arrays must be released at the same time. The permanent entities are intersections and roads.

These EVERY statements set up their attribute arrays which will be further described through DEFINE statements later in the preamble. Each entity of the type INTERSECTION has only two attributes, signal and capacity, and maintains a set called a QUEUE which stores cars

```
PERMANENT ENTITIES.....
    EVERY ONE.INTERSECTION HAS A F.SIGNAL,
    AND A F.CAPACITY, AND OWNS A QUEUE
    EVERY TWO.INTERSECTION HAS A S.SIGNAL, A S.CAPACITY,
    AND OWNS AN ORG.QUEUE, AND AN ALT.QUEUE
    EVERY ROAD HAS SOME CONDITIONS, A LOAD, A DISTANCE,
    A SPEED, A LOW.DENSITY IN ARRAY 1,
    A LOW.DENSITY.VALUE IN ARRAY 2, A HIGH.DENSITY IN
    ARRAY 3, A HIGH.DENSITY.VALUE IN ARRAY 4,
    A LOW.BUNCHING IN ARRAY 5, A LOW.BUNCHING:VALUE
    IN ARRAY 6, A HIGH.BUNCHING IN ARRAY 7, AND A
    HIGH.BUNCHING.VALUE IN ARRAY 8
```

before the INTERSECTION. Specifically ONE.INTERSECTION allows only cars on the local segment to form a queue. In the case of the second intersection there are queues depending on the approach, either through the original or alternate segments, forming ORG.QUEUE and ALT.QUEUE, respectively.

For the sake of readable English the attribute list is preceded by the indication of ownership, the word HAS, and the individual attributes by the articles A, AN, THE, or SOME. In addition the INTERSECTION is part of a set which is named QUEUE. This is established by the word OWNS. The entities ROAD—the route segment of the model—have attributes CONDITIONS, LOAD, DISTANCE, SPEED, and a number of DENSITY and BUNCHING functions. These later may be specifically defined to specific locations by the IN ARRAY n statement. The delimiters to set off a phrase or word are spaces. Therefore periods can be used to improve readability as in LOW.DENSITY. Values for the attributes may be data either read in from cards or generated within the model. The segment distance is a fixed input item while the number of cars on the road, load, will vary during the running of the model.

Temporary Entities—Program Elements with Named Attributes

In the case of vehicles moving through the model they need only temporary allocations of storage for their attributes data structures. The storage is returned to the available pool when the entity is DESTROYED.

```
TEMPORARY ENTITIES
    EVERY CAR HAS AN IDENTIFIER, SOME CARS.BEFORE,
    AN ENTRY.TIME, A ROUTE, AND MAY BELONG TO A QUEUE,
    AN ALT.QUEUE, AND AN ORG.QUEUE
```

As before, the structure of the entity-attribute-set relationship is established with the EVERY statement. The entity CAR has several named attributes: CARS.BEFORE, ENTRY.TIME, and ROUTE. These provide a tally of the number of cars waiting at the intersection, the time the car entered the model, and the car's current route, respectively.

The model is described by a structure based on its entities. These provide the cars, roads, and intersections. Their attributes fill out the model description and provide an organized structure. Set relationships enable selections to be made by an even broader generalized concept. Together these provide the static definition of the model. Now it is necessary to dynamically activate the static elements through a series of invitations to action, event notices. Later these event notices will be initialized and move through the model to set the various logical routines into action. The words MAY BELONG TO establish that a CAR may be part of the sets named QUEUE, ORG.QUEUE, and ALT.QUEUE.

Event Notices

```
EVENT NOTICES INCLUDE DRIVING.CONDITIONS, DEPARTURE,
OUR.CAR, ARRAY.VALUE, FIRST.SIGNAL.CONTROL, AND
SECOND.SIGNAL.CONTROL
```

The event notices will trigger a wide variety of actions. DRIVING.-CONDITIONS will operate only to vary road conditions. DEPARTURE will turn into a cascade of vehicles moving over all the routes. OUR.CAR will start us on the drive to the office at the local segment. ARRAY.VALUE will change the pointer used to establish the level of traffic according to the current time in the model, TIME.V. The signals operating in the intersections will be started by FIRST.SIGNAL.CONTROL and SECOND.SIGNAL.CONTROL. The EVENT NOTICE has five special attributes:

```
TIME.A--the simulated time when the event is to
occur.
EUNIT.A--the code indicating whether the event was
caused internally or externally.
P.EV., S.EV.S, and M.EV.S--set attributes for timing
routine to keep track of scheduled events.
```

```
The set used for timing has the name EV.S.
```

The EVENT NOTICE may also include information about or to be used by the event that is to be triggered. When this occurs an EVERY statement follows the EVENT NOTICE. In this case the ARRIVAL at the intersection and the TRAVEL.ON intersection require additional attributes.

```
EVERY ARRIVAL HAS AN AUTO IN WORD 6,
   A PLACE IN WORD 7,
EVERY TRAVEL.ON HAS AN AUTO IN WORD 6,
   A PLACE IN WORD 7,
```

Both the ARRIVAL and TRAVEL.ON event notices attributes are identical: AUTO and PLACE. These attributes will tie the event notice to a particular car at a particular location. They are additional elements in the array and called out as words six and seven. Note that CAR is an entity so that AUTO is an attribute of these event notices. The positive location of the data at a particular site avoids the possibility of ambiguity, which could happen if the compiler were free to establish locations.

During simulation, the order of precedence of event execution depends on the sequence of EVENT statements in the PREAMBLE. When another order of priority is desired, the statement PRIORITY ORDER IS enables the compiler to establish rules for events of different classes scheduled to occur at the same time. Events of the same class can be separated through the BREAK ARRIVAL TIES BY HIGH attribute value statement. Likewise, low value ties could be used to separate the events.

Additional Declarations

The preamble also contains instructions to the compiler which extend throughout the model as functions, sets, global variables, define to mean, and statistics-gathering statements. The function is used to determine one value from a set of data. An INTEGER function has only whole number values. The REAL function can have any numeric value within the limitations of the computer. It is used for precision or when fractional values are needed. Functions and variables are not limited to numeric values. There is also the ALPHA mode when strings of alphanumeric characters can be requested. These character strings may be stored and manipulated logically, as in the case of WITH VALUE NOT EQUAL "*END*" which could be used to locate the end of input data. The alpha expression, a literal, is enclosed in quotation marks.

Global VARIABLES are defined in the preamble and apply in every routine throughout the model. Sometimes it is not desirable for a variable to have the same value in all routines. For example, numerous local routines might use the same names for variables to simplify follow-

ing the model. If they are local variables, their values are different
in each case although the functions they perform are similar. A local
VARIABLE is defined in and pertains only to that routine.

A global VARIABLE will have a common meaning throughout the
model. In practice the same storage location is used. Therefore it must
appear in the preamble through a DEFINE statement. The distinction
between real and integer terms must be positively made if the default
to integer mode is to be avoided.

```
    DEFINE ENTRY.TIME AS A REAL VARIABLE
    DEFINE DENSITY AND BUNCH AS REAL FUNCTIONS
```

These statements set up DENSITY and BUNCH as functions to be
real valued for accuracy. These functions will interpolate traffic modification factors depending on the current route status. A convenience
in establishing a structure for a list of attributes is the SYSTEM
ATTRIBUTE.

```
NORMALLY,DIMENSION=0
THE SYSTEM HAS A PATH ''RETAINED MEMORY OF ROUTE USED
THE SYSTEM HAS A K ''INDEX TO OBTAIN COMPETING TRAFFIC
    VALUE
NORMALLY,DIMENSION=2
THE SYSTEM HAS A SIGNAL.DELAY
DEFINE TRAFFIC.VALUE AND SIGNAL.DELAY AS INTEGER,
    2-DIMENSIONAL VARIABLES
DEFINE CHANGE.SEQUENCE AS AN INTEGER,
    1-DIMENSIONAL ARRAY
```

The words VARIABLE and ARRAY may be used interchangeably
to label and structure data in lists and tables. The compiler is informed
of the array size and form through the dimension term. A one-dimen-
sional array is simply a list. The subscript for each element locates
that element in the list. Two-dimensional arrays require a pair of sub-
scripts to identify the element by row and column. The dimensionality
of the array must be stated in the preamble with a follow-up RESERVE
statement later in the deck to set up the actual storage allocation. For
single-point variables NORMALLY, DIMENSION = 0

The DEFINE TO MEAN statement instructs the compiler to substitute either a value or a string of items. In this case the term INITIAL
is translated into unity wherever it appears. One capability of the DE-
FINE TO MEAN statement allows the user to simplify his programming,
as in the statement: DEFINE BUMP TO MEAN ADD 1 TO. This

```
DEFINE INITIAL TO MEAN 1
DEFINE ORIGINAL TO MEAN 2
DEFINE ALTERNATE TO MEAN 3
DEFINE FINAL TO MEAN 4
DEFINE GREEN TO MEAN 1
DEFINE RED TO MEAN 2
DEFINE UNUSED TO MEAN 1
DEFINE FULL TO MEAN 2
DEFINE GOOD TO MEAN 1
DEFINE BAD TO MEAN 2
DEFINE BUMP TO MEAN ADD 1 TO
```

allows the use of the statements BUMP I and BUMP Y(I) and thereby increment a counter. There is also the provision to overrule a DEFINE TO MEAN statement through the SUPPRESS SUBSTITUTION and RESUME SUBSTITUTION statements. When used with a local DEFINE TO MEAN statement, these allow local freedom in that subroutine. Therefore a DEFINE statement can appear any where in the model.

```
DEFINE HOUR TO MEAN DAYS
DEFINE HOURS TO MEAN DAYS
DEFINE MINUTES TO MEAN /60 DAYS
DEFINE MINUTE TO MEAN /60 DAYS
DEFINE SECONDS TO MEAN /3600 DAYS
DEFINE SECOND TO MEAN /3600 DAYS
```

The above DEFINE statements represent a subterfuge to work around the lack of a simulated second interval in SIMSCRIPT II. The range from seconds to days can generate integer numbers too large for some computers. Now DAYS is the basis for the time interval. From this reference HOUR(S) can be substituted and MINUTE(S) and SECOND(S) have their appropriate factors.

```
DEFINE MTIME AS A REAL VARIABLE
DEFINE SUMO AS A REAL VARIABLE
DEFINE SUMSQO AS A REAL VARIABLE
DEFINE TOTALO AS AN INTEGER VARIABLE
DEFINE SUMA AS A REAL VARIABLE
DEFINE SUMSQA AS A REAL VARIABLE
DEFINE TOTALA AS AN INTEGER VARIABLE
DEFINE DURTO AS A REAL, 1-DIMENSIONAL ARRAY
DEFINE DURTA AS A REAL, 1-DIMENSIONAL ARRAY
```

This group of DEFINE statements are needed to calculate statistics for the drive duration. Since most of the variables are integers, these have to be specifically and globally defined as real since they are each used in more than one routine and require accuracy.

These are some of the possible preamble statements. In all probability the programmer would not have constructed them first. However, they are required by SIMSCIPT II in this order for the compiler. And they will be debugged first to enable the compiler to progress to the routines. Later the errors in the preamble will appear as bugs in the routines. Obviously the final version of the preamble is the result of iterations.

Main Program

The structure of SIMSCRIPT II like many programming languages depends on specific subroutines or subprograms to perform the actions of the program. These subprograms are self-contained and called by a symbolic name. The MAIN program provides the starting program to call only one subprogram which will in turn pass control on to another subprogram and so on until a STOP statement is executed. The tasks performed by the MAIN program are initialization, transfer of control to the internal timing routine, and bringing the simulation to a halt.

```
MAIN
NOW INITIALIZE RELEASE INITIALIZE''NOT ON MVT
   OPERATING SYSTEM FOR IBM SYSTEM 360''
LET BETWEEN.V = 'TRACE.PRINT'
START SIMULATION
STOP
END
```

The compulsory name MAIN defines the program. NOW INITIAL-IZE is one of many forms to call a routine INITIALIZE. Perform and call could also have been used. Since storage is limited, those routines used to set up the model are probably of no further use once the simulation has started. The release statement can be used to return the space used by the INITIALIZE routine to a free storage pool.

BETWEEN.V is a debugging statement. It is used to check if a trace is required every time a next event selection is made, but before the event routine has control.

The START SIMULATION statement is the next MAIN program statement. Since the compiler has now completed its actions, there are routines ready to take over as they are called. The START SIMULA-TION statement performs this function by allowing the interval timing routine to remove the first event from the events set and transfer control to the routine called by the event. The simulation is now under way until it runs out of events, reaches an END.OF.SIMULATION event, or a STOP statement. The STOP routine signals either the end or reinitialization for the next run.

Subprograms

The first subprogram reads input data and sets up global values. These include: initial conditions for intersections, roads, and traffic signals; storage for input data in the desired array structure; and event schedules for all the independent events which will start the model.

```
ROUTINE TO INITIALIZE

CREATE EACH ONE.INTERSECTION(1)
LET ONE.INTERSECTION = 1
LET F.SIGNAL=RED

LET F.CAPACITY=UNUSED

CREATE EACH TWO.INTERSECTION(1)
LET TWO.INTERSECTION = 1
LET S.SIGNAL=RED

LET S.CAPACITY=UNUSED
```

The spacing of the free format allows any identation for easy readability. The first statement names the subprogram, INITIALIZE, and has either the word TO or FOR added for readability. The subprogram was called from the MAIN program as described above. The EVERY statement in the preamble defined the permanent entity ONE.INTER-SECTION; the CREATE statement allocated storage for the array of entity attributes, F.SIGNAL and F.CAPACITY. The number of inter-sections is specified by the argument (1). The attribute values are initialized through the LET statements which set for each intersection the initial value of 1 and process for the attribute F.SIGNAL, the value for RED, and insert it in the proper array location. Since RED was defined by a global variable as 2, this is the number for the initial condition. Likewise F.CAPACITY is set to 1 through the variable UN-

USED. A similar process establishes the initial conditions for TWO.INTERSECTION.

```
CREATE EACH ROAD(4)
FOR EACH ROAD, DO
LET LOAD=O
READ SPEED(ROAD),DISTANCE(ROAD),CONDITIONS(ROAD),
   LOW DENSITY(ROAD), LOW.DENSITY.VALUE (ROAD)
   HIGH.DENSITY(ROAD), HIGH.DENSITY.VALUE (ROAD)
   LOW.BUNCHING(ROAD), LOW.BUNCHING.VALUE (ROAD)
   HIGH.BUNCHING(ROAD), HIGH.BUNCHING.VALUE (ROAD)
LOOP
```

The establishment of the entities ROADS with their speed and various factors is similar to INTERSECTION. For SPEED, CONDITIONS, DISTANCE, and the remaining density and bunching items, the data must be read in from input cards for each of the four route segments. The READ statement accomplishes this by filling in the eleven attribute values for each road segment. One set of statements performs these functions through the DO LOOP arrangement of statements. These data appear on the data card or cards as individual numbers in free format with blanks to separate each value. The data cards are separate from the model deck. All variables initially start on zero. The setting of LOAD to zero is merely good practice in anticipation of multiple runs.

In the preamble there were declarations of global variables. Where an array was involved, it is necessary to specify and allocate, in effect reserve, storage required for each array. The RESERVE statement can appear anywhere in the model except the preamble. Until a RESERVE statement is executed storage is not allocated for an array and therefore its element values are not accessible.

```
RESERVE CHANGE.SEQUENCE(*) AS 2
   LET CHANGE.SEQUENCE(GREEN)=RED
   LET CHANGE.SEQUENCE (RED)=GREEN
RESERVE SIGNAL.DELAY(*,*) AS 2 BY 2
   READ SIGNAL.DELAY
RESERVE TRAFFIC.VALUE(*,*) AS 3 BY 10

   READ TRAFFIC.VALUE
RESERVE DURTO(*) AS 60
RESERVE DURTA(*) AS 60
```

The RESERVE statement used above is the long form with (*) to indicate a one dimension array for CHANGE.SEQUENCE. Since this has already been clearly defined, a short form, RESERVE CHANGE.SEQUENCE AS 2, could have been used. The RESERVE SIGNAL.DELAY statement specifies the dimensionality of the array and its size through the words 2 BY 2. The additional statements insert data into the arrays by using the values for RED and GREEN, and by reading into arrays data for SIGNAL.DELAY, the duration of each signal condition, and TRAFFIC.VALUE, number of competing vehicles.

```
''SCHEDULE INITIALIZING EVENTS
SCHEDULE A DRIVING.CONDITIONS AFTER 6 HOURS''SET
  DRIVING CONDITIONS
SCHEDULE A DEPARTURE AFTER 7 HOURS''START COMPETING
  TRAFFIC
SCHEDULE AN OUR.CAR AFTER 7.5 HOURS''START OUR CAR
SCHEDULE AN ARRAY.VALUE IN 7 HOURS
SCHEDULE A FIRST.SIGNAL.CONTROL IN 7 HOURS
SCHEDULE A SECOND.SIGNAL.CONTROL IN 7 HOURS
RETURN
END
```

The last purpose of the initialization routine is to SCHEDULE the first events. The SCHEDULE statement creates an event notice and files it until the appropriate simulated time. Therefore the above statements will cause DRIVING.CONDITIONS to be established six hours and DEPARTURES for competing traffic seven hours after the initiation of model execution. Similarly the other events are generated and stored for future use when the simulated clock reaches that time. At this point the initialization is complete—storage has been allocated, initial values inserted, data read in, and the starting events await the progress of the simulation clock to trigger them into action. The RETURN statement signifies the end of the initialization. Control now passes back to the MAIN routine where the INITIAL routine is released. The START SIMULATION statement is now executed. It will select the first event from the list either on the basis of clock time or, when several events are scheduled for the same time, the order is the order of the SCHEDULE statements in the deck.

Events and Routines

The event DEPARTURE creates a competing car temporary entity which is processed through the routine until it is either passed on to another routine or destroyed. The event notice itself either is saved to reappear later or passes out of the model.

```
''GENERATES MODEL'S TRAFFIC PATTERN

EVENT DEPARTURE SAVING THE EVENT NOTICE
    DEFINE TIME AS A REAL VARIABLE
DEFINE N AND J AS INTEGER VARIABLES
''GENERATE COMPETING TRAFFIC
IF MOD.F(TIME.V,24) IS GREATER THAN 9.5,
  RESCHEDULE THIS DEPARTURE IN 21.5 HOURS
    RETURN
OTHERWISE.....
RESCHEDULE THIS DEPARTURE IN 3 MINUTES
FOR J=1 TO 3, DO THE FOLLOWING
    LET N=TRAFFIC.VALUE(J , K)
    ALSO FOR I=1 TO N, DO THIS
      CREATE A CAR
      LET IDENTIFIER=9
      LET ROUTE=J
      LET ENTRY.TIME=TIME.V
      ADD 1 TO LOAD(J)
    LET TIME = UNIFORM.F(0.0,0.05,1)
      +(DISTANCE(J)/SPEED(J))*
DENSITY(J)*BUNCH (J)*CONDITIONS(J)
CAUSE AN ARRIVAL IN TIME HOURS
      LET AUTO(ARRIVAL)=CAR
      LET PLACE(ARRIVAL)=ROUTE
    LOOP
RETURN
END
```

This routine is triggered by the event DEPARTURE, which is retained by SAVING THE EVENT NOTICE. Next local real and integer variables, time, N, and J are established. These variables are used only within this routine to determine the amount of traffic, N, and for which route segment, J, and the time an arrival event is to occur.

The competing traffic will be generated until 9:30 at which time a DEPARTURE event will be scheduled, reusing the event notice for the next day, $21\frac{1}{2}$ hours later and control is returned to the events list. If TIME.V has not yet reached 9:30, the IF statement is completed by OTHERWISE, ELSE, ALWAYS, or REGARDLESS. This aspect of the routine schedules another DEPARTURE for three minutes later to establish the next batch of competing traffic.

Details of this batch of traffic are specified starting with the **FOR** statement. The number of cars, N, is determined from input data TRAFFIC.VALUE(J,K) where it is the route segment and K is the number of this quarter hour. The value of K, a global constant, is obtained from the ARRAY.VALUE routine.

```
''SET SYSTEM VARIABLE TO THE CURRENT FIFTEEN MINUTE
  PERIOD
EVENT ARRAY.VALUE SAVING THE EVENT NOTICE
    IF MOD.F(TIME.V,24) IS LESS THAN 9.50
    LET K = K + 1
    RESCHEDULE THIS ARRAY.VALUE IN 15 MINUTES
RETURN
ELSE
    SCHEDULE THIS ARRAY.VALUE IN 21.5 HOURS
    LET K = 0
RETURN
END
```

Each car is established as a temporary entity by the **CREATE A CAR** statement. Attributes for the entity CAR must also be established. These are IDENTIFIER, ROUTE, and ENTRY.TIME. The values for each attribute are determined from the LET statements as 9, J, and TIME.V, respectively. Since the cars traverse roads there is a change to the attribute of the permanent entity ROAD; namely, increase by one the LOAD on that route segment.

The next step is to determine when each car will arrive at an intersection. This ARRIVAL event notice will occur according to the evaluation of the CAUSE AN ARRIVAL IN statement.

```
CAUSE AN ARRIVAL IN UNIFORM.F(0.0,0.05,1) +
DISTANCE(J)/SPEED(J))*DENSITY/(J)*BUNCH(J)*
CONDITIONS(J) HOURS
```

Two factors contribute to determining the **ARRIVAL** time. Since the DEPARTURE events which trigger the generation of each car occur every three minutes, UNIFORM.F(0.0,0.05,1) uniformly delays the cars according to random number stream 1 between 0 and 3 minutes. The remainder of the expression calculates how long it will take under current conditions to reach the intersection.

The ARRIVAL event notice has attributes AUTO and PLACE. These are obtained for each car by

```
LET AUTO(ARRIVAL)=CAR
LET PLACE(ARRIVAL)=ROUTE
```

Through the LOOP statement the process is repeated until there are ARRIVAL event notices to represent all the cars generated during this three-minute interval for all routes.

A similar routine establishes our car's entry into the model at the initial route segment. In this case the event notice is OUR.CAR and the IDENTIFIER is set to 1. Since the route is over the initial segment, the expression to generate an ARRIVAL is similarly derived but without any randomizing factor.

```
'' GENERATE OUR CAR DAILY

EVENT OUR.CAR SAVING THE EVENT NOTICE
   DEFINE TOTAL AS AN INTEGER SAVED VARIABLE
      CREATE A CAR
      LET IDENTIFIER=1
      LET ROUTE=INITIAL
      LET ENTRY.TIME=TIME.V
      ADD 1 TO LOAD(INITIAL)
      CAUSE AN ARRIVAL IN (DISTANCE(INITIAL)/SPEED
         (INITIAL))*DENSITY(INITIAL)*CONDITIONS(INITIAL)
         HOURS
      LET AUTO(ARRIVAL)=CAR
      LET PLACE(ARRIVAL)=INITIAL
         RESCHEDULE THIS OUR.CAR IN 24 HOURS
      BUMP TOTAL
PRINT 1 LINE WITH TOTAL THUS....
      TOTAL = ***
      IF TOTAL>100, NOW REPORT     STOP     ELSE....
RETURN
END
```

To assist in debugging each time our car is generated a line is printed indicating the total times this occurred. There remains only to determine when to halt the simulation. Since TOTAL is incremented at the beginning of each day's trip when it reaches 101 it is time to produce the report by branching to the REPORT routine, NOW REPORT which will in turn lead to STOP. If the trip is not the last one, the event is closed and the control returns to the timing routine. It will select the next event, update the simulated clock time and the simulation continues.

The ARRIVAL and TRAVEL.ON routines provide the flow of information for arrival of a car at an intersection and its progress through the intersection. Additional separate routines trigger the signal changes, and vary driving conditions.

The ARRIVAL routine encompasses logic to determine at which intersection the car has arrived, first or second, and for the second whether over the original or alternate routes. In addition once OUR.CAR completes the final segment the statistics have to be compiled.

This routine is triggered by the processing of an ARRIVAL event notice which has been previously scheduled by either the DEPARTURE or OUR.CAR routines. The event notice is saved to retain attributes for the TRAVEL.ON. For this routine there are a series of local variables that have to be defined and the load is reduced by one as a car arrives at the intersection. Then the logic determines that the car has not reached the final route segment. When the final route is completed by OUR.CAR it is necessary to compute how long the trip took by the statement LET MTIME = TIME.V — ENTRY.TIME. Furthermore separate records are kept if the trip took either the original, TOTALO, or the alternate route, TOTALA, through the test of the system variable IF PATH = ORIGINAL,GO TO OCOMP. Separate calculations are made for the trip statistics, mean and standard deviation. Either way once the data for the entire trip have been recorded there is no further use for this entity and it is removed from the system through the statements 'AFTCOMP' DESTROY THIS CAR DESTROY THIS ARRIVAL RETURN.

For the earlier cases where the trip is not over, there is need to sense which intersection, first or second, and for the second which queue, either ORG.QUEUE or ALT.QUEUE. The decision at the first intersection selects the original or alternate routing which in turn may depend on the number of temporary entities, cars, contained in the set, queue—LET CARS.BEFORE = N.QUEUE. The condition of the intersection is established through a boolean expression, IF QUEUE IS EMPTY AND F.SIGNAL = GREEN AND F.CAPACITY = UNUSED, all three conditions being met, then SCHEDULE A TRAVEL.ON IN 2 SECONDS for the specific case of LET ROUTE = ORIGINAL and set attributes of LET AUTO(TRAVEL.ON) = CAR,LET PLACE(TRAVEL.ON) = ROUTE. This will provide for the TRAVEL.ON event to appear on the events list 2 seconds later, which is the constant time for a car to travel the distance across the intersection. When a car does enter the intersection only one car is allowed in the area through the statement LET F.CAPACITY = FULL which prevents another vehicle from satisfying the boolean expression until the intersection is again clear. Any other logical situation caused by other cars in the queue, red signal light, or a car already in the intersection will follow the OTHERWISE statement FILE THIS CAR IN QUEUE. The system variable PATH must be set by OUR.CAR. Since the second intersection

```
''A CAR ARRIVES AT AN INTERSECTION VIA SOME ROAD
EVENT ARRIVAL SAVING THE EVENT NOTICE
    DEFINE ENTRY AS AN INTEGER VARIABLE
    LET CAR=AUTO(ARRIVAL)
    LET ROUTE=PLACE(ARRIVAL)
    SUBTRACT 1 FROM LOAD(ROUTE)
IF ROUTE IS NOT EQUAL TO FINAL, GO TO WHICH
ELSE
IF IDENTIFIER IS NOT EQUAL TO 1, GO TO AFTCOMP
OTHERWISE
    LET MTIME=TIME.V-ENTRY.TIME
    LET ENTRY=TRUNC.F(MTIME*60)
PRINT 1 LINE WITH MTIME, ENTRY, AND PATH THUS
    MTIME IS **.***** ENTRY **.**** PATH IS **
    IF PATH IS EQUAL TO ORIGINAL, GO TO OCOMP
    ELSE GO TO ACOMP
'OCOMP' LET DURTO(ENTRY)=DURTO(ENTRY)+1
    LET TOTALO=TOTALO+1
    LET SUMO=SUMO+ENTRY
    LET SUMSQO=SUMSQO+ENTRY * ENTRY
    GO TO AFTCOMP
'ACOMP' LET DURTA(ENTRY) = DURTA(ENTRY) + 1
    LET TOTALA=TOTALA+1
    LET SUMA=SUMA+ENTRY
    LET SUMSQA=SUMSQA+ENTRY * ENTRY
'AFTCOMP' DESTROY THIS CAR
GO TO DESTROY
'WHICH' IF ROUTE IS NOT EQUAL TO INITIAL GO TO
  NEXT.INTERSECTION
ELSE.....
'FIRST' LET CARS.BEFORE=N.QUEUE
    IF QUEUE IS EMPTY AND F.SIGNAL
      =GREEN AND F.CAPACITY=UNUSED,
    SCHEDULE A TRAVEL.ON IN 2 SECONDS
    LET ROUTE=ORIGINAL
    LET F.CAPACITY - FULL
GO TO CAR
    OTHERWISE.....
    FILE THIS CAR IN QUEUE
LET CARS.BEFORE - N.QUEUE
```

```
      IF CARS.BEFORE IS GREATER THAN 4,
          LET ROUTE=ALTERNATE
GO TO MORE.CARS
          ELSE.....
          LET ROUTE=ORIGINAL
GO TO MORE.CARS
'AAA' LET AUTO(TRAVEL.ON)=CAR
      LET PLACE(TRAVEL.ON)=ROUTE
'DESTROY' DESTROY THIS ARRIVAL
RETURN
'NEXT.INTERSECTION' IF ROUTE EQUALS ALTERNATE GO TO
   ALT
ELSE
      IF ORG.QUEUE IS EMPTY AND S.SIGNAL
        =GREEN AND S.CAPACITY=UNUSED,
      SCHEDULE A TRAVEL.ON IN 2 SECONDS
     LET S.CAPACITY=FULL
LET ROUTE=FINAL GO TO AAA
          OTHERWISE.....
LET ROUTE=FINAL
      FILE THIS CAR IN ORG.QUEUE
      GO TO DESTROY
'ALT' IF ORG.QUEUE IS EMPTY AND S.SIGNAL=RED AND
  S.CAPACITY=UNUSED,
      SCHEDULE A TRAVEL.ON IN 2 SECONDS
     LET S.CAPACITY=FULL
LET ROUTE = FINAL GO TO AAA
          OTHERWISE.....
LET ROUTE=FINAL
       FILE THIS CAR IN ALT.QUEUE
       GO TO DESTROY
'CAR' IF IDENTIFIER = 1 LET PATH = ROUTE GO TO AAA
     OTHERWISE GO TO AAA
'MORE.CARS' IF IDENTIFIER = 1 LET PATH
   = ROUTE GO TO DESTROY
ELSE GO TO DESTROY
END
```

has two possible queues, this logic set is repeated twice. The ARRIVAL event has fulfilled its function either by triggering a TRAVEL.ON event to occur later, or when the car is filed in a queue; therefore the 'DESTROY' DESTROY THIS ARRIVAL statement.

The TRAVEL.ON routine was triggered either through the ARRIVAL event which got the car to the intersection or a signal change. Now at a later time the intersection will again be able to handle another car. This routine starting from EVENT TRAVEL.ON SAVING THE EVENT NOTICE will provide the logic to accept the next car. Attributes of the TRAVEL.ON may be needed for another TRAVEL.ON or ARRIVAL. LET CAR = AUTO(TRAVEL.ON),LET ROUTE = PLACE(TRAVEL.ON) determine which car to remove from the queue and, likewise, transmit the attributes to the next TRAVEL.ON

```
''A CAR COMPLETES PASSAGE THROUGH AN INTERSECTION
EVENT TRAVEL.ON SAVING THE EVENT NOTICE
      DEFINE TIME AS A REAL VARIABLE
      LET CAR=AUTO(TRAVEL.ON)
      LET ROUTE=PLACE(TRAVEL.ON)
      IF ROUTE - FINAL, GO TO FINAL
   ELSE
'FIRST' LET F.CAPACITY=UNUSED
IF F.SIGNAL IS NOT EQUAL TO GREEN
      GO TO WAIT
ELSE
      IF QUEUE IS NOT EMPTY
      REMOVE FIRST CAR FROM QUEUE
      GO TO A
      ELSE
      GO TO C
'WAIT' DESTROY THIS TRAVEL.ON
      RETURN
   'A'    SCHEDULE THIS TRAVEL.ON IN 2 SECONDS
         LET AUTO(TRAVEL.ON)=CAR
         LET PLACE(TRAVEL.ON)=ROUTE
IF IDENTIFIER IS NOT EQUAL TO 1, DESTROY THIS CAR
         RETURN
         ELSE
         GO TO CARRY.ON
'B' RETURN
'FINAL'LET S.CAPACITY=UNUSED
IF S.SIGNAL IS NOT EQUAL TO GREEN GO TO RED
```

```
ELSE
      IF ORG.QUEUE IS NOT EMPTY
      REMOVE FIRST CAR FROM ORG. QUEUE
      GO TO A
      ELSE
'C' IF IDENTIFIER IS NOT EQUAL TO 1
                              DESTROY THIS CAR
                              DESTROY THIS TRAVEL.ON
                              RETURN
      ELSE
      GO TO CARRY.ON
'RED' IF ALT.QUEUE IS NOT EMPTY
      REMOVE FIRST CAR FROM ALT.QUEUE
GO TO A
ELSE
      GO TO C
'CARRY.ON' LET TIME=DISTANCE(ROUTE)/SPEED(ROUTE)*
  DENSITY(ROUTE)*
      BUNCH(ROUTE)*CONDITIONS(ROUTE)
      SCHEDULE AN ARRIVAL IN TIME HOURS
PRINT 1 LINE WITH TIME AND ROUTE THUS
AT CARRY.ON THE NEXT ARRIVAL IS *.****
  HOURS FOR ROUTE **
      ADD 1 TO LOAD(ROUTE)
      LET AUTO(ARRIVAL)=CAR
      LET PLACE(ARRIVAL)=ROUTE
    IF ROUTE = FINAL GO TO B
    ELSE
    LET PATH = ROUTE
    GO TO B
END
```

LET AUTO(TRAVEL.ON) = CAR, LET PLACE(TRAVEL.ON) = ROUTE. Since there is no further need for the competing traffic at this point, IF IDENTIFIER IS NOT EQUAL TO 1, DESTROY THIS CAR DESTROY THIS TRAVEL.ON RETURN ELSE GO TO CARRY.ON. The next event to be scheduled will be an arrival from the original, alternate, or final routes. 'CARRY.ON', LET TIME = DISTANCE (ROUTE)/SPEED(ROUTE-))*DENSITY(ROUTE)*BUNCH (ROUTE)*CONDITIONS(ROUTE) SCHEDULE AN ARRIVAL IN TIME HOURS. Add 1 TO LOAD(ROUTE) LET AUTO(ARRIVAL =

CAR LET PLACE(ARRIVAL) = ROUTE. Finally for OUR.CAR set
the system variable PATH to the route which will be used.

Road conditions can change during the day. This routine schedules
that capability through the DRIVING.CONDITIONS event.

```
' ' DRIVING CONDITIONS CHANGE WITH TIME OF DAY
EVENT DRIVING.CONDITIONS SAVING TIME EVENT NOTICE
IF MOD.F(TIME.V,24)=6, LET VISIBILITY=GOOD LET TIME=14
    GO SET  ELSE
    LET VISIBILITY=BAD  LET TIME=10
' SET' FOR J=1 TO 4, LET CONDITIONS(J)=VISIBILITY
RESCHEDULE THIS DRIVING.CONDITIONS AT TIME.V + TIME
RETURN
END
```

This routine sets up a scenario of driving conditions with the initial
conditions as good. The fixed scenario starts with the first event arriving after 6 hours. This is equal to the fixed value of 6 so that visibility

```
''SIGNALS CHANGE INDEPENDENTLY OF TRAFFIC FIRST
    INTERSECTION
EVENT FIRST.SIGNAL.CONTROL SAVING THE EVENT NOTICE
''IF TIME IS RIGHT SHUT OFF SIGNAL UNTIL NEXT DAY
IF MOD.F(TIME.V, 24) =10.5, RESCHEDULE THIS
FIRST.SIGNAL.CONTROL AT TRUNC.F(TIME.V)+21
    LET F.SIGNAL = RED
    GO TO A
ELSE...
LET F.SIGNAL = CHANGE.SEQUENCE(F.SIGNAL)
    RESCHEDULE THIS FIRST.SIGNAL.CONTROL IN
    SIGNAL.DELAY(1,F.SIGNAL) SECONDS
'A' IF F.CAPACITY EQUALS FULL RETURN
ELSE
IF QUEUE IS EMPTY RETURN
ELSE
REMOVE FIRST CAR FROM QUEUE
'B' SCHEDULE A TRAVEL.ON IN 2 SECONDS
    LET AUTO(TRAVEL.ON) = CAR
    LET PLACE(TRAVEL.ON) = ROUTE
    LET F.CAPACITY = FULL
RETURN
END
```

```
''SIGNALS CHANGE INDEPENDENTLY OF TRAFFIC SECOND
   INTERSECTION
EVENT SECOND.SIGNAL.CONTROL SAVING THE EVENT NOTICE
''IF TIME IS RIGHT SHUT OFF SIGNAL UNTIL NEXT DAY
IF MOD.F(TIME.V, 24) =10.5, RESCHEDULE THIS
    SECOND.SIGNAL.CONTROL AT TRUNC.F(TIME.V)+21
    LET S.SIGNAL = RED
    GO TO A
ELSE...
LET S.SIGNAL = CHANGE.SEQUENCE(S.SIGNAL)
    RESCHEDULE THIS SECOND.SIGNAL.CONTROL IN
    SIGNAL.DELAY(2,S.SIGNAL) SECONDS
'A' IF S.CAPACITY EQUALS FULL RETURN
ELSE
IF PATH = ALTERNATE GO TO C
ELSE
IF ORG.QUEUE IS EMPTY RETURN
ELSE
REMOVE FIRST CAR FROM ORG.QUEUE
'B' SCHEDULE A TRAVEL.ON IN 2 SECONDS
    LET AUTO(TRAVEL.ON) = CAR
    LET PLACE(TRAVEL.ON) = ROUTE
    LET S.CAPACITY = FULL
RETURN
'C' IF ALT.QUEUE IS EMPTY RETURN
ELSE
REMOVE FIRST CAR FROM ALT.QUEUE
    GO TO B
END
```

is good. The next event is scheduled to occur at 20 hours. At this time the modulo division remainder is not 6, so the visibility becomes BAD and the next event is scheduled for 10 hours later, the thirtieth hour, when the remainder is again 6. Then the visibility reverts to GOOD and the process starts over.

The traffic signals are independent routines. Because of the two queues at the second intersection there are two routines, first and second signal control. They can be allowed to continue all day or be turned off after our car has reached the office until the next day. Obviously there is nothing to be gained from running them while no events of interest go on.

The signal control event is provided by the initialization routine at 7 hours after the start. The signals are to run until 10:30, after which they are rescheduled for the next morning. The function TRUNC.F takes the current simulated time, TIME.V, and obtains from it only the integer value of time. To this value is added the number of hours until the signal control is needed the next morning. As a last bit of housekeeping the signal is left in the red condition. Since the queue should be empty at this time, it is good practice to introduce a safety check 'debug' which prints out characteristics of any cars still in the queue.

When the signal is being operated the sequence of control is established through the DEFINE RED TO MEAN 2 in the preamble and the LET F.SIGNAL = RED in the initialization routine. Therefore LET F.SIG-NAL = CHANGE.SEQUENCE(F.SIGNAL) changes the signal control status. The time duration for each state is established through RE-SCHEDULE THIS FIRST.SIGNAL.CONTROL, IN F.SIGNAL.DE-LAY(1,F.SIGNAL) SECONDS and is based on data read in during initialization. Since the signals affect the queue status, the statements IF QUEUE IS EMPTY RETURN ELSE REMOVE FIRST CAR FROM QUEUE evaluates the changes caused by the new light condition. When a car is not waiting the control passes back to the timing routine. When there is a car waiting in the queue and light changes to green an event TRAVEL.ON is triggered for 2 seconds later, the time used for each car to leave the queue and cross the intersection. The capacity of the intersection is changed to FULL to prevent other cars from going through the intersection without joining the queue. The second intersection is similar except that cars have to be removed from either the original or alternate queues.

The routines for the execution of the simulation are now complete. The remaining routines provide output so that the results are made available and mathematical expressions which could have been located anywhere among the routines. The output in SIMSCRIPT II is developed through the use of PRINT statements and the formating features of the language. There is flexibility to make the report appear as the programmer desires.

The REPORT routine is entered only once in the simulation from the routine to generate the traffic pattern which checked for 100 days of simulation. Once this function has been completed the next declaration is to STOP. In this routine the data that are to be presented are gathered and displayed. The print statement defines the number and content of the format lines which follow. PRINT 2 LINES WITH MO, SO, MA, SA THUS . . . Indicates that the variables for mean

```
ROUTINE TO REPORT
IF TOTALO = 0 ,GO TO NO.TOTALO ELSE
LET MO = SUMO / TOTALO
LET SO = SQRT.F((SUMSQO-((SUMO*SUMO)/TOTALO))/
   (TOTALO-1))
'NO.TOTALO' IF TOTALA = 0 GO TO NO.TOTALA ELSE
LET MA = SUMA / TOTALA
LET SA - SQRT.F((SUMSQA-((SUMA*SUMA)/TOTALA))/
   (TOTALA-1))
'NO.TOTALA' PRINT 2 LINES WITH MO, SO, MA, SA THUS
   TRIP TIME USING ORIGINAL ROUTE... AVG= **.**
      STD.DEV= **.**
   TRIP TIME USING ALTERNATE ROUTE... AVG= **.**
      STD.DEV.= **.**
SKIP 3 OUTPUT LINES
IF TOTALO = 0, GO TO NO.HISTO ELSE
PRINT 1 LINE THUS
   HISTOGRAM OF TRIP TIMES OVER ORIGINAL ROUTE
PERFORM HISTOGRAM GIVEN DURTO(*)
SKIP 3 OUTPUT LINES
'NO.HISTO' IF TOTALA = 0, GO TO NO.HISTA ELSE
PRINT 1 LINE THUS
   HISTOGRAM OF TRIP TIMES OVER ALTERNATE ROUTE
PERFORM HISTOGRAM GIVEN DURTA(*)
'NO.HISTA' RETURN
END
```

travel time and standard deviations are to be printed out in a specific order, original and then alternate routes.

```
TRIP TIME USING ORIGINAL ROUTE....AVG= **.**
   STD.DEV.=**.**
TRIP TIME USING ALTERNATE ROUTE...AVG= **.**
   STD.DEV.=**.**

**.** defines the decimal form of the printout.
```

The format of the output is not limited to text. **Histogram form** of output is available for global variables and for attributes of permanent entities. Since the variables DUTO and DUTA were defined in the preamble they are global variables and may be used to provide histogram output. The size of the histogram and its format were defined in the preamble.

The remaining mathematical routines define data which are used to define the particular values of the DENSITY and BUNCH functions which were used to calculate the progress of the traffic according to the load on that route segment.

```
ROUTINE DENSITY(ROAD)DEFINE ROAD AS AN INTERGER
 VARIABLE
IF LOAD ( LOW.DENSITY,RETURN WITH LOW.DENSITY,VALUE
 ELSE
IF LOAD ) HIGH.DENSITY,RETURN WITH HIGH,DENSITY,VALUE
 ELSE
RETURN WITH (LOAD/HIGH.DENSITY)*(HIGH.DENSITY.VALUE-
 LOW.DENSITY.VALUE)+LOW.DENSITY.VALUE
ROUTINE BUNCH(ROAD) DEFINE ROAD AS AN INTEGER VARIABLE
IF LOAD ( LOW.BUNCHING RETURN WITH LOW.BUNCHING.VALUE
 ELSE
IF LOAD ) HIGH.BUNCHING RETURN WITH HIGH.BUNCHING.
 VALUE ELSE
RETURN WITH(LOAD/HIGH.BUNCHING)*(HIGH.BUNCHING.VALUE-
 LOW.BUNCHING.VALUE)+LOW.BUNCHING.VALUE
```

An integral aspect of SIMSCRIPT II is the need to insert debugging routines in the body of the model. The programmer must create his own debugging structure. Some basic means of following the simulation process must be available to pin down a bug. In SIMSCRIPT II, as part of the inherent internal processing betweeen each event, there is the option to have access to the event through the BETWEEN.V statement. In the MAIN routine there is the statement LET BETWEEN.V = TRACE.PRINT. Therefore between each event the TRACE.PRINT routine will be entered. Since the event is identified by EVENT.V a particular print statement may be associated with each possible event by GO TO LABEL1, LABEL2, LABEL3, LABEL4, LABEL5, LABEL6, LABEL7, LABEL8, PER EVENT.V

The printout during a debugging process may be quite confused by the large number of lines of print as each event travels through the model. However, once the routine is operational, as might be the case for the signal controls which produce considerable output, then a statement such as 'LABEL5' PRINT 1 LINE WITH TIME.V AS FOLLOWS FIRST.SIGNAL.CONTROL TIME.V IS ****.** is changed to 'LABEL5' RETURN.

Unfortunately most bugs are not found through the limited inherent SIMSCRIPT II debugging fractures. Instead debugging is based on

```
ROUTINE TRACE.PRINT
GO TO LABEL1, LABEL2, LABEL3, LABEL4, LABEL5, LABEL6,
   LABEL7,LABEL8
PER EVENT.V
'LABEL1' PRINT 1 LINE WITH TIME.V THUS
TIME-V DRIVING CONDITIONS IS *****.****
RETURN
'LABEL2' PRINT 1 LINE WITH TIME.V AS FOLLOWS
DEPARTURE TIME-V IS *******.****
RETURN
'LABEL3' PRINT 1 LINE WITH TIME.V AS FOLLOWS
OUR.CAR TIME-V IS ******.****
RETURN
'LABEL4' PRINT 1 LINE WITH TIME.V AS FOLLOWS
ARRAY.VALUE TIME-V IS ****.****
RETURN
'LABEL5' RETURN ''FIRST SIGNAL CONTROL
'LABEL6' RETURN ''SECOND SIGNAL CONTROL
'LABEL7' RETURN ''TRAVEL.ON
'LABEL8' RETURN ''ARRIVAL
     END
```

a process of being informed that the simulation has terminated with
an error message and a limited set of current data.

```
                    AT LOCATION 00001A08
                    CALLED FROM 00003DB2
                    CALLED FROM 0000037A
                    CALLED FROM 00002C64


TIME.V =          7.1400      READ.V = 5      RCOLUMN.V =  25      EOF.V = 0
EVENT.V =    3                WRITE.V = 3     WCOLUMN.V =   0

********** ERROR NUMBER 206 **********
MANY ERRORS POSSIBLE. SEE MESSAGE FOR COMPLETION CODE OC6 IN IBM MANUAL
```

From this information the rough location of the error can be determined
from the hexadecimal value, 1A08. This address indicates the error
between 1910 and 1A48, RBSQUEUE, as shown in the following listings.
Likewise the immediate call to this routine came from 3DB2, RAR-
RIVAL. This is consistent with the value of EVENT.V =8, also an
arrival situation. Neither of the other locations help since 37A is
RTIMESR and 2664 is MAIN. Therefore the debugging must at this
point wait for another SIMSCRIPT II run with whatever additional
diagnostics are considered to be of greatest value. Obviously the debugging process will require considerable effort since the most probable

MODULE MAP

CONTROL SECTICN

NAME	ORIGIN	LENGTH
PRMA	00	140

NAME	ORIGIN	LENGTH
PRMA1	140	14C
RTIME$R	290	284
RAALTQ	518	134
RCALTQ	650	148
RBALTQ	798	134
RCALTQ	800	148
RXALTQ	A18	114
RYALTQ	B30	114
RZALTQ	C48	13C
RD$CAR	D88	98
RAORGQ	F20	134
RCORGQ	F58	148
RBORGQ	10A0	134
RCORGQ	11D8	148
RXORGQ	1320	114
RYORGQ	1438	114
RZORGQ	1550	13C
RA$QUEUE	1690	134
RC$QUEUE	17C8	148
RB$QUEUE	1910	134
RC$QUEUE	1A48	148

ENTRY

NAME	LOCATION	NAME	LOCATION	NAME	LOCATION	NAME	LOCATION
GDEPARTU	00	GTRACE	4	GTRAVEL$	8	GCAR	C
GIDENTIF	10	GCAR$BE	14	GENTRY$T	18	GROUTE	1C
GSALTQ	20	GPALTQ	24	GMALTQ	28	GSORGQ	2C
GPORGQ	30	GMORGQ	34	GS$QUEUE	38	GP$SQUEUE	3C
GM$QUEUEF	40	GSECOND$	44	GEV$E	48	GDRIVING	4C
GOUR$CAR	50	GFIRST$S	54	GARRAY$V	58	GARRIVAL	5C
GPATH	60	GK	64	GSIGNAL$	68	GMTIME	6C
G$UMO	70	GSUM$QA	74	GSUM$QO	78	GTRAFFIC	7C
GNDNEI	80	GNTWOI	84	GDURTA	88	GD$URTO	8C
GTWO$INT	90	GS$SIGNA	94	GS$CAPAC	98	GFALTQ	9C
GLALTQ	A0	GNALTQ	A4	GFORGQ	A8	GLORGQ	AC
GNORGQ	B0	GTOTALA	B4	GTOTALO	B8	GCHANGE$	BC
GROAD	C0	GCONDITI	C4	GLOAD	C8	GDISTANC	CC
G$PEED	D0	GONE$INT	D4	GF$SIGNA	D8	GF$CAPAC	DC
GF$QUEUE	E0	GL$QUEUE	E4	GN$QUEUE	E8	GN$ROAD	EC
G$UMA	F0	GW$DEPAR	F4	GW$TRACE	F8	GW$TRAVE	FC
GW$CAR	100	GW$SECON	104	GWEVE	108	GW$DRIVI	10C
GWOURC	110	GW$FIRST	114	GW$ARRAY	118	GW$ARRIV	11C
GI$DRIVI	120	GI$DEPAR	124	GIOURC	128	GI$ARRAY	12C
GI$FIRST	130	GI$SECON	134	GI$TRAVE	138	GI$ARRIV	13C
GI$TRACE	280	GEVENTS$	284	GSIZE$V	288		

186

NAME	ORIGIN	LENGTH		NAME	LOCATION	NAME	LOCATION	NAME	LOCATION	NAME	LOCATION	NAME	LOCATION	NAME	LOCATION
RX$QUEUE	1B90	114													
RY$QUEUE	1CA8	114													
RZ$QUEUE	1DC0	13C													
RCTWOI	1F00	1F4													
RC$ROAD	20F8	26C													
RCONEI	2368	154													
RINITIAL	24C0	75C													
MAIN	2C20	60													
RDEPARTU	2C80	3C4													
RQUP$CAR	3048	26C													
RFIRST$S	3288	298													
RSECOND$	3550	304													
RARPIVAL	3858	AC8													
RTRAVEL$	4320	564													
RDRIVING	4888	174													
RTPACE	4A00	70													
RREPORT	4A70	3A8													
RCENSITY	4E18	108													
RBUNCH	4F20	108													
RARRAY$V	5028	D4													
RTRACE$P	5100	30C													
LCES$F *	5410	4C													
LERR$F *	5460	C													
RAEVS *	5470	114													
RCRE$F *	5588	164													

RERR$R *	56F0	2C
RRES$R *	5720	26E
RRFI$R *	5990	50
RTIM1$R *	59E0	190
RTIM2$R *	5B70	9C
RTIM3$R *	5C10	200
RDIM$F *	5E10	3C
RHISTOGR*	5E50	180
RWTA$R *	6000	EC
RWTI$R *	60F0	190
RWTR$R *	6280	1B2
RWTS$R *	6438	95
RUSE$R *	64D0	4CC
RWTE$R *	69A0	2C4
RWTP$R *	6C68	6E
XCLP *	6CD8	26
XERP *	6D00	DC
XFLT *	6DE0	48
RRDR$R *	6E28	256
XCVT *	7080	1D8
XDCB *	7258	5EC
XEHR *	7848	E8
XFPE *	7930	8A
XOPU *	79C0	4C
XRND *	7A10	8E

GETO 689A UIBS 6940

XLAR 6E2E XEOF 6E9E

NAME		ORIGIN	LENGTH
XCLS	*	7AA0	196
XTRA		7C38	2C6
XWTC	*	7F00	104
RWTC$R	*	8008	100
RWTD$R		8108	338
RWTT$R	*	8440	59
RRMD$F	*	84A0	64
RSQRT$F	*	8508	C0
RTRUNC$F	*	85C8	60
RUNIFORM	*	8628	5C
XMAS	*	8688	1000
REFIELD$	*	9688	8C
RMODE$F	*	9718	160
RRDI$R	*	9878	1C0
RRFA$R	*	9A38	64
RRFD$R	*	9AA0	6C
RSFIELD$	*	9B10	7C
RSKIP$R	*	9B90	8E
XIDE	*	9C20	7C
XSE1	*	9CA0	114

NAME	LOCATION	NAME	LOCATION	NAME	LOCATION
XREC	9CCC	XRET	9D6C		

NAME		ORIGIN	LENGTH
XSPH	*	9DB8	BE
RRANDOM$	*	9E78	A4
RRDA$R	*	9F20	D4
RRDE$R	*	9FF8	340

NAME	LOCATION
RRDD$R	9FF8

NAME		ORIGIN	LENGTH
RRDT$R	*	A338	64
XFIX	*	A3A0	B2
XINR	*	A458	7C
XLV	*	A4D8	1804
XSE2	*	BCE0	190
XSLA	*	BE70	EE
XSTP	*	BF60	4C
XCVI	*	BFB0	21C
XSPI	*	C1D0	46

ENTRY ADDRESS   2C20
TOTAL LENGTH   C218

***GO   DOES NOT EXIST BUT HAS BEEN ADDED TO DATA SET

189

response to finding a bug is to add diagnostics for this unique situation and then try the model again.

The listing of the module map shows the location of the global variable and functions as part of the preamble and followed by the starting addresses and lengths for both interval SIMSCRIPT II routines and those developed as part of the model.

Frequently there is inadequate information obtained for debugging from the BETWEEN.V and MODULE MAP traceback features. Under these circumstances a much more elaborate debugging structure has to be added to the model. One particular limitation of BE-TWEEN.V is that the event has not yet been processed. Therefore if we want to know what caused this event to occur, an arrival for example, it is desirable to print the values for ENTRY.TIME, ROUTE, and IDENTIFIER as well as the current time, TIME.V. Since the LET CAR = AUTO (ARRIVAL) type statements have not yet been executed the required data are not available. This problem is overcome by inserting after LET ROUTE = PLACE(ARRIVAL) statement an additional print statement:

```
PRINT 1 LINE WITH ENTRY.TIME, ROUTE, IDENTIFIER,
AND TIME.V THUS ARRIVAL ENTRY.TIME ***.**** ROUTE *
IDENTIFIER * TIME.V ***.****
```

This will result in a line of output every time this routine is entered in the following format:

```
ARRIVAL ENTRY.TIME 7.5000 ROUTE 4 IDENTIFIER 1
    TIME.V 7.8363
```

Input Data

The data for the model appear as items in a separate section of the composite card deck. A card can contain many items in a free form structure, identified as such by blanks. The order of the data is the same as the READ statements. The cards are consecutive. The first four provide data describing each road segment with speed, distance, conditions, density, and bunching values.

```
25 1 1 10 1 25 2 1 1 1 1
35 3 1 15 1 50 2 5 1 10 2
55 5 1 20 1 75 2 1 1 1 1
15 1 1 1 1 1 1 1 1 1 1
```

```
''   SIMULATION OF A DRIVE TO THE OFFICE
''     A SIMSCRIPT II EXAMPLE


PREAMBLE

NORMALLY, MODE IS INTEGER..........
PERMANENT ENTITIES.....
   EVERY ONE.INTERSECTION HAS A F.SIGNAL, A F.CAPACITY ,
      AND OWNS A QUEUE
   EVERY TWO.INTERSECTION HAS A S.SIGNAL, A S.CAPACITY, AND
                    OWNS AN ORG.QUEUE, AND AN ALT.QUEUE
      EVERY ROAD HAS SOME CONDITIONS, A LOAD, A DISTANCE, A SPEED,
         A LOW.DENSITY IN ARRAY 1, A LOW.DENSITY.VALUE IN ARRAY 2,
         A HIGH.DENSITY IN ARRAY 3, A HIGH.DENSITY.VALUE IN ARRAY 4,
         A LOW.BUNCHING IN ARRAY 5, A LOW.BUNCHING.VALUE IN ARRAY 6,
         A HIGH.BUNCHING IN ARRAY 7 AND A HIGH.BUNCHING.VALUE IN ARRAY 8


TEMPORARY ENTITIES.....
         EVERY CAR HAS AN IDENTIFIER, SOME CARS.BEFORE, AN ENTRY.TIME, A ROUTE,
AND MAY BELONG TO A QUEUE, AN ALT.QUEUE, AND AN ORG.QUEUE

EVENT NOTICES INCLUDE DRIVING.CONDITIONS, DEPARTURE, OUR.CAR,
         ARRAY.VALUE, FIRST.SIGNAL.CONTROL, AND SECOND.SIGNAL.CONTROL
      EVERY TRAVEL.ON HAS AN AUTO IN WORD 6, A PLACE IN WORD 7
      EVERY ARRIVAL HAS AN AUTO IN WORD 6, A PLACE IN WORD 7

''   DATA COLLECTION AND ANALYSIS SPECIFICATIONS

   EXTERNAL EVENTS ARE TRACE '' FOR CAPABILITY TO TRIGGER BETWEEN.V
''    ADDITIONAL DECLARATIONS

DEFINE DENSITY AND BUNCH AS REAL FUNCTIONS
DEFINE ENTRY.TIME AS A REAL VARIABLE
THE SYSTEM HAS A PATH
THE SYSTEM HAS A K    ''INDEX TO OBTAIN COMPETING TRAFFIC VALUE
NORMALLY, DIMENSION=2
THE SYSTEM HAS A SIGNAL.DELAY
DEFINE TRAFFIC.VALUE AND SIGNAL.DELAY AS INTEGER, 2-DIMENSIONAL VARIABLES
DEFINE CHANGE.SEQUENCE AS AN INTEGER, 1-DIMENSIONAL ARRAY
NORMALLY, DIMENSION=0
DEFINE INITIAL TO MEAN 1
DEFINE ORIGINAL TO MEAN 2
DEFINE ALTERNATE TO MEAN 3
DEFINE FINAL TO MEAN 4
DEFINE GREEN TO MEAN 1
DEFINE RED TO MEAN 2
DEFINE UNUSED TO MEAN 1
DEFINE FULL TO MEAN 2
DEFINE GOOD TO MEAN 1
DEFINE BAD TO MEAN 2
DEFINE BUMP TO MEAN ADD 1 TO
DEFINE HOUR TO MEAN DAYS
DEFINE HOURS TO MEAN DAYS

             DEFINE MINUTE  TO MEAN /60 DAYS
             DEFINE MINUTES TO MEAN /60 DAYS
             DEFINE SECOND  TO MEAN /3600 DAYS
             DEFINE SECONDS TO MEAN /3600 DAYS
             DEFINE MTIME    AS A REAL VARIABLE
             DEFINE SUMO AS A REAL VARIABLE
             DEFINE SUMSQO AS A REAL VARIABLE
             DEFINE TOTALO AS AN INTEGER VARIABLE
             DEFINE SUMA AS A REAL VARIABLE
             DEFINE SUMSQA AS A REAL VARIABLE
             DEFINE TOTALA AS AN INTEGER VARIABLE
             DEFINE DURTO AS AN INTEGER 1-DIMENSIONAL ARRAY
             DEFINE DURTA AS AN INTEGER 1-DIMENSIONAL ARRAY

             END
```

```
ROUTINE TO INITIALIZE
    CREATE EACH ONE.INTERSECTION(1)
    LET ONE.INTERSECTION = 1
            LET F.SIGNAL = RED
            LET F.CAPACITY = UNUSED
       CREATE EACH TWO.INTERSECTION(1)
            LET    TWO.INTERSECTION = 1
            LET S.SIGNAL = RED
            LET S.CAPACITY = UNUSED

CREATE EACH ROAD(4)
FOR EACH ROAD, DO
    LET LOAD=0
READ  SPEED(ROAD), DISTANCE(ROAD), CONDITIONS(ROAD),
    LOW.DENSITY(ROAD),  LOW.DENSITY.VALUE(ROAD),
    HIGH.DENSITY(ROAD),  HIGH.DENSITY.VALUE(ROAD),
    LOW.BUNCHING(ROAD),  LOW.BUNCHING.VALUE(ROAD),
    HIGH.BUNCHING(ROAD),  HIGH.BUNCHING.VALUE(ROAD)
LOOP

RESERVE DURTO(*) AS 60
RESERVE DURTA(*) AS 60
RESERVE CHANGE.SEQUENCE(*) AS 2
    LET CHANGE.SEQUENCE(GREEN)=RED
    LET CHANGE.SEQUENCE(RED)=GREEN
RESERVE SIGNAL.DELAY(*,*) AS 2 BY 2
    READ SIGNAL.DELAY
RESERVE TRAFFIC.VALUE(*,*) AS 3 BY 10
    READ TRAFFIC.VALUE

''    SCHEDULE INITIALIZING EVENTS
SCHEDULE A DRIVING.CONDITIONS IN 6 HOURS    ''SET DRIVING CONDITIONS
SCHEDULE A FIRST.SIGNAL.CONTROL IN 7    HOURS
SCHEDULE A SECOND.SIGNAL.CONTROL IN 7    HOURS
SCHEDULE A DEPARTURE IN 7.1   HOURS ''START COMPETING TRAFFIC
SCHEDULE AN OUR.CAR  IN 7.5 HOURS    '' START OUR CAR
SCHEDULE AN ARRAY.VALUE IN 7 HOURS
RETURN
END

    MAIN
    NOW INITIALIZE    ''RELEASE INITIALIZE    THIS STATEMENT NOT ALLOWED UNDER MVT''
    LET   BETWEEN.V = 'TRACE.PRINT'
    START SIMULATION
    STOP
    END

    '' GENERATES MODEL'S TRAFFIC PATTERN

    EVENT DEPARTURE SAVING THE EVENT NOTICE
        DEFINE TIME AS A REAL VARIABLE
    DEFINE N AND J AS INTEGER VARIABLES
    ''GENERATE COMPETING TRAFFIC
    IF MOD.F(TIME.V,24) IS GREATER THAN 9.5, RESCHEDULE THIS DEPARTURE IN 21.5 HOURS
        RETURN
    OTHERWISE.....
    RESCHEDULE THIS DEPARTURE IN 3 MINUTES
    FOR J=1 TO 3, DO THE FOLLOWING
        LET N = TRAFFIC.VALUE(J , K)
        ALSO FOR I=1 TO N, DO THIS
            CREATE A CAR
            LET IDENTIFIER=9
            LET ROUTE=J
            LET ENTRY.TIME=TIME.V
            ADD 1 TO LOAD(J)
        LET TIME   =    UNIFORM.F(0.0,0.05,1)+(DISTANCE(J)/SPEED(J))*
        DENSITY(J)*BUNCH  (J)*CONDITIONS(J)
        CAUSE AN ARRIVAL IN  TIME HOURS
            LET AUTO(ARRIVAL)=CAR
            LET PLACE(ARRIVAL)=ROUTE
        LOOP
    RETURN
    END
```

192

```
''  GENERATE OUR CAR DAILY

EVENT OUR.CAR SAVING THE EVENT NOTICE
    DEFINE TOTAL AS AN INTEGER SAVED VARIABLE
       CREATE A CAR
       LET IDENTIFIER=1
       LET ROUTE=INITIAL
       LET ENTRY.TIME=TIME.V
       ADD 1 TO LOAD(INITIAL)
       CAUSE AN ARRIVAL IN (DISTANCE(INITIAL)/SPEED(INITIAL))*
           DENSITY(INITIAL)*CONDITIONS(INITIAL) HOURS
       LET AUTO(ARRIVAL)=CAR
       LET PLACE(ARRIVAL)=INITIAL
          RESCHEDULE THIS OUR.CAR IN 24 HOURS
       BUMP TOTAL
PRINT 1 LINE WITH TOTAL THUS....
       TOTAL = ***
       IF TOTAL    2, NOW REPORT        STOP       ELSE....
RETURN
END


''  SIGNALS CHANGE INDEPENDENTLY OF TRAFFIC FIRST INTERSECTION
EVENT FIRST.SIGNAL.CONTROL SAVING THE EVENT NOTICE
''  IF TIME IS RIGHT SHUT OFF SIGNAL UNTIL NEXT DAY
IF MOD.F(TIME.V, 24) =10.5, RESCHEDULE THIS
FIRST.SIGNAL.CONTROL        AT TRUNC.F(TIME.V)+21
    LET F.SIGNAL = RED
    GO TO A
ELSE...
LET F.SIGNAL = CHANGE.SEQUENCE(F.SIGNAL)
        RESCHEDULE THIS FIRST.SIGNAL.CONTROL IN
             SIGNAL.DELAY(1,F.SIGNAL) SECONDS
'A' IF F.CAPACITY EQUALS FULL RETURN
ELSE
IF    QUEUE IS EMPTY RETURN
ELSE
REMOVE FIRST CAR FROM    QUEUE
'B' SCHEDULE A TRAVEL.ON IN 2 SECONDS
    LET AUTO(TRAVEL.ON) = CAR
    LET PLACE(TRAVEL.ON) = ROUTE
    LET F.CAPACITY = FULL
RETURN
END

''  SIGNALS CHANGE INDEPENDENTLY OF TRAFFIC SECOND INTERSECTION
EVENT SECOND.SIGNAL.CONTROL SAVING THE EVENT NOTICE
''  IF TIME IS RIGHT SHUT OFF SIGNAL UNTIL NEXT DAY
IF MOD.F(TIME.V, 24) =10.5, RESCHEDULE THIS
   SECOND.SIGNAL.CONTROL     AT TRUNC.F(TIME.V)+21
    LET S.SIGNAL = RED
    GO TO A
ELSE...
LET S.SIGNAL = CHANGE.SEQUENCE(S.SIGNAL)
        RESCHEDULE THIS SECOND.SIGNAL.CONTROL IN
             SIGNAL.DELAY(2,S.SIGNAL) SECONDS
'A' IF S.CAPACITY EQUALS FULL RETURN
ELSE
IF PATH = ALTERNATE GO TO C
ELSE
IF ORG.QUEUE IS EMPTY RETURN
ELSE
REMOVE FIRST CAR FROM ORG.QUEUE
'B' SCHEDULE A TRAVEL.ON IN 2 SECONDS
    LET AUTO(TRAVEL.ON) = CAR
    LET PLACE(TRAVEL.ON) = ROUTE
    LET S.CAPACITY = FULL
RETURN
'C'     IF ALT.QUEUE IS EMPTY  RETURN
ELSE
REMOVE FIRST CAR FROM  ALT.QUEUE
   GO TO B
END
```

```
''  A CAR ARRIVES AT AN INTERSECTION VIA SOME ROAD
EVENT ARRIVAL SAVING THE EVENT NOTICE
    DEFINE ENTRY AS AN INTEGER VARIABLE
      LET CAR=AUTO(ARRIVAL)
      LET ROUTE=PLACE(ARRIVAL)
      SUBTRACT 1 FROM LOAD(ROUTE)
IF ROUTE IS NOT EQUAL TO FINAL, GO TO WHICH
ELSE
IF IDENTIFIER IS NOT EQUAL TO 1, GO TO AFTCOMP
OTHERWISE
      LET MTIME=TIME.V-ENTRY.TIME
      LET ENTRY=TRUNC.F(MTIME*60)
PRINT 1 LINE WITH MTIME, ENTRY, AND PATH THUS
    MTIME IS **.***** ENTRY **.**** PATH IS **
      IF PATH IS EQUAL TO ORIGINAL, GO TO OCOMP
      ELSE GO TO ACOMP
'OCOMP'   LET DURTO(ENTRY)=DURTO(ENTRY)+1
          LET TOTALO=TOTALO+1
          LET SUMO=SUMO+ENTRY
          LET SUMSQO=SUMSQO+ENTRY * ENTRY
          GO TO AFTCOMP
'ACOMP'   LET DURTA(ENTRY) = DURTA(ENTRY) + 1
          LET TOTALA=TOTALA+1
          LET SUMA=SUMA+ENTRY
          LET SUMSQA=SUMSQA+ENTRY * ENTRY
'AFTCOMP'      DESTROY THIS CAR
GO TO DESTROY
'WHICH' IF ROUTE IS NOT EQUAL TO INITIAL GO TO NEXT.INTERSECTION
ELSE.....
'FIRST'   LET CARS.BEFORE=N.QUEUE
          IF QUEUE IS EMPTY AND F.SIGNAL=GREEN AND F.CAPACITY=UNUSED,
          SCHEDULE A TRAVEL.ON IN 2 SECONDS
          LET ROUTE=ORIGINAL
      LET F.CAPACITY = FULL
GO TO CAR
          OTHERWISE.....
          FILE THIS CAR IN QUEUE
    LET CARS.BEFORE = N.QUEUE

    IF CARS.BEFORE IS GREATER THAN 4, LET ROUTE=ALTERNATE
GO TO MORE.CARS
          ELSE.....
          LET ROUTE=ORIGINAL
GO TO MORE.CARS
'AAA'     LET AUTO(TRAVEL.ON)=CAR
          LET PLACE(TRAVEL.ON)=ROUTE
'DESTROY'    DESTROY THIS ARRIVAL
RETURN
'NEXT.INTERSECTION' IF ROUTE EQUALS ALTERNATE GO TO ALT
ELSE
        IF ORG.QUEUE IS EMPTY AND S.SIGNAL=GREEN AND S.CAPACITY=UNUSED,
        SCHEDULE A TRAVEL.ON IN 2 SECONDS
    LET S.CAPACITY=FULL
LET ROUTE=FINAL   GO TO AAA


        OTHERWISE.....
  LET ROUTE=FINAL
        FILE THIS CAR IN ORG.QUEUE
        GO TO DESTROY
'ALT'   IF ORG.QUEUE IS EMPTY AND S.SIGNAL=RED   AND S.CAPACITY=UNUSED,
        SCHEDULE A TRAVEL.ON IN 2 SECONDS
    LET S.CAPACITY=FULL
  LET ROUTE = FINAL          GO TO AAA
        OTHERWISE.....
  LET ROUTE=FINAL
        FILE THIS CAR IN ALT.QUEUE
        GO TO DESTROY
'CAR' IF IDENTIFIER = 1 LET PATH   = ROUTE    GO TO AAA
  OTHERWISE GO TO AAA
'MORE.CARS' IF IDENTIFIER = 1 LET PATH = ROUTE   GO TO DESTROY
  ELSE GO TO DESTROY
END
```

`''` A CAR COMPLETES PASSAGE THROUGH AN INTERSECTION


```
EVENT TRAVEL.ON SAVING THE EVENT NOTICE
        DEFINE TIME AS A REAL VARIABLE
    LET CAR=AUTO(TRAVEL.ON)
    LET ROUTE=PLACE(TRAVEL.ON)
     IF ROUTE = FINAL, GO TO FINAL
  ELSE
'FIRST'  LET F.CAPACITY=UNUSED
IF F.SIGNAL IS NOT EQUAL TO GREEN
 GO TO WAIT
ELSE
   IF QUEUE IS NOT EMPTY
   REMOVE FIRST CAR FROM QUEUE
   GO TO A
   ELSE
   GO TO C
'WAIT'  DESTROY THIS TRAVEL.ON
   RETURN
'A'     SCHEDULE THIS TRAVEL.ON IN 2 SECONDS
        LET    AUTO(TRAVEL.ON)=CAR
        LET    PLACE(TRAVEL.ON)=ROUTE
IF IDENTIFIER IS NOT EQUAL TO 1, DESTROY THIS CAR
        RETURN
     ELSE
        GO TO CARRY.ON
'B' RETURN
'FINAL'LET S.CAPACITY=UNUSED
IF S.SIGNAL IS NOT EQUAL TO GREEN GO TO RED
ELSE
        IF ORG.QUEUE IS NOT EMPTY
          REMOVE FIRST CAR FROM ORG.QUEUE
      GO TO A
        ELSE
'C'  IF IDENTIFIER IS NOT EQUAL TO 1         DESTROY THIS CAR
                                             DESTROY THIS TRAVEL.ON
                                             RETURN

        ELSE
        GO TO CARRY.ON
 'RED' IF ALT.QUEUE IS NOT EMPTY
         REMOVE FIRST CAR FROM ALT.QUEUE
   GO TO A
    ELSE
        GO TO C
'CARRY.ON'    LET TIME=DISTANCE(ROUTE)/SPEED(ROUTE)*DENSITY(ROUTE)*
              BUNCH(ROUTE)*CONDITIONS(ROUTE)
        SCHEDULE AN ARRIVAL IN TIME HOURS
PRINT 1 LINE WITH TIME AND  ROUTE THUS
AT CARRY.ON THE NEXT ARRIVAL IS *.**** HOURS FOR  ROUTE **
     ADD 1 TO LOAD(ROUTE)
     LET AUTO(ARRIVAL)=CAR


     LET PLACE(ARRIVAL)=ROUTE
  IF ROUTE = FINAL GO TO B
  ELSE
   LET PATH = ROUTE
  GO TO B
END
```

`''` DRIVING CONDITIONS CHANGE WITH TIME OF DAY

```
EVENT DRIVING.CONDITIONS SAVING THE EVENT NOTICE
IF MOD.F(TIME.V,24)=6, LET VISIBILITY=GOOD  LET TIME=14   GO SET
ELSE  LET VISIBILITY=BAD  LET TIME=10
'SET'  FOR J=1 TO 4, LET CONDITIONS(J)=VISIBILITY
RESCHEDULE THIS DRIVING.CONDITIONS AT TIME.V + TIME
PRINT 1 LINE WITH VISIBILITY AND TIME THUS
    VISIBILITY = ****        TIME = ****
RETURN
END
```

195

```
   EVENT TRACE
LET BETWEEN.V = 'TRACE.PRINT'
 RETURN    END




 ROUTINE TO REPORT

  IF TOTALO = 0 ,GO TO NO.TOTALO      ELSE
  LET MO = SUMO / TOTALO
  LET SO = SQRT.F((SUMSOO-((SUMO*SUMO)/TOTALO))/(TOTALO-1))
  'NO.TOTALO' IF TOTALA = 0 GO TO NO.TOTALA ELSE
  LET MA = SUMA / TOTALA
  LET SA = SQRT.F((SUMSQA-((SUMA*SUMA)/TOTALA))/(TOTALA-1))

  'NO.TOTALA' PRINT 2 LINES WITH MO, SO, MA, SA THUS
     TRIP TIME USING ORIGINAL ROUTE...  AVG= **.**   STD.DEV= **.**
     TRIP TIME USING ALTERNATE ROUTE... AVG= **.**   STD.DEV.= **.**
SKIP 3 OUTPUT LINES
 IF TOTALO = 0, GO TO NO.HISTO ELSE
PRINT 1 LINE THUS
     HISTOGRAM OF TRIP TIMES OVER ORIGINAL ROUTE
PERFORM HISTOGRAM GIVEN DURTO(*)
SKIP 3 OUTPUT LINES
 'NO.HISTO' IF TOTALA = 0, GO TO NO.HISTA  ELSE
PRINT 1 LINE THUS
     HISTOGRAM OF TRIP TIMES OVER ALTERNATE ROUTE
PERFORM HISTOGRAM GIVEN DURTA(*)
 'NO.HISTA' RETURN
END




ROUTINE DENSITY(ROAD)
IF LOAD ¤ LOW.DENSITY, RETURN WITH LOW.DENSITY.VALUE  ELSE
IF LOAD   HIGH.DENSITY,  RETURN WITH HIGH.DENSITY.VALUE  ELSE
RETURN WITH (LOAD/HIGH.DENSITY)*(HIGH.DENSITY.VALUE-LOW.DENSITY.VALUE)+
LOW.DENSITY.VALUE
END




ROUTINE BUNCH(ROAD)
IF LOAD ¤ LOW.BUNCHING,  RETURN WITH LOW.BUNCHING.VALUE  ELSE
IF LOAD   HIGH.BUNCHING,  RETURN WITH HIGH.BUNCHING.VALUE    ELSE
RETURN WITH (LOAD/HIGH.BUNCHING)*(HIGH.BUNCHING.VALUE-LOW.BUNCHING.VALUE)+
LOW.BUNCHING.VALUE
END




'' SET SYSTEM VARIABLE TO THE CURRENT FIFTEEN MINUTE PERIOD

EVENT ARRAY.VALUE SAVING THE EVENT NOTICE
   IF MOD.F(TIME.V,24) IS LESS THAN 9.50
     LET K = K+ 1
     RESCHEDULE THIS ARRAY.VALUE IN 15 MINUTES
RETURN
ELSE
     SCHEDULE THIS ARRAY.VALUE IN 21.5 HOURS
        LET K = 0
RETURN
END
```

```
ROUTINE TRACE.PRINT
GO TO LABEL1, LABEL2, LABEL3, LABEL4, LABFL5, LABEL6,LABEL7,LABEL8
PER EVENT.V
'LABEL1'  PRINT  1 LINE WITH TIME.V THUS
TIME-V DRIVING CONDITIONS IS *****.****
RETURN
'LABEL2'  PRINT 1 LINE WITH TIME.V AS FOLLOWS
DEPARTURE TIME-V IS *******.****
RETURN
'LABEL3'  PRINT 1 LINE WITH TIME.V AS FOLLOWS
OUR.CAR  TIME-V IS ******.****
FETURN
'LABEL4'  PRINT 1 LINE WITH TIME.V AS FOLLOWS
ARRAY.VALUE TIME-V IS ****.****
RETURN
'LABEL5'  RETURN  ''FIRST SIGNAL CONTROL
'LABEL6'  RETURN  '' SECOND SIGNAL CONTROL
'LABEL7'  RETURN  ''TRAVEL.ON
'LABEL8'  RETURN  ''ARRIVAL
          END
```

The fifth card provides timing sequences for the traffic lights.

$$60\ \ 60\ \ 45\ \ 60$$

The amount of competing traffic for any particular time is determined from the array of input data set up indicate traffic levels for ten 15-minute intervals. The above statements remove the first seven hours from the traffic so that only a 10 valued array is necessary on cards six through eight.

```
2 10 10 10 10 10  5 3 2 1
3 15 15 15 15 15  7 4 3 2
4 20 20 20 20 20 10 5 4 3
```

The SIMSCRIPT II version of the drive to the office example does the same things as the GPSS version. There are radical differences in how the simulation languages operate. The understanding of the language is different. The readability in English language-like statements is contrasted to the block diagram statement form. The results are equivalent. The amount of effort is considerably different. Computer running time itself is not the critical resource, but rather the length of time it takes an individual to develop a working program. The listing of the SIMSCRIPT II program starting on page 191 gives an overview after the parts have been covered.

PROBLEMS

1. The approach used for the drive to the office follows the simple and direct approach of the initial GPSS treatment of Chapter 2. Consider instead the alternate, generalized approach attempted in Chap-

ter 6. Now for the SIMSCRIPT II implementation introduce the following concepts:

    (a)   Generalized number of routes and intersections

    (b)   Record the performance of particular trips—best, worst, chronological—and only those over some routes.

2. Describe the debugging superstructure that should be added to the basic model to facilitate the debugging process. Include:

    (a)   Ability to trace sequences of events.

    (b)   Quantify records of traffic volumes.

    (c)   Indicate the current degree of progress when the bug occurred.

    (d)   Restrict the amount of output to provide a limited history rather than the full detail of every event.

3. The hardest areas to model are those with extensive simultaneous interaction. Questions such as when to schedule overtime for a machine shop, how to schedule traffic signals over an extensive route structure, and the adequacy of numerous services provided by a time—shared computer installation may have to be considered. These require extensive analysis as part of the modeling process. Outline the key elements and their interactions to show the structure of general models for the listed areas.

BIBLIOGRAPHY

1. *The SIMSCRIPT II Programming Language*, P. J. Kiviat, R. Villanueva, and H. M. Markowitz, Prentice-Hall, Englewood Cliffs, N.J., 1969.
2. *The SIMSCRIPT II Programming Language Reference Manual*, P. J. Kiviat and R. Villanueva, Prentice-Hall, Englewood Cliffs, N.J., 1969.
3. *The SIMSCRIPT II Programming Language: IBM 360 Implementation*, P. V. Kiviat, H. J. Shukiar, J. B. Urman, and R. Villanueva, RM-5777-PR, The RAND Corporation, Santa Monica, Calif., 1960.

Part II

APPLICATIONS

This section reviews a series of simulation applications. The difficulty lies not in the particular applications selected; but in reducing each application to a chapter rather than a book. Since, each chapter must communicate a feeling of confidence that other dissimilar applications can also be simulated effectively. Obviously actually developing a simulation is the best way to obtain confidence in the technique. Reading what has been done is only second best.

The different problem areas selected are resource management, scheduling, and the comparison of systems from both developmental and operational viewpoints. These broad topics cannot be looked at abstractly. Instead specific applications must be used to show the customer's need and intent. Each application is presented in the problem language. The simulation approach is emphasized with regard to goals and level of detail. The actual details are not included, because a particular implementation depends on the experience and preferences of the modeler. As yet model construction is an art rather than a science. If it has been made to work in a reasonable time, it is considered an adequate model.

Criteria applied to the selection of each model include:

How was the problem defined?
Where were the input data obtained?
How did the model evolve?
How should the results be presented?
To what degree are the results valid?

The basic question of whether or not simulation achieved the goals set or the model was quite obviously avoided by selecting only successful applications.

199

Some of the references compare simulation and analytic methods. In general these applications, in their full state, are beyond the capability of the analytic method, at least under the problem definition used. If the systems are partitioned, the subsystems might well be approached analytically, but that changes the problem.

The language used in each application is not emphasized. While it is true that the effort to obtain the results might vary considerably depending on the language, these applications may be simulated in a number of ways. Discussion of those applications of greater complexity is found in Chapter 16.

8

Illustrative Example I—
Prediction of Passenger
Railroad System Performance

The economic justifications for railroad developments are based on a number of factors, including vehicle capacity and speed, passenger demand, trip profiles, schedules and right-of-way speed, and headway restrictions. The effects of new vehicles in the highly constrained environment of intercity railroad passenger service represents one simulation example from this problem class. The goal of simulation is to predict vehicle utilization and required fleet size. From these data the capital and operating costs can be determined, enabling a prediction of profitability.

This chapter describes some simplified approaches to railroad scheduling that preceded the development of the simulation approach and then considers the general ground transportation model with its many interacting factors—schedules, vehicle characteristics, passengers, and routes. This model is constrained to one type of transportation rather than the general problem of getting people from home to destination and back over a variety of transportation modes. The model developed, however, is typical of transportation models which could be used for any mode of transportation. This model represents one of the more complex transportation systems since it involves different sizes for trains and their numerous storage yard locations. The model could be made to interface with models representing other modes of transportation. In this manner a total transportation system model would be aggregated.

Basic Scheduling Considerations

A fundamental question in determining the efficiency of a given schedule from the operator's point of view is, "how many vehicles do I need to meet this given schedule?" The answer to this question yields infor-

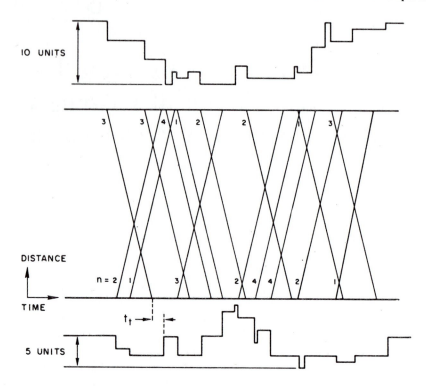

TOTAL VEHICLE REQUIREMENT = 10 + 5 = 15 UNITS

Figure 8.1   Determination of vehicle requirements.

mation on the average vehicle utilization, which affects average unit costs. The problem is complicated by the ability to vary train lengths only at a limited number of locations and the need to schedule departure intervals according to varying demand.

Instead of starting with the full ramifications of the complex system, let us first consider the simple case presented in Figure 8.1, vehicles providing transportation between two points. The vertical dimension between the two horizontal straight lines represents the distance between the two stations, while the horizontal lines mark the time of day. The diagonal lines, therefore, represent the movement of vehicles between the two terminals. The slopes of these lines vary with vehicle speed. As a preliminary to the scheduling process a number is noted at the origin of each train, representing the number of vehicles for that train.

Traces above and below the horizontal lines show the instantaneous vehicles inventory at each terminal, starting with some arbitrary level at the beginning of the daily cycle. As vehicles leave a terminal, the inventory trace makes a step-down equal to the number of departing vehicles. As another train arrives at the same terminal a step upward equal to the arriving train size is made after a turnaround time, t_1. When the number of departures is equalized between the terminals, both inventories will return at the end of the cycle to their original value. The minimum number of vehicles needed to keep this schedule can then be determined. The answer indicated in Figure 8.1, 15 vehicles, could not have been determined by considering any *peak* in the traffic or the total number of vehicles in motion at any time. Thus there are no obvious simple analytical solutions or shortcuts to develop the answer for even this simple case.

The simplest possible case of vehicle flow in the two-station system is shown in Figure 8.2. Here the schedule is continuous, with a constant headway between departures, and has a departure at the same time from each station. Furthermore the sizes of the transport units are assumed so that each train is made up of just one unit. By inspection the number of vehicles required to satisfy this schedule is given by determining

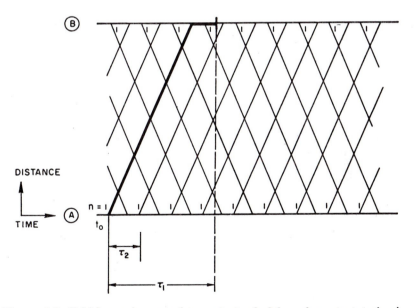

Figure 8.2 Vehicle requirements for constant schedule and constant train size.

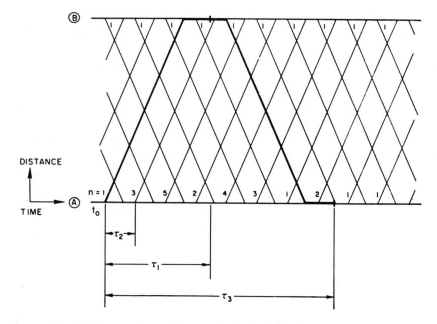

Figure 8.3   Vehicle requirements for constant schedule and constant train size from one terminal.

the departure requirements within the time equal to the time required for one trip and one turn around. After this time span all other departures can be satisfied by vehicles already counted.

A slightly more realistic case is one in which there is a continuous schedule with constant departure headway and single-unit vehicles leaving one terminal, while the total number of vehicle departures from the other terminal is a maximum. In this case, shown in Figure 8.3, the maximum number of vehicles can be determined by considering the departures during a complete cycle of time, τ_3. By inspection it may be determined that if the placement of τ_3 is chosen so that the summation of all vehicles leaving station A within the time increment τ_3 is a maximum, then all future departure requirements can be met by vehicles already counted during τ_3. Assuming the increment τ_3 is properly selected so as to maximize the summation

$$\sum_{\tau_0}^{(t_0+\tau_3)} (n_1) \tag{1}$$

where n_i = number of vehicles in trains leaving Station A, then the number of vehicles required is given by

$$N = \sum_{t_0}^{(t_0+\tau_1)} (n_1) + \sum_{(t_0+\tau_1)}^{(t_0+\tau_3)} (n_1 - 1) + \frac{<\tau_1}{\tau_2>}. \tag{2}$$

The time increments τ_3 and τ_1 are related to the variables describing the system; thus

$$\tau_3 - \tau_1 = \frac{D}{V_B} + \frac{N l_s}{60} + \frac{t_t}{60} \tag{3}$$

taken to the next higher integral multiple of departure headway, τ_3, and

$$\tau_1 = \frac{D}{V_B} + \frac{(N+1)t_s}{60} + \frac{t_t}{60} \text{ exactly.} \tag{4}$$

Let D = distance between terminals, miles.

V = average block velocity, mph.

N = number of intermediate stops between terminals

t_s = time spent at each station, min.

t_1 = turn-around time (including in-station time at departure station), min.

Equation (2) gives a simple method of estimating the minimum inventory requirements for the simple type of case shown in Figure 8.3 without developing a complicated scheduling diagram and considering yard inventories throughout the day. This approach is useful for simple cases because the passenger flow between any pair of cities tends to peak in one direction in the morning and the other direction in the afternoon. If the smallest unit of transportation is sufficiently large, as in the case of unit trains, the flow opposite to the peak flow during the peak hours can actually be handled by a succession of minimum-sized, single-vehicle trains. Thus the vehicle requirements for these simple cases can be determined by simple inspection of departure demands.

This simplified type of analysis of vehicle requirements is unsatisfactory for determining the efficient utilization of railroad equipment. First, transportation between only two cities is unrealistic, other stations must be included. Also, trains do not always maintain a constant train length from origin to destination. Cars are often dropped at intermediate stations where a significant percentage of the traffic terminates. In these cases an exceptionally poor load factor would be encountered by maintaining constant train size. There would be little opportunity to turn around some vehicles for a second trip during a peak period.

New or special rolling stock should be operated on regular, frequent schedules at the highest practical rates of utilization. To handle these factors properly, one needs more sophisticated analytical techniques than have been indicated so far.[1-5]

Difficulties of the More General Problem

Determination of fleet size for a given schedule includes many factors. First, the size required for each train must be determined. This is a function of the maximum number of people simultaneously on board the train. This maximum load, in turn, is a function of those stations the train is to service, daily demand between those stations, distribution of demand with time of day, and schedule frequency. These factors are not independent of each other and may also depend on other factors. Thus daily demand and its distribution throughout the day between any two points will depend on the trip time, schedule frequency, and trip cost, as well as on the desirability of competing modes between those areas and the efficiency of connecting transportation available at each station.

The interaction of all these factors in determining the overall demand for transportation on a given system is quite complex, and no universally accepted analytical prediction techniques exist although, there are, in general form, *gravity* and *elasticity* models.[6,7] An interconnected set of simulation models could be developed to represent the transportation system. It would be composed of a series of models covering limited aspects of transportation. One such limited model could include the various factors governing fleet size assuming a given demand, its distribution, and an accommodating schedule. There is a relationship between demand levels and corresponding train sizes. Once train sizes are determined there is a straightforward relationship among block times, storage yards, turn-around times, and car requirements for a particular system. The model described below is used to explore these relationships for the multi-station case, while treating travel demand and its distribution as arbitrary, known functions.

The function of the model may be considered with reference to Figure 8.4, which indicates the elements of a three-station case. For a train departing from A and going to C the size of the train is determined by the maximum passenger load that can occur either between A and B or B and C. The number of passengers boarding each vehicle is determined by departure time from station A, and the number of passengers accumulated according to the distribution since the last departure time, Δt.

The movement of individual trains through the simulated system de-

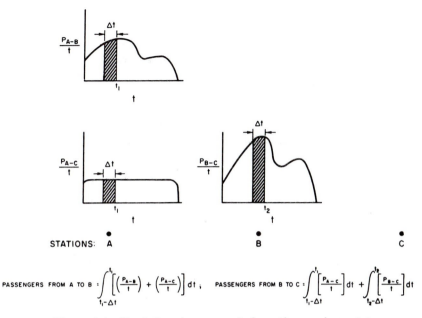

Figure 8.4 Simulation of passenger train seating requirements.

termines the required train size according to the point where the maximum number of passengers are on board each train. Trains originate and terminate at selected yard locations. For each yard the model keeps a running record of the number of vehicles stored during the day as shown in Figure 8.1. For the situation in which the minimum value for each yard reaches zero, the starting inventory represents the number of vehicles to be supplied by that yard. The total fleet size is the summation of the contributions of all yards.

The simulation model provides insight into the operation of the transportation system by representing individual trains of varying sizes and different passenger loads. The model requires a description of the transportation system as input and, in turn, provides output statistics representative of actual operations. Individual actions are represented with adequate detail to provide meaningful indications of system performance.

The model is summarized in the following key inputs and outputs:

Inputs

1. Total daily demand on each pair of stations on a transportation network.
2. Demand distribution with time of day.
3. Departure schedule from each originating point.

    4. Train route description and schedule.

    5. Distance between stations.

    6. Average train speed.

    7. Turn-around time.

    8. Minimum headway between trains.

Outputs

    1. Hourly system statistics.

    2. Vehicle fleet size.

    3. Waiting time mean and distribution.

    4. Number of people boarding at each station.

    5. Cumulative load factor.

    6. Statistics for each train.

    7. Times of origination and termination.

    8. Train number, size, and average speed.

    9. Peak passenger load and where it occurred.

  10. Total number of passengers hauled.

  11. Passenger service index, such as percentage of total travel time spent in motion.

In addition the safety aspect is simulated by ensuring that no trains ever approach one another within a specified block distance.

Overall Model Structure

This simulation model is general in nature and can be used to simulate ground transportation systems having up to 50 stations. The model

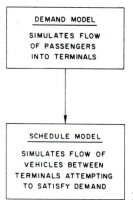

DEMAND MODEL

SIMULATES FLOW
OF PASSENGERS
INTO TERMINALS

SCHEDULE MODEL

SIMULATES FLOW OF
VEHICLES BETWEEN
TERMINALS ATTEMPTING
TO SATISFY DEMAND

Figure 8.5 Demand and schedule model.

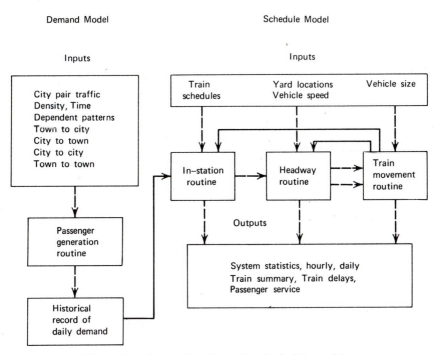

Figure 8.6   Interacting demand and schedule models.

was originally developed using GPSS III and later updated with features of GPSS 360.

There are two principal aspects to any transportation system; first, passengers arriving at the various stations and intending to travel from one place to another and second, vehicles traveling between stations attempting to satisfy these demands.   As shown in Figure 8.5, the logic model consists of two interacting submodels, the Demand Model which simulates the flow of passengers into the various stations, and the Schedule Model which simulates the flow of vehicles along the rail lines connecting the terminals.

These models interact as described in Figure 8.6.   As shown the Schedule Model consists of subroutines in which trains enter specified stations according to defined schedules, pick up and discharge passengers, depart, and proceed to the next station.

The Demand Model

At intervals throughout the simulated 24-hour day, groups of passengers are created for each city pair.   Each group size is computed, based

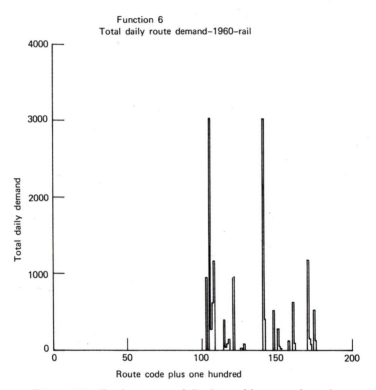

Figure 8.7   Total passenger daily demand between city pairs.

both on the total passengers per day who travel between that particular pair of cities, as shown in Figure 8.7, and on the percentage of that total occurring at that particular time of day. A different time distribution is used for each direction of travel between the paired cities. Typical time distribution curves for passenger demand are shown in Figures 8.8*a* and *b*. These represent daily inbound and outbound distributions of passengers traveling between a small community and a large city. There is a very sharp peak in demand during the morning for service toward the large city. In the afternoon there is a similar, but less sharp, demand peak from the large city. Long-haul traffic has less well-defined morning and afternoon peaks. The model uses these graphic relationships to establish the characteristics of the service required to meet passenger demand.

The daily passenger demand data, consisting of group size, origin, destination, and time of day, are recorded on magnetic tape and used as a standard input to the Schedule Model. Each alternative system is evaluated using this same standard passenger demand data. Alternative systems may then be compared on a relative basis. As the simulated time of day progresses in the Schedule Model, this magnetic tape is examined for passengers desiring service at that particular time. From these demand data the passengers join queues representing the north- or southbound platforms of the various stations, where they wait for the arrival of a train scheduled to stop at their destination.

The demand model is based on the passenger activity expected during three-minute intervals throughout the 24-hour day. The input data are obtained by taking the total daily passenger traffic between two cities and, in accordance with the hourly passenger demand pattern, dividing this number by 20 for the passenger-arrival rate over the three-minute intervals. These data were subjected to randomizing influences by multiplying the resultant data by random number ratios that could either increase or decrease the number. To reduce the possibility of having

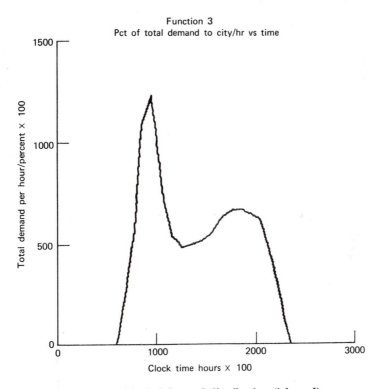

Figure 8.8a Typical demand distribution (inbound).

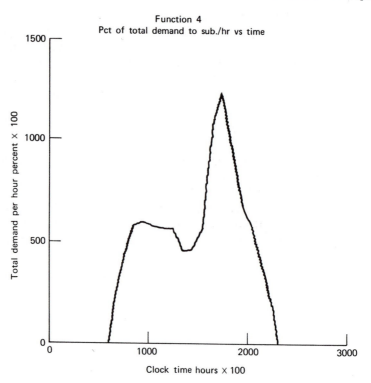

Figure 8.8b Typical demand distribution (outbound).

events occur at the same time, a random delay was introduced, giving an equal probability of passengers arriving at any time within the three-minute interval. The minimum interval between two groups—the time resolution of the model—was 1/100 of an hour.

Transfer of these data to magnetic tape requires certain changes to accommodate the GPSS tape record structures. The main difficulty is if the daily number of passengers exceeds the storage capacity of a reel of magnetic tape. This occurs because GPSS uses the maximum possible space, 100 parameters, on the magnetic tape even if the transaction has needed only one parameter. This difficulty was overcome by gathering the data from individual transactions to fill up all 100 parameters of the transaction and then storing the record on tape. The result was to temporarily store the number of people traveling in the Demand Model until the gathering transaction made its rounds on a three-minute schedule. This process added to the Schedule Model the requirements to take each of these composite transactions apart before the passengers could join the queues at the station.

The Schedule Model

The Schedule model simulates the flow of vehicles along the rail lines connecting the terminals and obtains statistical data on the entire system as well as on each train. The rail system is assumed to be a 2-track system with the stations located on sidings at each side of the main tracks. As such a through train can bypass a local train that is in a station. The bypass capability can be deleted to simulate restricted right-of-way transportation systems, such as tunnels. Similarly additional sidings can be added to permit between-station passing, if desired.

The Schedule model must be provided with input information:

1. Departure time of each successive train from each yard. Figure 8.9 shows the plot of typical input data.

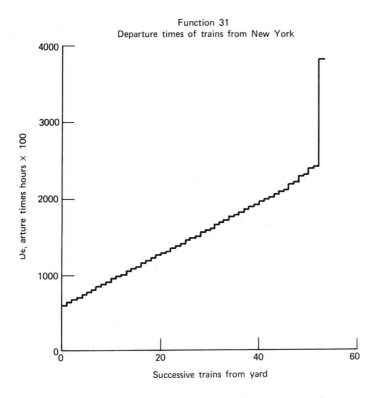

Figure 8.9 Typical train departure schedule input function.

2. Stations serviced by each successive train from each yard. Figure 8.10 indicates stations skipped in plotted form.

3. Distance between each pair of stations.

4. Specific average vehicle speed permitted between each pair of successive stations.

5. Time spent in each station.

6. Number of seats per car.

7. Length of track section that will be reserved for only one train at a time.

8. Turnaround time required to ready cars for reuse.

The Schedule model is used in a two-phase regime; that is, the model must run twice. In the first phase the trains are assumed to be able to seat all the passengers provided by the demand model. The simulation, operating under this constraint, determines the *minimum* size for

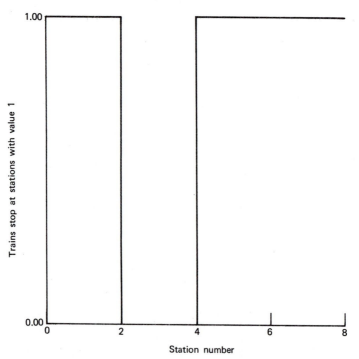

Figure 8.10   Typical input train function describing stations serviced.

each train, between storage yards. This procedure continues for the full 24-hour period to determine sizes for all trains. The second phase of the model uses for *input* the number of vehicles determined in phase one as required for each train and then proceeds to determine the fleet size.

The train movements are scheduled so that each yard has the same number of vehicles at the beginning and end of each day. The statistics concerning passenger delays, train performance, and system conditions are determined during the running of appropriate parts of the model.

The overview of the simulation model shown in Figure 8.11 begins with the train's departure from the storage yard according to the established schedule. A transaction representing the train is labeled with the following: originating yard, route stops, anticipated speed characteristic, and final destination. At this point, phase one, the eventual number of vehicles has yet to be determined.

Transactions that represent trains and transactions that represent passengers interact during the "in-station subroutine." The Demand Model has provided the passengers for this day's operation, and the Schedule Model has provided the trains and processed the passengers into station queues. The Schedule Model boards passengers bound for destinations serviced by a particular train or by a connecting train. Likewise passengers detrain at this station because they have reached their destination or wish to board a connecting train. A current record of train population is maintained to permit later calculation of passenger miles. As the train proceeds to other stations a record is maintained of the maximum train population. These data will be used after the first run to determine the train size.

There is a nominal time that the train will spend at the station platform before departing for the next station. The sections of track between adjacent stations are divided into blocks of specified length. Succeeding trains will not be permitted to enter a block occupied by another train. Based on the train speed and block length, the train will in time advance from one track block to the next.

Upon arrival at its next stop a computation is made of the passenger miles and seat miles accumulated by both the total system and that particular train. The passengers destined for this station are detrained, and the train population is accordingly decremented. As before, the platform is examined for people waiting for this train, and the appropriate people are entrained. The train population is incremented accordingly, and a test is made to determine if the train population exceeds any previous maximum. If so, the value and the station number at which this occurred are stored. Again, the train spends a selected time

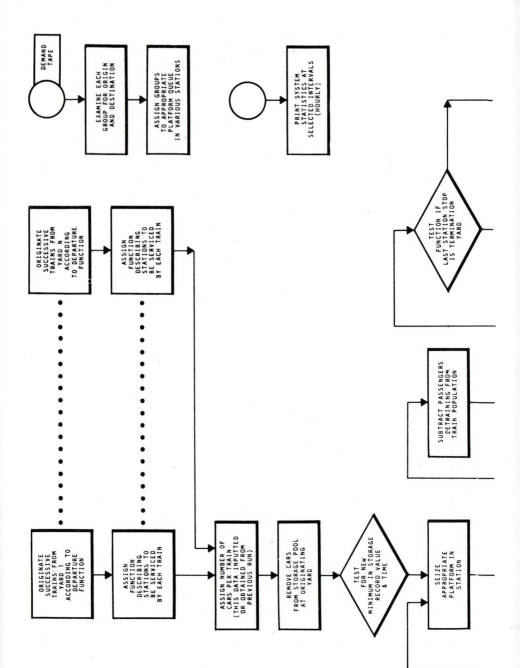

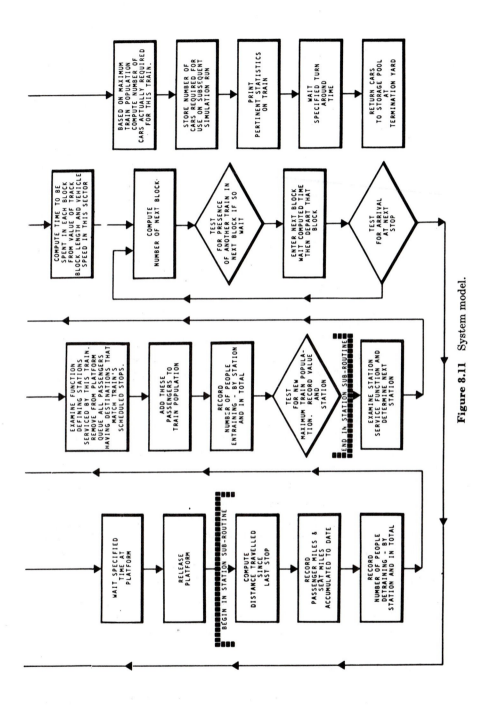

Figure 8.11 System model.

217

at the station platform and then departs. The above process is repeated for each train at each of its scheduled stations until the train arrives at its termination station.

A single subroutine is used to process the movements of trains and passengers in the stations. This "In-Station Subroutine" exists only once within the model. It is structured with indirect addressing so that each entering transaction carries with it the needed labels to describe the actual station number, train characteristics, and passenger demand.

After discharging its passengers at the station platform, the train waits an amount of time equal to the specified turnaround time before returning the cars to the storage pool for subsequent trains originating from that yard. The average speed of the train from its origination, including platform time, is computed and recorded.

At the final station the remaining passengers are detrained and the stored value of maximum train population is examined. From the maximum value the number of cars (or units) required is computed and recorded. At the end of the first phase of the simulation run, the number of cars required for each train has been determined.

The minimum fleet size is determined from the collection of data listing the maximum number of cars required from each yard. Since the number of cars for each yard is not known at the start of the second phase, a large nominal value is assigned as the yard population. The maximum number of cars removed from the yard at any time is the required information. When each train leaves the yard, the remaining inventory is compared with the previous minimum. If this is now a new minimum, the value and time are retained as historical results. After the daily peak traffic has subsided, the vehicles are restored to yards in order to balance the system for the next day. Abnormal traffic patterns could be accommodated by this balancing routine.

The simulation output is the hourly system statistics and a series of text statements describing the characteristics of each train as it terminates. A sample of the GPSS III output is shown on page 219–221.

An interesting aspect of the Schedule model is the lack of any non-deterministic element. There are probabilistic factors in the Demand Model. If weather conditions and breakdowns were to be included in the determination of fleet size by including a reserve for contingencies, then probabilistic elements could appear in the model.

Some Typical Results

During the development of the model, some sample runs were made of typical transportation situations. One model run concerned the flow

SAVEX LOCATION 220 CONTAINS THE TRAIN IDENTIFICATION NUMBER--ORIGIN*1000 PLUS DIRECTION *100 PLUS ITH TRAIN FROM YARD
SAVEX LOCATION 221 CONTAINS TIME TRAIN TERMINATED
SAVEX LOCATION 222 CONTAINS TIME TRAIN ORIGINATED
SAVEX LOCATION 223 CONTAINS NUMBER OF CARS ON TRAIN
SAVEX LOCATION 224 CONTIANS AVERAGE SPEED OF TRAIN
SAVEX LOCATION 225 CONTAINS TOTAL MILES TRAVELLED
SAVEX LOCATION 226 CONTAINS TOTAL PASSENGERS CARRIED
SAVEX LOCATION 227 CONTAINS MAX. PASSENGERS AT ANY TIME
SAVEX LOCATION 228 CONTAINS STATION WHERE MAX LOAD OCCURRED
SAVEX LOCATION 229 CONTAINS TOTAL PASSENGER MILES/ONE HUNDRED
SAVEX LOCATION 230 CONTAINS NUMBER OF TIMES TRAIN DELAYED BECAUSE OF INSUFFICIENT HEADWAY
SAVEX LOCATION 231 CONTAINS AVERAGE LOAD FACTOR EQUAL TO 1000*PASSENGER MILES DIVIDED BY SEAT MILES

SAVEX NR,	VALUE	NR,	VALUE	NR,	VALUE	NR,	VALUE	NR,	VALUE
220	1010	221	1208	222	1050	223	5	224	75
225	91	226	326	227	326	228	2	229	265
231	728								

SAVEX NR,	VALUE	NR,	VALUE	NR,	VALUE	NR,	VALUE	NR,	VALUE
220	5109	221	1208	222	1050	223	8	224	75
225	91	226	634	227	634	228	3	229	530
231	910								

SAVEX NR,	VALUE	NR,	VALUE	NR,	VALUE	NR,	VALUE	NR,	VALUE
220	1007	221	1226	222	900	223	8	224	75
225	226	226	716	227	616	228	2	229	1126
231	778								

SAVEX NR,	VALUE	NR,	VALUE	NR,	VALUE	NR,	VALUE	NR,	VALUE
220	8104	221	1226	222	900	223	7	224	75
225	226	226	628	227	524	228	5	229	1051
231	830								

TIME OF DAY IN HOURS TIMES ONE HUNDRED

CAR STORAGE STATISTICS

1200

STORAGE	CAPACITY	AVERAGE CONTENTS	AVERAGE UTILIZATION	ENTRIES	AVERAGE TIME/TRANS	CURRENT CONTENTS	MAXIMUM CONTENTS
NUMBER							
1	32767	192.57	.0059	249	928.06	187	200
5	32767	190.81	.0058	210	1090.34	167	200
8	32767	190.70	.0058	217	1054.57	173	200

STATISTICS ON PASSENGER QUEUES AT EACH STATION PLATFORM-NORTHBOUND PLATFORMS CARRY STATION NUMBER PLUS TEN

QUEUE	MAXIMUM CONTENTS	AVERAGE CONTENTS	TOTAL ENTRIES	ZERO ENTRIES	PERCENT ZEROS	AVERAGE TIME/TRANS	$AVERAGE TIME/TRANS	TABLE NUMBER	CURRENT CONTENTS
NUMBER									
1	616	177.77	3730	356	9.5	57.19	63.23	0	34
2	264	88.18	682	0	.0	155.16	155.16	0	44
3	118	53.02	216	0	.0	294.56	294.56	0	98
5	96	39.19	410	0	.0	114.71	114.71	0	34
6	70	25.16	118	0	.0	255.83	255.83	0	4
13	222	57.48	820	0	.0	84.12	84.12	0	82
15	594	84.15	2988	266	8.9	33.79	37.10	0	46
16	242	95.66	472	0	.0	243.20	243.20	0	64
17	146	59.32	606	0	.0	117.48	117.48	0	28
18	486	140.95	1820	72	4.0	92.94	96.77	0	74

$AVERAGE TIME/TRANS = AVERAGE TIME/TRANS EXCLUDING ZERO ENTRIES

MINIMUM NUMBER OF CARS IN STORAGE AND TIME WHEN MINIMUM OCCURRED IS GIVEN IN SAVEX LOCATIONS 440-450. MINIMUM STORAGE VA

LUE IS GIVEN IN SAVEX LOCATION 440 PLUS YARD NUMBER.TIME THAT MINIMUM OCCURRED IS GIVEN IN SAVEX LOC.441 PLUS YARD NO.

SAVEX NR,	VALUE	NR,	VALUE	NR,	VALUE	NR,	VALUE	NR,	VALUE
441	174	442	900	445	167	446	1150	448	173
449	1100								

SUMMARY OF PEOPLE BOARDING TRAINS.SAVEX LOCATIONS 421-430 CONTAIN NUMBER OF PEOPLE WHO HAVE ENTRAINED AT EACH STATION

SAVEX LOCATION 431 CONTAINS GRAND TOTAL OF PEOPLE WHO HAVE ENTRAINED AT ALL STATIONS

SAVEX NR,	VALUE	NR,	VALUE	NR,	VALUE	NR,	VALUE	NR,	VALUE
421	3696	422	638	423	856	425	3318	426	522
427	578	428	1746	431	11354				

SUMMARY OF PEOPLE DEBOARDING TRAINS.SAVEX LOCATIONS 401-410CONTAIN NUMBER OF PEOPLE WHO HAVE DETRAINED AT EACH STATION

SAVEX LOCATION 411 CONTAINS GRAND TOTAL OF PEOPLE WHO HAVE DETRAINED AT ALL STATIONS

SAVEX NR,	VALUE	NR,	VALUE	NR,	VALUE	NR,	VALUE	NR,	VALUE
401	3416	402	564	403	482	405	1762	406	456
407	326	408	660	411	7666				

SUMMARY OF PASSENGER MILES AND SEAT MILES.SAVEX LOC 451 CONTAINS TOTAL PASS. MILES.SAVEX LOC 452 LISTS TOTAL SEAT MILES

SAVEX LOC 453 GIVES SYSTEM LOAD FACTOR EQUAL TO 1000*PASSENGER MILES DIVIDED BY SEAT MILES

SAVEX NR,	VALUE	NR,	VALUE	NR,	VALUE
451	1059902	452	1444640	453	733

SUMMARY OF TRAIN STATISTICS OBTAINED WHEN EACH TRAIN TERMINATES of railroad passenger cars in intercity traffic between New York City and Washington, D.C. Traffic between New York, Newark, Trenton, Philadelphia, Wilmington, Baltimore, and Washington was considered. The demand for transportation between these cities corresponds to the level of average daily demand experienced in 1960. A schedule was formulated with trains running between Philadelphia and New York every hour in either direction and with hourly service between New York and Washington in each direction. Since the Washington trains stop at Baltimore, Wilmington, and Philadelphia, half-hourly service between New York and Philadelphia is thereby provided.

If service were provided by a pool of interchangeable, self-powered vehicles with a unit seating capacity of 80 people, Figure 8.12 indicates the number of cars required to satisfy the passenger demand as a function of the average vehicle speed between stations. These numbers represent an absolute minimum inventory; some cars would have to be added to allow for routine maintenance. Since the number of car miles traveled in each case does not vary significantly with speed, the

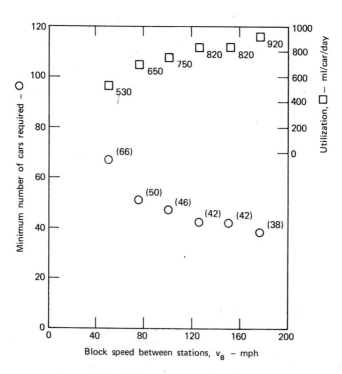

Figure 8.12 Effect of car speed on inventory.

average vehicle utilization in miles per day is easily computed from the number of cars required and the distances involved. This value is also shown in Figure 8.12. Due to the effects of station stops, turnaround requirements, and demand variations, the utilization does not increase linearly with speed. Both relationships shown in Figure 8.12 are actually step curves. Since finding the precise locations of the steps would involve taking data at many train speeds, curves have not been drawn through the discrete data points presented. Some additional runs were made to assess the effects of turnaround time required on the vehicle inventory. Figure 8.13 indicates the effect of increasing turnaround time from 10 minutes to 1 hour. The inventory requirements for the schedule considered are increased by about 20% for a 1-hour turnaround time.

Other information yielded by these computer runs is system load factor, which is important in determining the expected ticket price and average train length, which is important in determining crew costs. Having statistics on the operation of whole system permits calculation

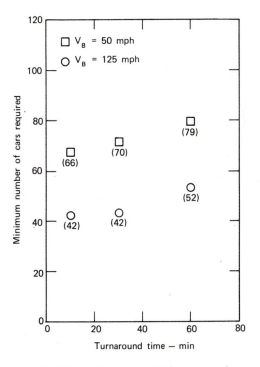

Figure 8.13 Effect of turnaround time on car inventory.

of car costs in dollars per car mile directly, rather than estimation of cost of operation on a fleet basis.

Comments

This model illustrates how a model has areas of potential extension within and beyond the practical constraints and boundaries already established. It also shows the ability of a model to be used for more than one purporse.

Large complex systems such as rail transportation can be modeled by handling different parts of the problem and then tying the pieces into a unified total system. This combination of models used to determine fleet size is only a part of the total transportation system. For example the Demand Model considers as fixed data the number of passengers and the times they require service. This is actually not the case, because faster and more frequent service could increase the demand, especially if the cost did not increase. The model of the mode or modes of travel selected by the public were reduced for this example to a constant condition.

The model of right-of-way usage was also simplified. Freight trains were not considered. Therefore the interaction of passenger trains with others of considerably slower speed could not be obtained readily, although it would be relatively simple to introduce this complication into the model structure.

Finally the model was not used as a tool for schedule development. This is an area of great potential use. Different schedules can be introduced as input data and the simulation used to evaluate their effectiveness. Graphic output to show the potential gains from varying the departure times of specific trains would be helpful in this area.

This model indicates the potential of the modeling technique in the area of rail system analysis. It should be noted, however, that transportation problems in general have been approached using the simulation technique. There exist many large and complex models serving ground and air transportation both in planning and operational roles.[8-10]

PROBLEMS

1. Set up a schedule with the size for each train to provide service among points A, B, and C. There are no further stops. The number of cars per train will differ to reflect the variation in service

demand as shown in Figure 8.8. Three hours are required to go from A to C. Between 7 a.m. and 7 p.m. provide hourly service in both directions. Balance the number of cars used each day. Storage yards are located at A and C. Twice as many passengers go from A to B as from A to C. Assume there are no right of way restrictions and there is an unobstructed track in each direction.

2. Estimate the additional effort required to convert the scheduling effort of (1) to represent a normal railroad environment.

3. Compare the railroad model with equivalent models for the scheduling of aircraft and buses. Is one general model a feasible approach? What is the influence of other factors on the schedules?

4. Indicate how the model would have to be changed if the demand were to be considered as being influenced by the service provided. How would this affect the evaluation of the simulation results?

5. Suggest an approach to use to simulate the entire transportation requirements of a region.

6. Flow chart the model structure to determine the maximum number of trains over a right-of-way during a fixed time period. In addition consider trains of different speeds adding stopping locations.

7. The operation of a fleet of taxicabs is to be simulated. Describe the ways to compare the performance of alternate concepts with regard to:
   (a) Territory covered.
   (b) Fleet size.
   (c) Dispatching rules.
   (d) Different demand patterns.

8. How can the factors of No. 7 be varied in a general simulation?

BIBLIOGRAPHY

1. Scheduling a Vehicle Between Origin and Destination to Maximize Traveler Satisfaction, D. Youn, *Proc. 22nd ACM Conf.*, Thompson, Washington, 233–245, 1967.

2. *Techniques of System Engineering*, S. M. Shinners, McGraw-Hill, New York, 1967.

3. Simulation of Railroad Operations, Railway Systems and Management Association, Chicago, Ill, 1966.

4. Operation of a Rapid Transit System—A Dynamic Programming Approach to the Solution of Optimum Train Schedules K. M. Dale, *Rec. IEEE Sys. Sci. Cyber. Conf.*, 85–92, 1968.

5. Application of Zone Theory to a Suburban Rail Transit Network, D. D. Eisle, *Traffic Quart.*, **22**, 1:49–67 (1968).

6. *Demand for Intercity Passenger Travel in the Washington-Boston Corridor,* Systems Analysis and Research Corporation, Boston, Mass., 1963.

7. *Multimode Assignment Models,* J. M. McLynn and R. H. Watkins, Davidson, Talbird, and McLynn, Bethesda, Md., 1965.

8. A Simulation for a Short Takeoff and Landing System Traffic Analysis, A. B. Newmann, *IEEE Trans. Sys. Sci. Cyber.,* **SSC-6,** 3:162–172 (1970).

9. Detailed Simulation of Military Aircraft Operations and Logistics, J. H. Keeney, *Proc. 4th Conf. Appl. Simul.,* 32–38, 1970.

10. Applying Simulation Techniques to Air Traffic Control Study, R. C. Baxter, J. Reitman, and D. Ingerman, *ibid.,* 39–44.

9

Illustrative Example II— Simulation as an Aid and Controller in Production

Scheduling—The Answer to the Proverbial "What If?"

Production management wants answers to four fundamental questions.

1. Where am I now?
2. What will happen if I do this?
3. How does this compare to what I did last time?
4. How will this change costs?

Data processing has provided answers to the first question. There are programs for inventory status, shipments to date, location of incoming items, and current status of purchase orders. The logical next extension is to go one step further into the management information system and provide answers to the questions

5. When will we ship this order?
6. Will the shipping date change if these orders are put on overtime?
7. Should we order one or two more facilities of this type?
8. What task priority keeps the shop most effectively utilized?

Questions of this nature are tied directly into the third area, what happened last time we did this, if there was a last time. Ability to answer all the questions sets up information to close the loop by understanding the interrelated cost considerations.

This chapter seeks to show how simulation techniques may be applied to a second category of questions. These are concerned with how to make the best use of available resources. Then, with historical cost data, there can be a projection of costs for alternative schedules. The primary effort is directed toward the scheduling rather than accounting aspects of the problem.

This application is an outgrowth of work originally sponsored by the Air Force[1] for predicting production rates impacted by conflicts for resources between established and developing types of integrated circuits. Solid-state production is a particularly appropriate area for computer usage, because there has been little accumulated experience for management to draw on. It is a dynamic field with new products constantly moving from research and development into production. It is both adding to and displacing existing products and technologies.

The production of integrated circuits consists of processing a batch of silcion slices through a number of separate processing and inspection steps. These slices are approximately $1\frac{1}{2}$ inches in diameter and eventually may yield up to 1000 separate electronic circuits. Figure **9.1** shows a magnified portion of the slice, showing only one circuit from several hundred circuits on the slice.

The production steps involve the use of various types of equipments: mechanical polishers, chemical etchers, diffusion furnaces, photoengraving apparatus, and testing devices. Different operator skills are required to use this equipment. The number of processing steps varies for the different device types, with 26 major steps the maximum required. Each major step may contain up to six minor steps.

Complications in the processing arise from the dynamic changes taking place in the technology. Circuits are rarely made the same way for 2 years, and most processes change within a year. The processing steps for a general circuit class may have branch points. Material moves along a sequence of common steps until a decision point is reached. After that point the slice becomes a unique circuit. Some process steps handle batches while others process the slices individually. The number of circuits on one slice is highly variable, with one good slice frequently yielding 1000 separate chips or circuits. For special and development circuit types this may satisfy an order. The yields are highly variable and generally rather low. Sometimes a complete batch, when processed together, may be faulty, but this is more characteristic of early steps, in which the value of the slice is still low. As the slice value increases it may become desirable to process only one slice at a time through some process steps.

Under these circumstances the manager seeks to know his current yield data and status of in-process inventory, and obtain and adhere to a production schedule. He is faced with a classical job shop problem and very little experience other than limited historical yield data to go on. Obviously simulation is one possible means of improving the situation.[2-21]

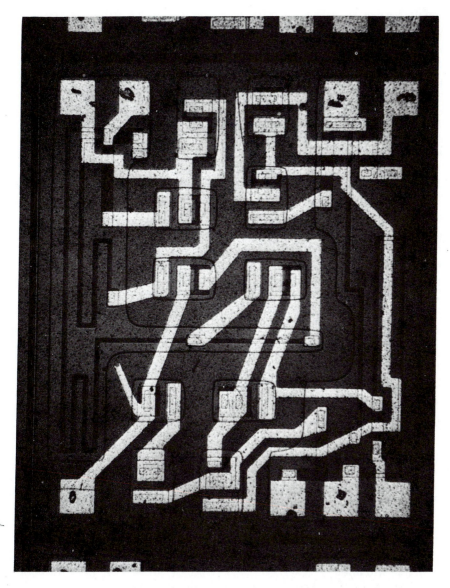

Figure 9.1 One example of an integrated circuit chip is master dice breadboard, which features a monolithic integrated layout.

Management Data Model Structure

The model described is the result of an evolutionary process. The original Air Force model in GPSS III considered the production process from a theoretical viewpoint in order to compare production times and costs between many short production runs of custom circuits versus a few long runs of standard circuits. This basic model has evolved through several stages into an everyday tool to assist management. The problem separates into two major and distinct areas: (1) inventory control, yield data, and prediction of eventual output, and (2) resource management, utilization, scheduling and costs, and all aspects of output prediction. The two are tied together since the first must provide the data base for the second. The first part could have been implemented in any language, but it is the requirements for the second that suggested the choice of a discrete event simulation language. Since this choice permitted all parts of the data processing system to be consistent with each other, a modified version of GPSS with permanent data storage on direct access devices was used. The logic and function performed in both the data processing and simulation aspects of the system are shown in Figure 9.2. The different elements in the system may be used for specific limited problems. For example the inventory control function needs only the blocks shown in dotted areas. Only for the full system is the entire flow chart called into use.

Total Inventory Control

To achieve inventory control a number of items must be handled by the data processing system. In addition there is a considerable amount of calculation to show current status and historical trends. The basic functions are the following:

- Slice movement for the previous day by step name.
- Yield data for each step of slices processed the previous day.
- Compare previous day's yield for that step with accumulated historical yield data.
- Project final number of slices using historical yield data.
- Locate all committed in-process slices by type, priority, and work order number.
- Stock uncommitted completed slices by type and quantity.
- Stock partially completed uncommitted slices available for differentiated processing.

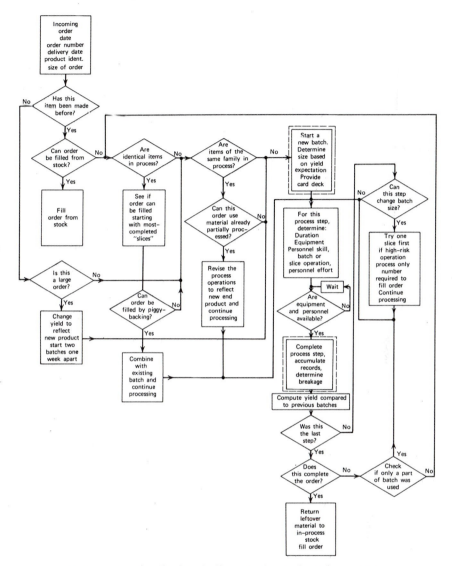

Figure 9.2 Simplified logic diagram of manufacturing system.

A mechanized system was instituted to obtain the data to follow each batch step by step. This was accomplished using a set of prepunched tabulating cards containing the common data of device name, work order number, and product code, and within the deck one card for each step number and name. This permitted reading of the card by both the

	RUN NUM	NUM OF SLICES	STEP NAME	PROD CODE	DEVICE NAME	STEP NUM
PRODUCTION CARD	06151		EMIT DIFF	4019	5VOLT SEC	13
CARD AFTER UPDATE KEYPUNCH	06151	0700	EMIT DIFF	4019	5VOLT REC	13

BATCH SIZE SLICES WHOLE · BROKEN SLICES

Figure 9.3 Card formats.

GPA MODE 4   CODE 5

AS OF 5.00 PM ON   728

		TOTAL SLICES AT STATION	TOTAL SLICES MOVED YESDAY	YIELD FOR YESDAY	STEP YIELD SINCE 701	CUM YIELD SINCE 701
1	CLEAN SLUG DIF	0.0	0.0	0	99	15.3
2	CLEAN SLUG OXI	42.0	0.0	0	94	15.5
3	MOAT PHOTO	252.5	21.0	100	91	16.5
4	MOAT ETCH	75.0	20.0	95	87	18.2
5	CLEAN ISO OXID	11.0	5.0	100	91	21.0
6	POLY GROWTH	27.0	16.0	84	94	23.1
7	GLASTRATE PROC	14.0	20.0	91	98	24.6
8	INSP POLISH	79.5	31.5	100	97	25.2
9	CLEAN BASE OX	43.5	11.0	100	95	26.0
10	BASE PHOTO	25.5	0.0	0	93	27.4
11	BASE DIFF	5.0	0.0	0	90	29.5
12	EMIT PHOTO	9.5	0.0	0	66	32.8
13	EMIT DIFF	12.5	8.0	61	84	49.8
14	1ST CONT PHOTO	35.0	0.0	0	93	59.4
15	GAIN ADJUST	0.0	0.0	0	100	63.9
16	CHEM OX REMOVE	10.0	0.0	0	96	63.9
17	ALUM ALLOY(1)	3.0	5.5	84	100	66.6
18	PHOTO INTERCON	0.0	10.5	100	96	66.6
19	ALUM ALLOY(2)	14.5	10.5	100	89	69.4
20	BACK ETCH	7.5	0.0	0	78	78.0
21	PRE EL PROBE	4.5	0.0	0	100	100.0
22	ELECT PROBE	0.0	0.0	0	100	100.0
23	INSP EL PROBE	0.0	0.0	0	100	100.0
24	SCRIBE & BREAK	0.0	0.0	0	100	0.0
25	INSP DICE	0.0	0.0	0	0	0.0
26	DICE	0.0	0.0	0	0	0.0

Figure 9.4 Inventory status for one item.

operator and the computer. The card format is shown in Figure 9.3. At each station the operator noted the quantity in pencil on the card. This item was keypunched at the end of each day. The institution and control of the data collection was probably the most difficult task in the entire project. Until complete confidence in the mechanized files was established, there was no actual aid to management.

The resultant data for different devices are shown in two ways.

1. The full listing, in Figure 9.4, of data pertaining to only one device for a particular month and day showing for each process step the total slices currently at the station, the slices that moved the day before, the preceding day's percentage yield for each step, and based on current experience what the final yield will be.

2. The graphic presentation, in Figure 9.5a and b, the current inventory and slice movement for another device type. This graphic presentation is designed to provide data for a quick appraisal of the situation.

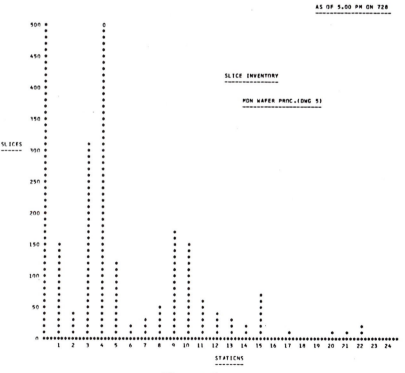

Figure 9.5a

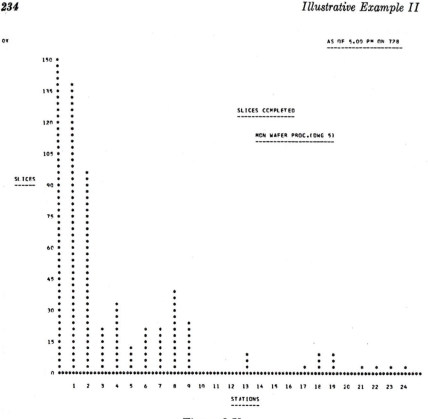

Figure 9.5b

One advantage gained from instituting the data collection system before the entire system became operational was to make available these useful inventory status reports. The benefit these reports gave to management provided additional impetus to see the full system implemented. The inventory status system showed management current areas or difficulty indicated by production trends, backlogs, yields, and overall activity.

Output Prediction Model

The output prediction model provides management with the ability to consider "what if" and make their decisions accordingly. To accomplish these aims the model can provide a variety of services, such as:

1. Schedule personnel and equipment for the next 24-hour period.
2. Predict the delivery dates for orders in process and compare them with scheduled delivery dates.

3. Predict delivery dates for new orders.

4. Predict the influence of different forecast order levels on delivery schedules.

5. Provide anticipated utilization for equipments and people.

6. Evaluate alternative priority systems among new orders and those partially processed.

7. Anticipate the impact of the introduction of new products and processes.

8. Determine the benefits obtained from improved yields for analysis with commensurate changes in costs.

9. Provide a full set of cost data relating to the process, equipments, and utilization.

Some of the above model outputs are routine whether daily, weekly, or aperiodic and are well defined. Other outputs used to provide the bases for special decisions require new formats and are the result of tailoring the basic model. The model is organized for routine output around a core of output from the production job shop submodel.

Job Shop Submodel

This submodel starts with the release of a work order in the form of a set of tabulating cards which define product, process steps to be used, and number of slices in the batch. The inventory control data provides the anticipated yield on the basis of historical experience. The new work order competes with existing work orders for equipment and people. All active work orders are in the model. The basis for the scheduling of work orders for the next process step is indicated directly by the controlling process drawing. The steps called out by the process drawing are stored for each type of product, including the detail of the individual process substeps together with required equipments and skills. Some types of devices require more substeps than others. The basic procedure for all is as follows:

1. *Determine equipment required.* The diffusion furnaces are generally used for one or possibly two steps. The photo-processing equipments are used between the different diffusion operations. There are also specific equipments for the process steps making interconnections on the slices and to provide bases for mounting the unit in its package. In addition there are test equipments depending on the process step.

2. *Determine the operator skill category.* Operators may have relatively wide capability to operate a variety of diffusion furnaces or test equipments. They may specialize in the use of one equipment, however,

for one step or aspect of photo processing. Characteristics are established for individual operators and their prime assignments.

3. *Establish the nominal time allocation for the process step.* Each step is classified according to whether there is a single time to collectively process the entire batch or whether each slice is processed individually. The nominal duration of the steps is obtained from the inventory-control data, which shows the number of batches processed through particular equipments over a long period of time and the daily records of operator assignments. This approach allows for set-up and clean-up times, as well as the actual operation time.

4. *Set priority among competing batches for equipment and operators.* The scheduling must take into account when, during the shift, tasks may begin, since some of the process steps are continuous for 8-hour periods. Shift structure may be varied from the basic 5-day, 8-hour shift. Within each priority level, the long-duration tasks are separated to start at the beginning of the shift. The remaining short-duration tasks of equal priority then compete for the remaining resources. Different priority schemes can be used. A simple one that appears to work well is to increase the priority of the batch in relation to the step it is waiting for. This tends to move batches through the higher-ordered tasks quickly. A priority discipline is needed because of the numerous photo steps competing for the same resources. In addition the predicted and promised delivery dates are compared with batches falling behind schedule getting higher priority. Management may also arbitrarily assign different priority levels to different products or individual orders.

5. *Determine according to the preset rules the next process step.* Slices which fail the tests are eliminated either individually or as an entire batch. At selected stages in the processing some types of wafers can be interconnected in different ways according to special requirements. Since the number of chips from one slice may be sufficient for an entire order, only one may be interconnected, with the remainder held in reserve or released for other sales orders. At any step the number of remaining slices may indicate that a new batch should be started.

The job ship submodel in addition to controlling and scheduling the processing of slices gathers statistics on the utilization of resources, total activity, and operating costs.

Forecast Submodel

The forecast submodel can determine whether an order can be filled from stock and can predict future production system requirements. Since the system contains all the current inventory data, it is readily apparent that incoming orders can be compared with finished and in-process stock. Where there is sufficient stock to fill the order, the choice is presented for approval. Similarly, but in a more complicated fashion, an order may be filled by completing the processing of some of the partially processed slices stock. Finally, for a small order, batches already in process are scanned to determine whether the order may be added. These opportunities for better use of available material depend on the number of slices required and the production codes for both basic and final materials. When the available material differs slightly from the desired slices it may still be used, but a rework step must be added.

The main function of the forecast submodel as a management tool is to predict the capability of a production system. The projection, until completion of the current batch, is not sufficient to determine the influence of new products, rising sales trends, and long lead times for equipment delivery and operator training. Management needs the ability to use various sets of anticipated orders and product mixes to aid in determining choices for expansion and reasonable delivery schedules. These tasks are accomplished through the use of a potential orders scenario or several scenarios based on varying sales forecasts. These include such factors as product mix and volume, delivery schedules, and rate of introduction of new products. The latter item is most important, because yields are the critical item in the production system and, characteristically, they start low and improve with experience.

The combination of the different submodels starts with current data and then makes projections into the future using sales forecast scenarios. From these submodels management may determine if the available facilities will be sufficient to meet desired delivery schedules and, more important, obtain advance insight into those areas where bottlenecks will develop. Plots of slice queues for particular process steps are shown in Figure 9.6. The plots are of slices in the queue and the time spent there rather than of queue size alone. This presentation of information indicates the size of the problem and the length of time it has existed. From presentations such as these the system provides the insight needed for effective operation. The extensive management information system provided by a discrete event simulation language for both routine data processing and simulation goes far beyond what could be obtained with conventional techniques alone.

The extension of the system representation to include a full-scale economic model is a small effort compared to what was needed to make the production system model work. There are cost data required for equipment and personnel. These are standard rates and are applied directly to model outputs such as diffusion furnace utilization and man

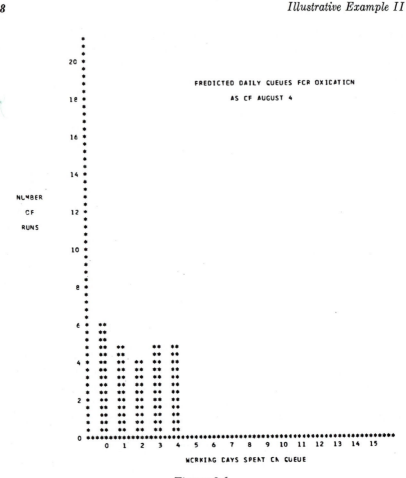

Figure 9.6

hours expended. One complex area not explored was the use of the forecast model to indicate predicted profit or loss and overhead rates for alternative forecast situations. This is within the state of the art from the simulation viewpoint, but it has a long way to go to gain management acceptance.

Description of GPSS Model

Since nearly 800 GPSS blocks were used to construct the various phases of this model, a fully detailed description of the model is beyond the scope of this chapter. However, certain features of GPSS which helped to make the model and limit the modeling effort should be identified.

The portion of the model simulating the production job shop was structured only once. Then, using indirect addressing, the same series of blocks was used to detail each of the process steps required. The use of indirect addressing made construction of the model easier, because the logic was developed only once. The catch is in the requirement for generality. Since each process step goes through the same series of blocks, a TEST block was inserted into the transaction stream to separate only those batches going through branching steps. These are the steps in the process when the device substrates are finished, enabling interconnection steps to begin. At this step the batches were sent through additional tests to determine how much of the batch should continue and how much should be delayed. One transaction represented the batch until slices were processed independently. Then a SPLIT block provided an additional transaction to represent that slice and the original transaction had its batch size decremented accordingly.

One of the variables investigated in the simulation was the differing priorities which batches could have for each step. Therefore part of the steup for each step—equipment required, personnel classification— also required the determination of the batch priority. When all batches had the same priority, the internal structure of GPSS sent a large number of batches through the first step and few got beyond. Therefore various priority schemes were tried out using the PRIORITY block. The one which appeared to push through the most orders was to increase the priority of the batch after it had been processed through the step or short series of steps.

The model sets up the interaction of incoming orders with the available material and automatically sets the step processes required to produce the material. Records are maintained of each order and the location of each batch. GPSS/360 was modified to store these matrix savevalue data on disk, thus enabling the inventory model to be run and updated each day, with the new day's data stored on disk rather than in the model. This technique avoided the use of the SAVE feature and made file maintenance and interaction with GPSS straightforward. With this flexibility it is possible to find any batch on any day and to determine its schedule. The model permits each item to be available for human judgment. In spite of this large amount of additional record keeping to control each order and slice, the running time for the model was about 10 weeks of simulated production for each minute of IBM

7040-7094II time using GPSS III and six minutes using GPSS/360 on a Model 50.

The input order submodel was related to the entire model through a magnetic tape record. The scenario covering the description and generation of incoming orders required only one GENERATE, 10 ASSIGN, one WRITE, and one TERMINATE block to set up incoming orders for 6 months into the future. Once recorded this scenario was used by each version of the production submodel. Only these few blocks were required to set up the incoming order sequence. If, however, specific orders from sales forecasts were to be used, this set of blocks would be required to describe each separate order. A GPSS macro card provides up to 10 descriptors for each order.

The incoming order processing submodel receives each incoming order in turn. The processing of these orders requires a considerable number of blocks. These are used to control logic: was the circuit made before, is there any finished stock or partially finished stock, and what size batch shall be started? This consideration requires maintaining historical records, and this may be done in GPSS through the use of "user chains" to achieve list processing. The code number file of all devices previously made is inspected to see if that item was made before. An UNLINK block is used to search the file for a particular code number and remove that transaction from the user chain. When the code number is not found, it means the device was not made before. Similar approaches were used to search the inventory to find out whether the order could be filled from stock, where the slices for a particular order currently were, and what was in the queue for a particular step.

Description of Model Outputs

The various outputs from the forecast simulation model provide answers in the areas of scheduling, equipment utilization, and production bottlenecks. Scheduled work for each day is shown in Figure 9.7 by step number, batch number, and type of device being processed. The lines of printout are in chronological order for each day. The system status is presented in terms of queues for resources shown in Figure 9.8, the list of orders awaiting processing.

When the processing of an order is finished a printout occurs. This printout contains the order number, the elapsed time to produce the order (tenths of hours), the actual time from the start of the simulation, and the calculated number of calendar days since the order was received.

From the management viewpoint, Figure 9.9a and b represent the exceptions to the orderly processing of the system, namely, those runs

```
                        DAILY JOB SHEET
            WEEK NUMBER   1              MONDAY

ITEM DEVICE RUN   STEP    PROCESS      INSPECTION
     TYPE   NO.   NO.   START  END    START   END

  1   2024  5081   24    800   830     830    842
  2   1012  3131   18    800   924     948   1000
  3   1033  6951   18    800   924    1000   1012
  4   2024  6011   18    800   924    1012   1024
  5   1033  7021   18    800   924    1024   1036
  6   1033  7821   18    800   940    1036   1048
  7   1033  9221   18    800   948    1048   1100
  8   1038  5531   18    800  1112    1112   1124
  9   1033  6621   15    800   930     930    942
 10   1038  4671   13    800   930     942    954
 11   1038  6481   11    800  1130    1324   1336
 12   1038  4431   10    800  1400    1600   1612
 13   1033  5731    6    800  1230    1336   1348
 14   1032  5681    5    800  1000    1012   1024
 15   1032  8861    4    800  1100    1624   1636
 16   1032  6371   24    830   900     900    912
 17   2024  5121   23    830  1000    1000   1012
 18   1033  6631   15    930  1100    1124   1136
 19   1038  5521   13    930  1100    1212   1224
 20   1033  6621   15    942  1112    1148   1200
 21   1038  5321   16    948  1154    1154   1206
 22   1032  5631   16    948  1154    1206   1218
 23   1032  6661   16    948  1154    1218   1230
 24   1032  6671   16    948  1154    1230   1242
 25   1033  6711   16    948  1154    1242   1254
 26   1032  6731   16    948  1154    1254   1306
 27   1032  6741   23   1000  1100    1100   1112
 28   1038  4671   13   1000  1130    1224   1236
 29   2024  5121   24   1100  1130    1136   1148
 30   2024  6871   17   1100  1412    1412   1424
 31   1032  6751   15   1100  1230    1236   1248
 32   2024  6521   13   1100  1230    1248   1300
 33   1033  6591    5   1112  1312    1348   1400
 34   1032  8881    4   1112  1412       0      0
 35   1032  6741   24   1130  1200    1200   1212
 36   1033  5231   11   1130  1500    1524   1536
 37   1033  6631   15   1136  1306    1312   1324
 38   1038  5531   18   1154  1342    1342   1354
 39   1032  6761   16   1154  1400    1400   1412
 40   1032 .5581   16   1154  1400    1412   1424
 41   1033  6611   16   1154  1400    1424   1436
 42   2024  7271   16   1154  1400    1436   1448
 43   2024  7301   16   1154  1400    1448   1500
 44   1032  5941   23   1200  1300    1300   1312
 45   1033  4761   15   1230  1400    1424   1436
 46   2024  6531   13   1230  1400    1500   1512
 47   2024  5541   23   1300  1400    1400   1412
 48   1032  6751   15   1306  1336    1448   1500
 49   1038  5521   13   1312  1442    1512   1524
 50   2024  7201   16   1342  1548    1548   1600
 51   1032  5941   24   1400  1430    1436   1448
```

Figure 9.7

which have not moved since a particular date. In this case Figure 9.9a shows where runs have been held up for the previous week and 9.9*b* shows how long they have been held up.

A record of each week's activity is printed out once each simulated week. The model output at the end of a desired period of time, either weekly or monthly, provides a full rundown of the use of the system. Figure 9.10 shows a summary of activity for the previous week.

The GPSS structure also produces certain standard reports. Among these there are two of interest: the utilization to date for all the personnel and equipments. These are calculated in standard GPSS on the basis of a 168-hour week. The standard output gives the utilization of the facility, the number of batches processed, the average time the batches occupied that facility, and what batch has the facility at the time the report is produced. Utilization of personnel is similarly summarized.

```
           QUEUES ON MONDAY  8.00 AM

        FACILITY QUEUES ( 1 EACH EXCEPT FOR 6 PHOTO)

          FACILITY              USED BY STEP(S)        QUEUE LENGTH

      SLUG DIFFUSION            1                          .
      OXIDATION                 2,5B,11                   45.
      PHOTO                     3,12,14,16,18,20           .
      ETCH                      4,22                      22.
      PYROLYTIC                 5B                        83.
      XTAL GROWTH               6                          .
      DEBURR                    7                          .
      BACK LAP - FRONT LAP      8,9                        .
      POLISH                    10                         .
      BASE DIFFUSION - STEP A   13                         .
      BASE DIFFUSION - STEP B   13                        11.
      EMITTER - STEP A          15                         .
      EMITTER - STEP B          15                        6.
      GAIN ADJUSTMENT           17                        3.
      ALUM EVAP                 19                         .
      INTERCONN ALLOY           21                         .
      GOLD EVAP                 22                         .
      GOLD ALLOY                24                         .

                      QUEUES FOR PEOPLE

       PERSONNEL SKILL        NUMBER TRAINED         QUEUE LENGTH

      DIFFUSION                 7                          .
      PHOTO                     6                         43.
      POLYCRYSTALLINE           1                        21.
      LAPPING & POLISHING       1                        76.
      BACKING                   1                        32.
```

Figure 9.8

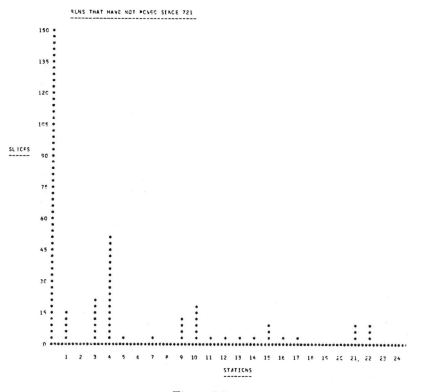

Figure 9.9a

Graphic Presentation of Results

An example of the use of graphic output techniques is the plot of how long each order took to be completed. This is shown in Figure 9.11a. The increase in time to complete orders at certain times can be attributed to the additional workload. Figure 9.11b shows the system unable to continue to deliver on schedule and the delivery dates become longer and longer.

Conclusion

The simulation technique provides a quick, low-cost means to gain insight into a complex manufacturing problem. This insight was gained in a new area where historical data are lacking. The results

	RUN NO.	DATE LAST MOVED	STATION NUMBER
1	28883	711	1
2	29233	717	1
3	29243	717	1
4	29253	721	1
5	29283	721	1
6	29293	721	1
7	29303	721	1
8	29313	721	1
9	29323	721	1
10	29333	721	1
11	29343	721	1
12	29353	721	1
13	29363	721	1
14	29373	721	1
15	29383	721	1
16	29393	721	1
17	29403	721	1
18	8861	628	3
19	8881	628	3
20	8901	628	3
21	9001	630	3
22	9041	629	3
23	28833	718	3
24	28843	718	3
25	28943	717	3
26	28953	717	3
27	28983	717	3
28	29003	717	3
29	29013	717	3
30	29023	717	3
31	29033	717	3
32	29073	717	3
33	29103	718	3
34	29123	718	3
35	29133	718	3
36	29163	718	3
37	29183	718	3
38	29193	718	3
39	29203	718	3
40	29223	718	3
41	6591	718	4
42	6601	718	4
43	6111	518	4
44	5801	518	4
45	6071	518	4
46	6081	518	4
47	5811	518	4
48	5821	518	4
49	6091	518	4

Figure 9.9b

ITEMS PROCESSED LAST WEEK

			SLICES	RUNS
STEP	1	SLUG DIFFUSION	.	.
STEP	2	OXIDATION	.	.
STEP	3	MOAT PHOTO	.	.
STEP	4	MOAT ETCH	42.	6.
STEP	5	POLY OXIDATION	.	.
STEP	6	XTAL GRWOTH	30.	5.
STEP	7	DE-BURR	.	.
STEP	8	BACK LAP	.	.
STEP	9	FRONT LAP	30.	5.
STEP	10	POLISH	30.	5.
STEP	11	OXIDATION-BASES	50.	10.
STEP	12	BASE PHOTO	90.	18.
STEP	13	BASE DIFFUSION	51.	17.
STEP	14	EMITTER PHOTO	75.	25.
STEP	15	EMITTER DIFFUSION	63.	21.
STEP	16	1ST CONT. PHOTO	126.	42.
STEP	17	GAIN ADJUST	30.	10.
STEP	18	2ND CONT. PHOTO	45.	15.
STEP	19	ALUM. EVAP	.	.
STEP	20	INTERCONN. PHOTO	.	.
STEP	21	INTERCONN. ALLOY	3.	1.
STEP	22	BACK ETCH	8.	4.
STEP	23	GOLD EVAP.	32.	16.
STEP	24	GOLD ALLOY	32.	16.
		TOTAL	737.	216.

Figure 9.10

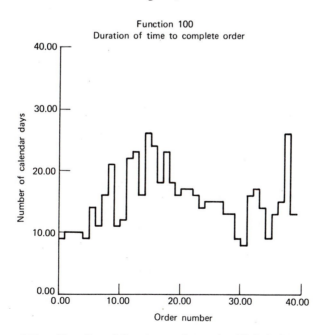

Figure 9.11a Duration of time to complete order (digital plotter output).

245

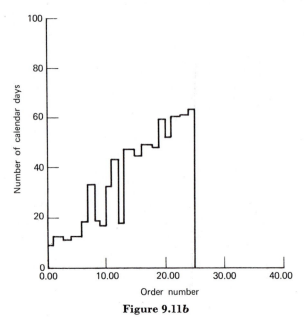

Figure 9.11b

from the model agreed with subsequent experience when the current factors were used as data. The freedom to try out new ideas and find out what might happen if they were used could only be accomplished through the use of computer techniques and a higher-order discrete event language.

PROBLEMS

1. Consider the extension of this model to provide the special needs of the following groups: production scheduling, sales, accounting, and new product development. List major areas of expansion, those eliminated, and those requiring change.

2. For only the MIS functions, but with live data, compare the advantages and disadvantages of using different higher-order programming languages: FORTRAN, COBOL, ALGOL, PL/1, GPSS, and SIMSCRIPT.

3. For the example in which the model is to be used by the planning group, list the special reports that the system should be able to provide. Describe one report in detail.

4. Indicate how the model can be modified to compare the performance of alternative assembly-line implementations. Consider, in addition to the assembly operations, the costs for material, scrap, labor, facilities, and operations. To compare system performance, suggest several figures of merit.

5. For a real-time MIS data bank the amount of data that could be entered can vary over a wide range. Show how to reduce the amount of data entered to a minimum. Also indicate the associated penalties.

6. Illustrate different ways of presenting results to the management level. Consider printer capability, plotter features, and CRT graphics and alphanumerics real-time display.

7. In the job shop the choice of priority scheme is critical to establish a high level of output. Indicate which schemes might work and how they would be tried, using the model.

8. Outline a program to train job shop production controllers through the use of the model and several variations of the basic shop.

9. List different industries in which variations of this basic model may be used with slight or moderate modification. Would you prefer to modify the existing model or create a new one for a different industry?

BIBLIOGRAPHY

1. *Developments of Techniques for Automatic Manufacture of Integral Circuits,* Vol. 1, F. S. Preston, A. Spitalny, and W. S. Mann, Norden Division of United Aircraft Corporation, Dayton, Ohio, AFML-TR 65-386, 1965, 61-61, E1-E28.

2. Heuristics in Job Shop Scheduling, W. S. Gere, Jr., *Manag. Sci.,* **13,** 3:167–190 (1966).

3. Job Shop Simulation of Orders That Are Networks, D. R. Trilling, *J. Ind. Eng.,* **17,** 2:59–71 (1966).

4. Linear Programming for Production Allocation, B. B. Henry and C. H. Jones, *J. Ind. Eng.,* **18,** 7:403–412 (1967).

5. Load Forecasting, Priority Scheduling, and Simulation in a Job Shop Control System, M. H. Bulkin, J. L. Colley, and M. W. Steinhoff, Jr., *Manag. Sci.,* **13,** 2:B29–B51 (1966).

6. Simulation Model of a Computer Controlled Automatic Warehouse, D. F. Thompson and J. S. Cnossen, *Simulation,* **10,** 6:297–304 (1968).

7. A Simulation Study of Dequencing in Batch Production, R. H. Hollier, *Operational Res. Quart.,* **19,** 4:389–407 (1968).

8. The Development of a Community Health Service System Simulation Model, F. D. Kennedy, *Rec. IEEE Sys. Sci. Cyber. Conf.,* 85–92 (1968).

9. Simulation Study of a Test Equipment Calibration and Certification System, S. K. Didis and C. E. Carpenter, *J. Ind. Eng.,* **17,** 8:437–441 (1966).

10. Production Scheduling Under Seasonal Demand, R. C. Vergin, *J. Ind. Eng.,* **17,** 5:260–266 (1966).

11. Simulation Model for Vocational Educational Facility Planning, D. A. Carter, A. Colker, and J. Leib, *J. Ind. Eng.*, **19**, 2:68–75 (1968).
12. Simulation: Dynamic Tool for Plant Expansion, P. Lipton, *J. Ind. Eng.*, **1**, 1:21–26 (1969).
13. The Application of a Simulation Rational for Facility Evaluation, T. P. Dunn and F. M. Smith, *Naval Eng. J.*, **81**, 1:79–87 (1969).
14. A Generalized Manufacturing Line Simulation System for Production, Equipment, and Manpower Planning, G. V. Raju and A. Bendazzi, *Proc. 3rd Conf. Appl. Simul.*, Los Angeles, 64–75, 1969.
15. An Order Picking and Shipping Model, L. W. Hillman, *ibid.*, 204–212.
16. Simulation of the Operation of the Coal Supply System at a 2000 MW Generating Station, R. D. Walkley and N. D. Hutson, *ibid.*, 213–225.
17. Simulating Scheduling Plans, C. E. Montagnon, *ibid.*, 350–362.
18. GASP II Simulation of Parts Inventory for Electronic Fuze Production, C. C. Peterson, *ibid.*, 454–469.
19. GPSS Study of Work-in-Process Inventory, H. L. Bowen, Jr., *Proc. 4th Conf. Appl. Simul.*, New York, 162–169, 1970.
20. Application of the GERTS II Simulator in the Industrial Environment, G. E. Whitehouse and K. I. Klein, *ibid.*, 170–177.
21. Simulation of a Distribution Center (Warehouse) Central Conveyor System, D. Grossman, *ibid.*, 303–312.

10

Illustrative Example III— System Effectiveness of a Weapons System

I'LL NEVER KNOW IF THE SYSTEM WILL WORK AND I HOPE NEVER TO FIND OUT.

The designer of a system may be faced with problems the solutions for which can almost never be verified. The example in this chapter— predicting the performance of a manned aircraft weapons system—is characteristic of the application of simulation to systems whose performance cannot be verified. Simulation provides a means of comparing the relative performance of alternative system designs before one design has been implemented. Iterated use of simulation contributes insight through each subsequent simulation version. From the iteration process there emerges guidance for new design approaches.

Models are helpful in various stages of system design, beginning with the initial concept formulation and extending to eventual comparison with actual experience.[1-3] The example of the manned aircraft weapons system illustrates the use of simulation for comparing the performances of different system design approaches after the concept formulation stage.

Model Structure

The weapons system simulation was used to predict the performance of a squadron of manned tactical attack aircraft under a variety of mission conditions. The model was developed using GPSS and has evolved during several different applications. Each new application has modified both the goals and details. One version of this particular model has as its prime function the comparison of different airborne sets of equipment to correctly identify targets.

The performance of manned tactical-attack aircraft varies considerably with the nature of the targets, surrounding terrain, enemy defenses, weather, and operational techniques. The model is set up to simulate anticipated situations within the spectrum of potential operational use.

The criteria of effectiveness for weapons system evaluation dictate a problem framework that extends far beyond the simple question, how often does the system eliminate the desired target? Together with that basic question is an entire list of technical, financial, and operational considerations that must be evaluated to make the performance of the simulated system coincide as closely as possible with eventual performance. These considerations require that the input data for the model include estimates of costs, support system, manpower skill, and the obvious items of technical capability. In addition the problem must be arbitrarily but judiciously bounded to avoid having the system interact with an infinite number of other systems. Since the system in real life does interact with numerous other systems the representation of the form and the degree of interaction must be controlled and specified. For example the model only applies when the air base is not itself affected by enemy action. This factor could be included, but on what terms? Independent, unrelated attacks would be simple to include but are relatively meaningless. Extension to an interactive war game considering the deployment of numerous squadrons would require considerably more effort and provide comparisons significantly different from those needed to evaluate the attack mission.

This example illustrates the use of an overall model. One interacting with several significant factors to restricted degrees of depth; and the level of detail is coarse. The important elements considered are the following:

1. *Operational regime.* The intended mission types for this weapons system consider target characteristics and enemy defenses, mission frequency, number of aircraft deployed, and aircraft characteristics.

2. *Effectiveness measure.* The criteria to measure the performance of the system, the number of targets eliminated, and the cost in men and material.

3. *Support concept.* The host of support activities, ranging from those directly tied to the aircraft rearming, refueling, and servicing, to the logistics pipeline with associated costs and time factors.

4. *Limited resources.* Solutions to this system design problem are obtained much more readily if the system is assumed to have everything necessary when needed. Unfortunately this is not the case. The num-

ber of aircraft in the squadron number and skill of support personnel, spares, and amount of support activities are all limited.

5.. *Tie-in with existing systems.* Previously existing systems will impose constraints. These could range from the skill of the flight crew, based on their level of training, to the logistics support concept employed—repair at the squadron level versus replacement.

The goal for the simulation is to provide a means of predicting—with increasing accuracy as the system becomes better defined—the behavior of the complex weapons system. The model should follow the development of the system starting after early concept formulation and developing with the system design. Initially the model is used to help identify the sensitive areas—those areas in which small changes have great influence on the results. Once these are known alternative solutions can be investigated until the overall system design is satisfactory. Later the model may be updated to represent the system as new problems are resolved. These modifications may require expanding part of the model into great detail, modifying the model for new developments or in response to test data, and, finally, representing the operational system. The risks in the development of the model lie in the need for skill to structure the model as a realistic representation, to evaluate its performance, and to avoid excessive costs and poor responsiveness through inclusion of irrelevant detail.

Model Details

This example consists of a series of interlocking submodels, shown from an overall viewpoint in Figure 10.1. The basic elements are the following:

1. Targets to be eliminated in the form of mission requests with the characteristics of anticipated mission profiles.

2. Maintanance of the aircraft by the squadron environment with replenishment activities and support resources.

3. The individual aircraft flights composed of penetration to the target, attack, and return to base.

4. Evaluation of mission—either elimination of target or the requirement for another mission to continue the attack.

Obviously, these five fundamental parts of the model and their subparts could be combined in an almost infinite number of ways. One

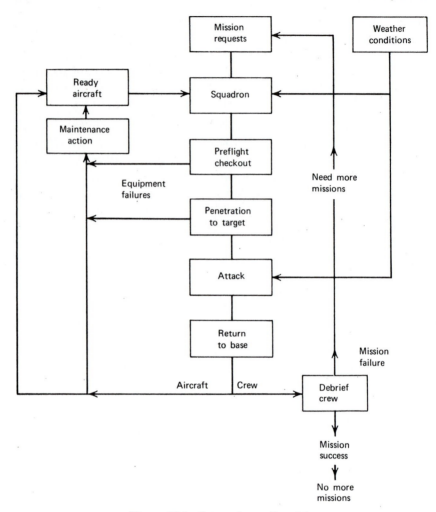

Figure 10.1 Scope of overall model.

way to restrict the number of actions and paths this combinatorial problem could follow is to use only one scenario. This scenario establishes the time and sequence of the targets to be eliminated by the squadron. Further scenario control is maintained by using the same sequence of random numbers to provide the characteristics for important situations. The specific behavior of the model is varied by changing the system characteristics. The scenario can be changed by using different random numbers. System effectiveness can be measured by the availability, total flight hours, cost, or number of losses. Since missions are flown until the target is eliminated, the number of missions required to eliminate the targets was used as the figure of merit.

Targets

This submodel creates the targets and their characteristics. Each target is a transaction generated with a specific time for the target request to be received by the squadron. This target request transaction contains the data in parameters describing the priority, location, size, indicated attack method, weapons to be used for each target, target defense capability, and an urgency factor. To provide a constant scenario the sequence of initial target requests is always the same. Different target characteristics can be used to anticipate system behavior against other targets.

When a target is eliminated during the first mission, no additional missions are flown against that target. For targets only partially destroyed another series of missions is undertaken. This process is repeated either until the target is eliminated or all available resources have been expended. Damage to the target reduces its ability to defend itself. Repair of the target's defenses, however, was not considered.

The urgency factor of the mission request provides the time sense that some targets must be destroyed as soon as possible after the squadron receives the request or the requirement will have changed. These targets have a time element associated with them—a time by which they are either eliminated or the system has failed to accomplish its task. A truck convoy moving along a relatively exposed stretch of road is an example. Even though the target has not been eliminated, no additional missions are flown against these targets after the time has passed.

Squadron

The squadron submodel represents the current condition of the squadron—that is the number of available aircraft, air crews, and ground resources. The mission requests are received by the squadron submodel. The interactions between mission requests and the squadron prior to takeoff are shown grossly in Figure 10.2. This submodel has provision for the squadron commander to determine extent of aircraft commitment, aircraft utilization, and squadron resources required for preflight preparation and aircraft checkout.

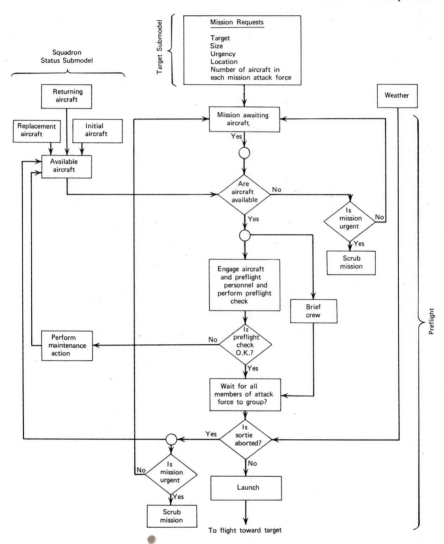

Figure 10.2　Mission and squadron interaction prior to take-off.

　　The mission request may call for all available aircraft or a smaller number.　To determine the number of available aircraft a file is maintained of transactions representing individual aircraft on available aircraft chain.　There are three additional files: aircraft still on missions, those not in operationing condition—whether waiting for repair, personnel, or spares, and those being refueled and rearmed.　The history of each aircraft is retained to be able to relate cause and effect. When the squadron receives the mission request the condition of each aircraft is interrogated and aircraft are selected. This approach preserves the interrelationships that would be lost if the aircraft were treated on a straight probabilistic basis. That is all aircraft form a pool and, based on statistics, a certain number from that pool are currently available.[4] The number of aircraft ready for a mission during a busy period would be controlled by the immediate past history, namely, limited resources available to refuel, rearm, or repair.[5]

The model is initialized with standard aircraft range and payload data. However, the squadron may impose its own modifications, because of the effect of local conditions on these data. Payload versus range characteristics of a particular aircraft are shown in Figure 10.3. This curve indicates a constant payload for a range up to 1000 miles, after which the range falls off linearly. The value used for the abscissa is absolute miles. Actually this is not true for most aircraft, because takeoff weight, runway length, temperature, flight speed, and altitude modify standard data. In the model a single index is used to modify the absolute miles according to the conditions anticipated for a particular mission.

The primary concern of the squadron submodel is the maintenance of records and their processing to indicate squadron availability, that is, its overall capability to react to mission requests. This is not the critical element in the simulation. The real figure of merit is the ability of the squadron to perform its assigned task—elimination of targets.

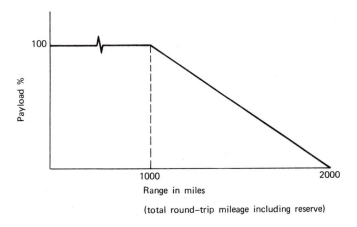

Figure 10.3 Typical range versus payload function.

Availability has become one of the factors frequently cited, so the model must also be used to obtain availability factors for comparison with other factors such as squadron performance and costs.

Missions

The largest aspect of the model concerns the missions. This part may be further subdivided into the phases of the mission: preflight, penetration, attack, return, landing, and mission evaluation. In turn each of these mission phases may be modeled in greater detail according to the needs of the system analysis.

Preflight

The preflight model briefs the aircrew and checks out the aircraft to ensure all equipments are operational. These squadron activities vie for the limited resources of men and equipment. If the preflight checkout is satisfactory, the aircraft awaits clearance for takeoff. Since a final check is made on the current weather forecast, the takeoff clearances may come for individual aircraft or for groups of any size. When the checkout fails, the decision is made whether to replace the faulty equipment and still have the aircraft ready for the mission, replace this aircraft with another, eliminate one aircraft from the mission, or scrub the mission.

Penetration

Penetration to the target is simulated as shown in overview in Figure 10.4. The distances to be flown over friendly and hostile territory, altitudes used, and airspeed are obtained from input data and can be changed to relate to particular missions. A mission may be aborted during penetration to the target either because of system failure or enemy action. Hostile interference is considered only from ground sources. If hostile enemy air action were to be considered, this would be implemented by introducing additional major subsections to the model to represent enemy air defense strategy, deployment of enemy air squadrons, and air-to-air combat. In effect this would have to become a different model for this phase.

The effectiveness of hostile elements is related to the altitude and speed of the aircraft. Higher altitude reduces the influence of small caliber ground fire, but increases risk from antiaircraft missiles. Similarly high speed reduces enemy effectiveness, but increases the risks of navigational error and collision with the ground during low-altitude

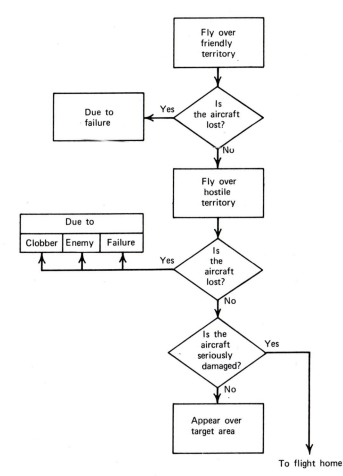

Figure 10.4 Flight toward target.

flight. The risks of aircraft failure, enemy action, and collision depend on terrain, aircraft equipment characteristics, and aircraft aerodynamic response. Not all of these factors may be required for simulating a particular mission. The model contains these elements for those missions that require them. When particular capabilities are not required that part of the model is bypassed.

Attack

The attack phase of the mission, shown in Figure 10.5, is only general, because the model is intended as a basic tool to study a wide variety

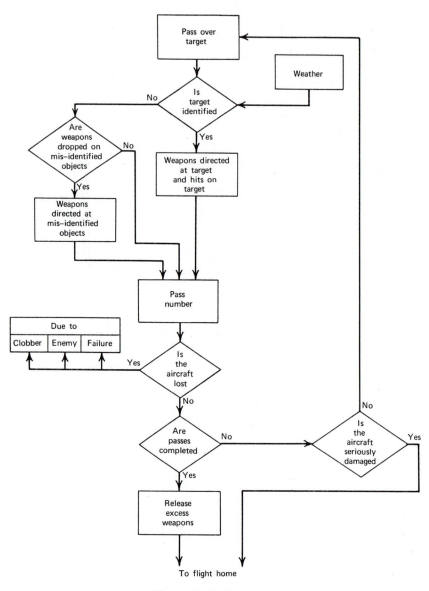

Figure 10.5 Attack.

of missions. In one particular application the model has been used to sense the influence on mission effectiveness of aircraft equipped with different capabilities for target identification. The simulated identification of the target depends on:

1. Target characteristics—type, size and surroundings.
2. Environmental conditions—terrain and weather.
3. Aircraft—speed and altitude.
4. Airborne equipment—component of weapons system under evaluation.
5. Tactical factors—number of passes over the target.

The attack phase of the model first considers target identification or misidentification and then determines, in the case of correct target identification, the extent of damage inflicted. The amount of damage is based on the combined influence of weapon type and size, aiming method, technique used to establish the weapon release point, aircraft airspeed, attitude, and altitude. Not all weapons need be used on the first pass. Some may be retained for another pass or a secondary target. The extent to which these factors are detailed depends on the purpose of the simulation. The broad overview model does not detail these factors, since they may be combined into a representation of overall accuracy. Similarly the tactics employed are not considered, merely the fact that the aircraft reached, identified, and attacked the target.

The target is considered destroyed in the model after the sum of weapon effects exceeds the level initially set to eliminate that target. An alternative method could establish a particular level of accuracy for a single attack to destroy the target. There would then be repeated attacks if the target were not completely eliminated during one attack. In either of these methods the target may be attacked by more than one aircraft and the effects are additive. Likewise if the target is eliminated and weapons remain, secondary targets are attacked.

The targets are not passive. During the attack they can damage or destroy attacking aircraft. The defensive capability of the target is considered to be affected by the amount and accuracy of the weapons used. Therefore as the target is repeatedly attacked, the number of weapons on target reduces the effectiveness of the defense. This lessened defensive capability is then applied to successive series of attacking aircraft. In an opposite manner a heavily defended target can reduce the capability of the squadron to continue to attack through attrition of aircraft.

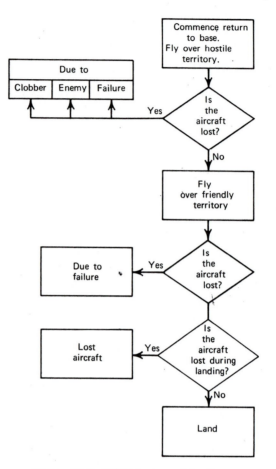

Figure 10.6   Flight home and landing.

Return

The aircraft-return phase of the mission, shown in Figure 10.6, is very much like the part of the model concerned with reaching the target—except in reverse order. Different values may be used for airspeed, altitude, and hostile-action risk. The landing is considered separately.

Damage sustained by the aircraft can cause it to be lost at any subsequent time. In addition to aircraft loss, the model considers the cumulative effects of damage incurred throughout the mission. In this case the determination of the extent of aircraft damage is made after return to the base when this factor helps establish the effort necessary to return the aircraft to service.

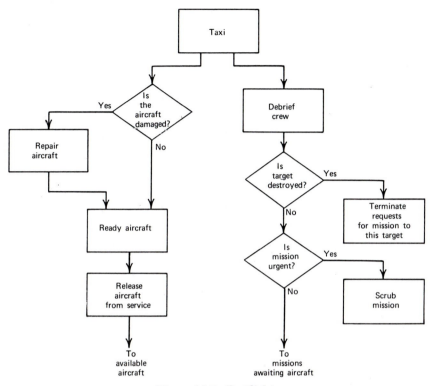

Figure 10.7 Postflight.

Postflight

Once the aircraft is on the ground two separate and parallel actions begin, as shown in Figure 10.7. Each aircraft is checked out. If serviceable it is refueled and rearmed. Damaged or failed aircraft are withdrawn from service until repairs are completed. Routine scheduled maintenance is not considered in this model. The nominal duration of servicing is taken from input data. The length of time the aircraft is out of service, will depend on competition for service personnel. In this version of the model, since no records are kept of individual spares, they are assumed to be available.

The crew debriefing—the remaining major postflight activity—assesses target damage. Based on target damage reported and on urgency, other missions may be flown against the same target. Timing for the later flights is dependent on whether the crew debriefing or subsequent reconnaissance flight is the source of the damage assessment.

System Effectiveness Factor

The purpose of interactivity relating targets and missions is to compare against a yardstick of ideal conditions—no misidentification and perfect weapons accuracy—the simulated maximum system success with more realistic capabilities. The simulation, therefore, shows an increase in the number of missions flown to accomplish the same results over the ideal mission. This provides one simple comparison among different weapons systems. The differences between ideal and realistic models are only in the equipment characteristics. The external factors—weather, terrain, enemy action, and even scenario of pseudorandom numbers—remain the same.

The running time for this model indicates the great amount of data obtained for relatively little computer cost. The model simulated 30 days of squadron behavior in 6 minutes using the IBM 360/50. This time should be regarded as an indication of the time required to obtain one point for reference. But 10, 20, or even 100 points are well within a practical economic range.

Output

This model could be used with a minimum of output, merely the system effectiveness factor. While this may satisfy some customers, more detail is needed to show how the results were obtained. The model contains numerous records of potential interest: historical record of each mission, accomplishments, and operating conditions. These records are kept in chronological order of mission completion on a FIFO chain Separate records are kept showing how each aircraft was lost. For each target records are kept indicating the extent of damage and how it was inflicted. From these individual records insight may be gained as to why something did or did not occur. Each aircraft may be followed through its missions. The factors contributing to cause a particular number of missions to eliminate a target—failure to identify this target, or an inaccurate weapons drop on that target—are available for further study.

A typical example of the data retained for later analysis is the record of each mission flown. There are 63 items of information contained in the record. Of these, 20 are pseudorandom numbers which provide the control of each nondeterministic element in the model. These are used to establish the percent effectiveness of the weapons or the probability of correct identification of the target. The total of items stored in the 63 parameters associated with the mission request transactions is shown on page 264 in a form included in the model listing.

The presentation of results from the model benefited from the ability to draw graphs using the high-speed printer. One measure used to compare mission success and effectiveness was the total number of passes required to eliminate the same targets. Factors, except for the accuracy of target identification, were held constant. Three separate cases were simulated: the ideal or no-target misidentification case, and two systems with different target identification capabilities. The summary of these three runs shows the number of attack passes for this scenario. The distribution of activity is not constant; there is more in the beginning than in the end. This is because of the scenario of mission requests, which is not constant, and the weather conditions during certain times, which degrade the system capability. There are two mission types intermixed: the urgent, which are flown only once, and the nonurgent, which are flown until the target is eliminated. There are two nonurgent targets to each urgent one.

The direct comparisons with the ideal case grows from 60 in the ideal case to 188 for one system and 267 for another.

The use of just one summary number is much too coarse to obtain insight into how the totals were obtained. A better way is to compare the day-to-day activity of the squadron over the period of the simulation (in this case, 25 days). The ideal case, shown in Figure 10.8a, represents the minimum number of missions to accomplish the requests considering weather and enemy action. The same requests are simulated for two different avionics systems in Figure 10.8b and d, for one system, and Figure 10.8c and e, for a poorer system. The number of attack passes has gone up and the daily distribution has changed, because there are now occasions when all aircraft are committed and there are still outstanding mission requests.

Still more data might be needed to understand the different system performances. One report was organized to show the limited data that would be characteristic of a weekly summary of squadron activities. This summary would provide a direct comparison of the three systems with regard to such items as mission requests, missions flown, targets destroyed, bombs used, and so forth. The more detailed form of the report is shown in Figure 10.9 for one week and the cumulative to-date report in a limited data form is shown in Figure 10.10. One additional measure of mission success and effectiveness as a function of target identification accuracy was the number of passes made by the aircraft to eliminate the same targets.

The GPSS report generator was used to create weekly summaries that

```
* THE MISSION REQUEST PARAMETERS ARE USED AS FOLLOWS
*
*1  RN INDEX          2·TAIL NUMBER       3  PRE FLGT TIME     4  FAILED UNIT
*5  TOTAL TIME        6  HOSTILE DIST      7  HOSTILE SPEED     8  HOSTILE ALT
*9  TARGET SIZE      10  URGENCY = 1      11  TARGET NUMBER    12  ATTACK DURATN
*13 SORTIE PHASE     14  PASS NUMBER -    15  PASS NUMBER +    16  IDENTIFD 1/99
*17 TARGET STATUS    18  LOSS CAUSE       19  LOST AIRCRAFT    20  RN INDEX
*21 TO PARAMETER 40 TWENTY RANDOM NUMBERS
*
*41 SUM OF RNS       42  TOTAL RISK       43  TRGT SORTIE NO   44  WEATHER 1-10
*45 FLIGHT LOG       46  FRIENDLY DIST    47  FRIENDLY SPEED   48  FRIENDLY ALT
*49 BOMB CEP         50  BOMBS DROPPED    51  MISSION TYPE     52  BOMBS ON TRGT
*53 BOMBS LEFT       54  INTNDD PASSES    55  BOMB LOAD        56  MISIDENTIFIED
*57 MISIDEN + HIT    58  PHASE RISK       59  24HR CAPS - 1    60  MISIDEN BOMBS
*61 SORTIE DELYD     62  INTENTION = 0    63  A/C DAMAGE
```

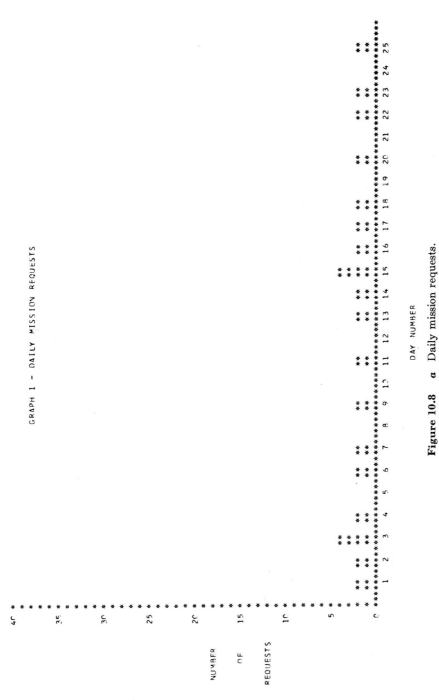

Figure 10.8 *a* Daily mission requests.

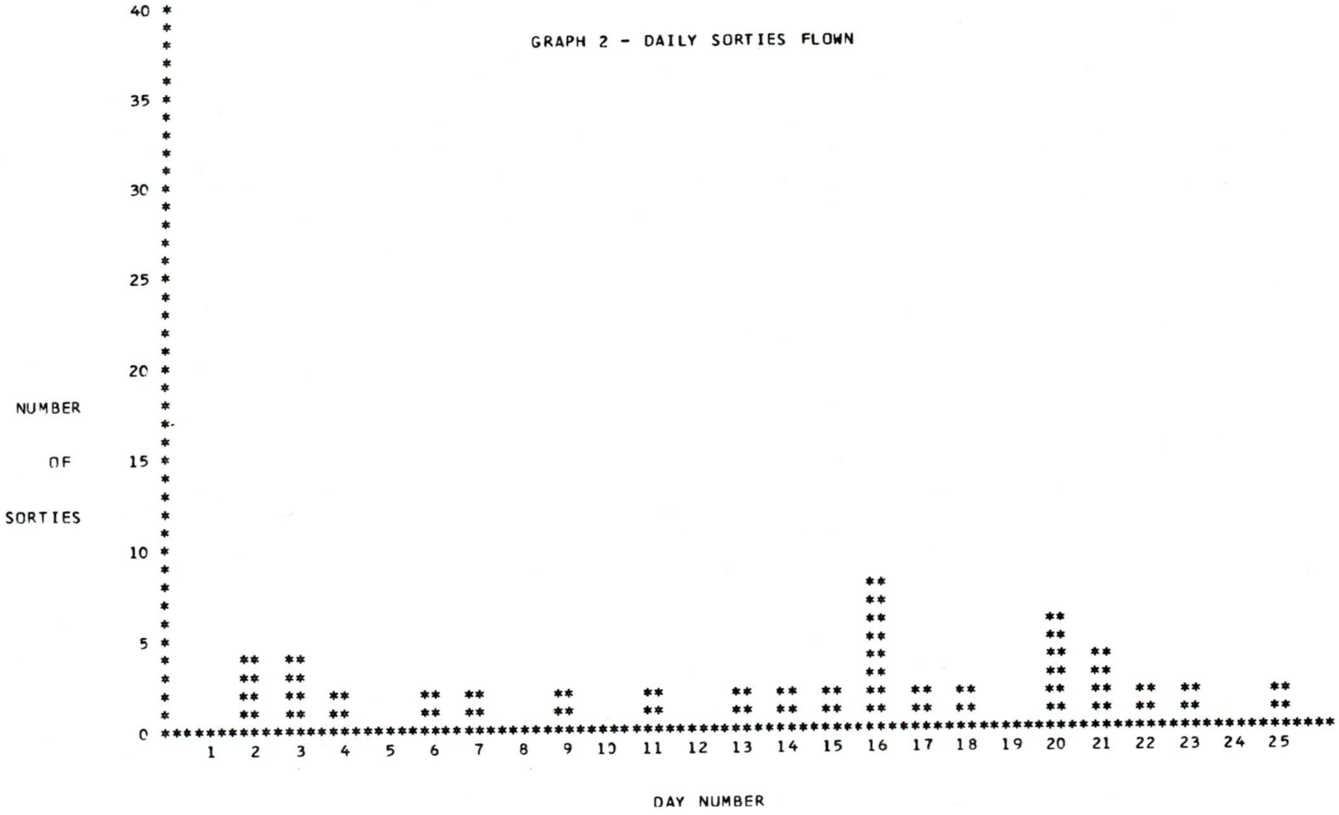

Figure 10.8b Daily sorties flown—improved system.

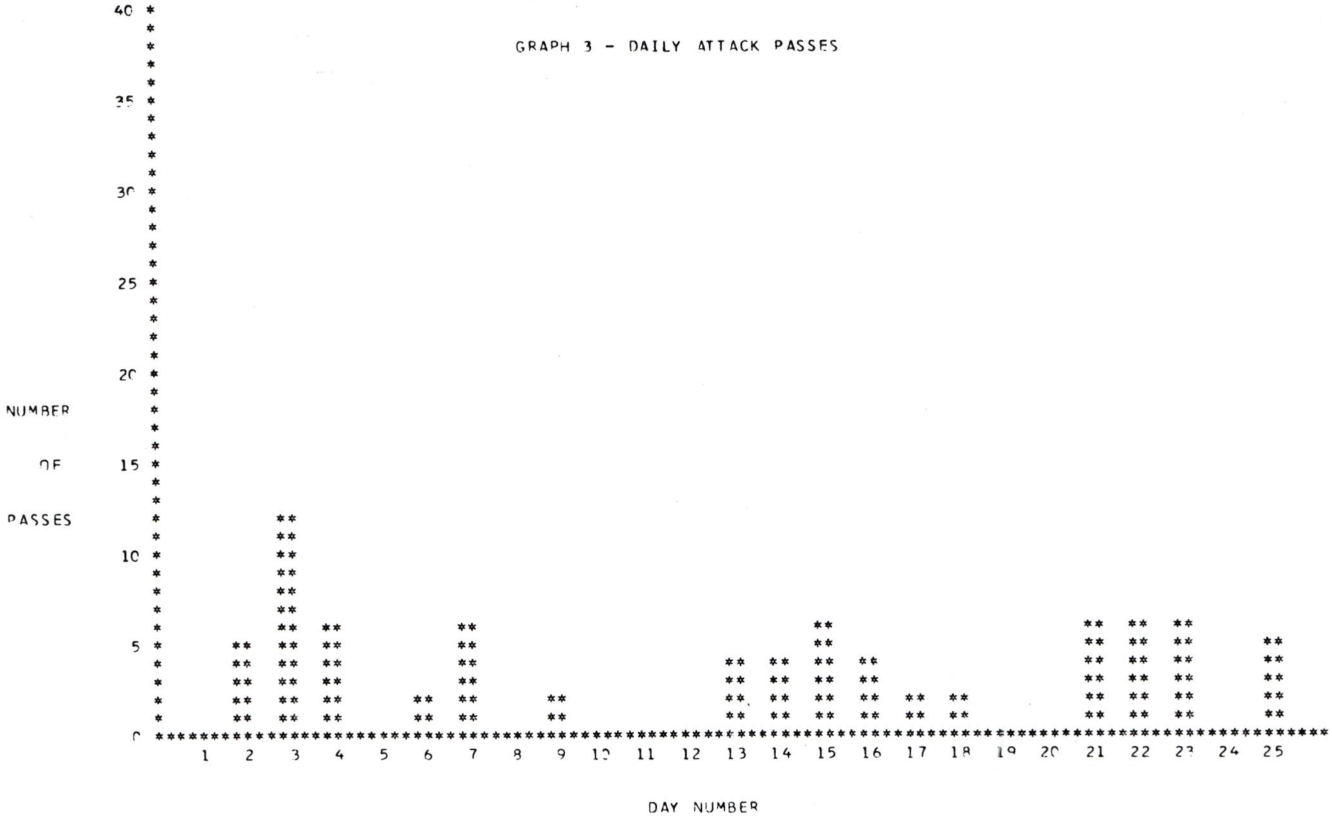

Figure 10.8c Daily attack passes—basic system.

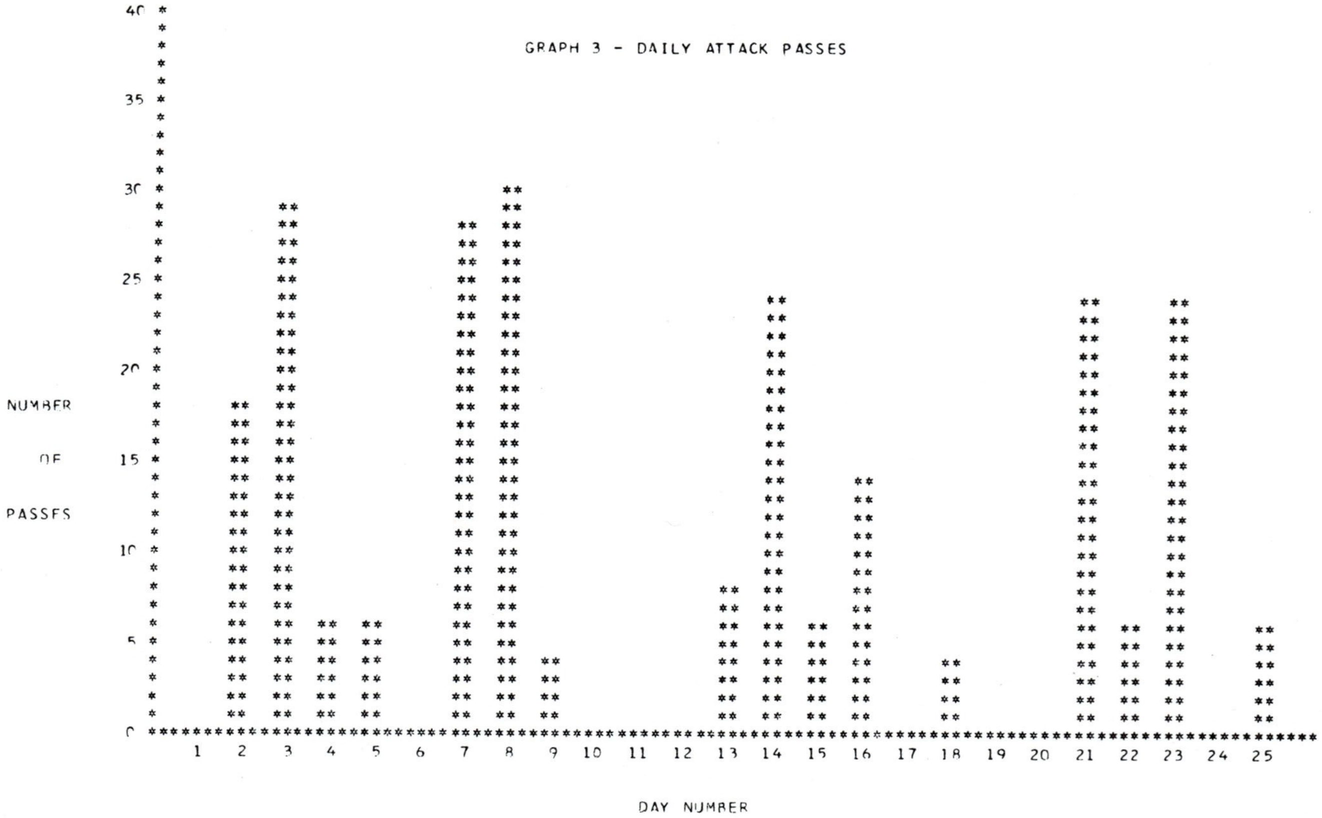

Figure 10.8*d* Daily attack passes—improved system.

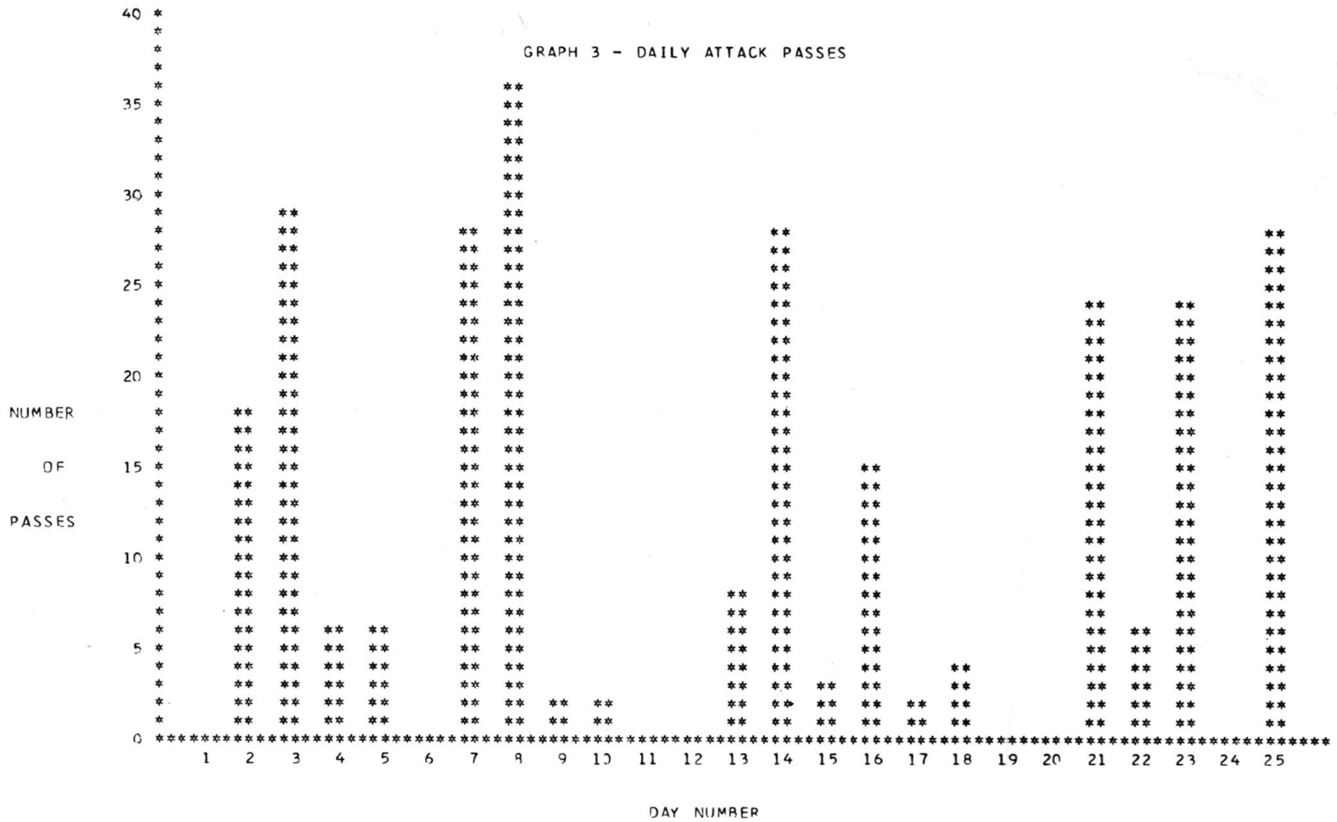

Figure 10.8e Daily attack passes—basic system.

WEEKLY SUMMARY OF FLIGHT OPERATIONS

WEEK NUMBER 4

THE PRINCIPAL STATISTICS ARE AS FOLLOWS-	CASE 1	CASE 2	CASE 3
NUMBER OF TARGETS	6	6	6
TARGETS DESTROYED	6	4	4
SORTIES REQUESTED	14	25	34
SORTIES FLOWN	14	25	34
SORTIES SCRUBBED			
URGENT SORTIES	4	4	4
TOTAL BOMBS EXPENDED	168	300	408
BOMBS DIRECTED AT TARGET-POUNDS	76000	126000	132000
BOMBS ON TARGET-POUNDS	76000	29200	28800
BOMBS DIRECTED AT MISIDENTIFIED TARGETS		24	102
BOMBS SCUTTLED	16	24	42
WEIGHT OF EACH BOMB	500	500	500
TOTAL PASSES MADE	32	55	80
NUMBER OF PASSES WITH			
CORRECT TARGET IDENTIFICATIONS	32	47	54
MISIDENTIFIED OBJECTS		8	26
MISIDENTIFIED OBJECTS ATTACKED		3	15

Figure 10.9

could be considered the limited data that would be widely distributed. The same data are available in a different form in the structure and format of the matrix savevalue. Fullword matrix savevalue one, Figure 10.11, shows the summary of activity to date by the target for the third case. These different presentations of simulation data indicate that special outputs are needed to provide the system design team with readily interpretable data for their use.

In addition to these results the usual squadron data may be obtained showing the number of hours flown per month, the number of missions flown, and the number of urgent missions canceled because of lack of aircraft or bad weather. The normal GPSS statistics can be used to give the percentage of utilization of the aircraft by the squadron and

WEEK NUMBER 4 (CONTINUED)

	CASE 1	CASE 2	CASE 3
TOTAL AIRCRAFT LOSSES	2	2	3
LOST ENROUTE TO TARGET			
OVER FRIENDLY TERRITORY			
OVER HOSTILE TERRITORY	1	1	1
LOST OVER TARGET	1		1
LOST ENROUTE FROM TARGET			
OVER HOSTILE TERRITORY		1	1
OVER FRIENDLY TERRITORY			

CUMULATIVE SUMMARY OF FLIGHT OPERATIONS

AFTER 4 WEEKS, THE PRINCIPAL RESULTS ARE

	CASE 1	CASE 2	CASE 3
TOTAL NUMBER OF TARGETS	23	23	23
TOTAL NUMBER OF TARGETS DESTROYED	22	16	17
TOTAL SORTIES REQUESTED	46	129	144
TOTAL SORTIES FLOWN	46	125	140
TOTAL SORTIES SCRUBBED	2	7	6
TOTAL AIRCRAFT LOST	4	7	6
SYSTEM COMPARISON FACTOR	100	35	31

Figure 10.10

other data on usage and historical records. If these data were to be widely used, their formats would also be tailored for the typical user.

Model Details

This model was originally implemented using the GPSS III language. Later it was used to test and evaluate new features of GPSS/360 and GPSS V. The original model required several HELP routines to simplify the transfer of data between two transactions. One transaction would represent the mission request, while the other would represent the aircraft. The data transferred between these two transactions could be the identity or tail number of the aircraft assigned for this mission,

CUMULATIVE SUMMARY OF FLIGHT OPERATIONS

AFTER 4 WEEKS THE PRINCIPAL RESULTS ARE

A TOTAL OF 144 SORTIES WERE REQUESTED

OF THESE 140 SORTIES WERE FLOWN AGAINST

21 TARGETS AND 6 SORTIES WERE SCRUBBED

THIS RESULTED IN THE DESTRUCTION OF 17 TARGETS AT A COST OF 6 AIRCRAFT

SUMMARY OF TARGET CHARACTERISTICS

TARGET NUMBER	NO.OF PASSES	NO.OF IDENTIFI-CATIONS	NO.OF MISIDENTI-FICATIONS	POUNDS OF BOMBS AT TARGET	POUNDS OF BOMBS ON TARGET	NO.OF BOMBS AT MISID.OBJ	NO.OF BOMBS SCUTTLED	NO.OF AIRCRAFT LOST	TARGET DESTROYED YES = 1
	1		3	4	5	6	7	8	9

MATRIX FULLWORD SAVEVALUE   1

COLUMN	1	2	3	4	5	6	7	8	9
ROW 1	13	2	11	12000	4840	60	84	2	1
2	16	8	8	24000	6110	30	18	0	1
3	0	0	0	0	0	0	0	0	0
4	18	12	6	24300	5250	18	6	0	1
5	12	7	5	15000	4840	18	0	0	1
6	1	0	1	0	0	12	0	0	0
7	63	15	48	37000	4320	142	50		1
8	4	1	3	5000	4360	24	12		1

Figure 10.11

272

the amount of time logged for this mission, or the transfer of the target number from the target request to the aircraft assigned to fly the mission. The set or group capability of GPSS V eliminated the need to use HELP subroutines for this function. Instead the transaction became a member of a group and the SCAN or ALTER blocks were used to transfer data.

There are numerous possibilities for the organization of the mission-requests scenario. In the GPSS III model one transaction was created representing each mission request and its input data. These transactions were recorded on magnetic tape and read into the main model using the JOBTAPE feature. The matrix savevalue structure of GPSS V permitted these data to be inserted into a fixed array that was part of the main model. The matrix savevalue used INITIAL cards to describe each mission, permitting a scenario of separately described and different missions. The mission request transaction obtained its data from one row of the matrix.

The indirect addressing ability of GPSS could have been used to create a compact model of few blocks. Then each phase of the model would be cycled through the same set of blocks. The indirect addressing would provide new data each time the aircraft advanced to the next phase of the mission. Such a compact model uses fewer blocks, but is less susceptible to bypassing levels of detail and is harder to debug. Therefore this model was constructed with each of the six active phases using its own routines. This approach is more suited to a general class of model since a subroutine such as "attack" could be replaced or extensively modified without affecting the rest of the model. The aircraft transaction included a label to indicate which routine it was currently in for historical and debugging purposes.

Record keeping in the model extended beyond what was needed for output reports and included dynamic lists, kept on chains, of the present assignment of each aircraft and historical records for a limited period of time activity of the squadron, missions flown, aircraft lost, missions awaiting aircraft assignment, and aircraft awaiting clearance for takeoff. This extensive use of lists and the processing of these files would not be necessary for specific production runs of the model. However, when using the model to determine the overall sensitivity of the system to particular factors, it is useful to be able to go back and point out the contributing elements that happened to provide the results obtained.

The model has almost 400 blocks. Of these, 150 are used to represent the mission, 100 to represent housekeeping for the squadron and mission status, 50 to represent the description of the target and the weather, and the rest to facilitate the output report.

Summary

This example is typical of the application of simulation to the design process.[6] The model developed is general, but only for a particular area. In turn this general model can be made specific to provide support for the design team. In fact a general model is only general during that part of the system design phase in which different concepts are being evaluated. Once the model is in use, it becomes the current tailored representation of the system and much less general.

The dynamic nature of the system design process requires other, more detailed uses for an established model. New data become available and are inserted. The model grows, but unevenly. Only those areas requiring further analysis become more developed. The rest may become less detailed or even represented by simpler relationships. Rarely is the model in a static state for long, which is as it should be for a useful tool.

Weapons system design, with its broad interlocking relationships, is probably the first area in which detailed simulation has become a part of the system design process. Such models are by no means limited to weapons system design—important as that is. They can be used to explore tactics and procedures, logistics requirements, and the application of already designed weapons systems in operational environments different from those for which they were originally designed.

PROBLEMS

1. Model generality is of great concern when several alternative systems are to be evaluated. Consider the evaluation of different approaches to the design of aircraft support systems in order to increase the probability that the aircraft will operate when needed.
   (a) Higher reliability components—fewer failures.
   (b) Easier maintenance—functional construction.
   (c) Quicker parts replacement—more accessible units.
   (d) More accurate failure diagnostics—built-in test.
   (e) More accurate failure diagnostics—improved support equipment.
   (f) Improved technical skill level, more effective training and greater incentives.
   (g) Limited funds—can trade-off between initial and operative costs.

Flow chart one model which will allow these factors to be changed—greatly, slightly, individually, or in combinations. Indicate how these changes may be accomplished by modification of input data.

2. Suggest a figure of merit and its constituent factors to compare the different approaches.

3. It is desired to control the actual servicing of existing aircraft using a model more detailed than the flow chart of No. 1. Add to the flow chart those details considered necessary to control both scheduled and unscheduled servicing.

4. Compare the modeling technique using the single multi-purpose model with special models designed for each approach.

5. One of the approaches for No. 1 is to be represented in an overall flight profile model. Reduce the flow chart to a coarse level of detail consistent with the evaluation of overall systems.

6. If, instead of aircraft, automobiles were to be the subject of the study considered in No. 1, to what extent could the model flow charted be used again?

BIBLIOGRAPHY

1. *Simulation Using Digital Computers*, G. W. Evans, G. F. Wallace, and G. L. Sutherland, Prentice-Hall, Englewood Cliffs, N.J., 1967.

2. A Tactical Warfare Simulation Program, J. B. Fain, H. W. Karr, and W. W. Fain, *Naval Res. Logis. Quart.*, **13**, 4:413–436 (1966).

3. Simulation and the Logistics System Laboratory, M. A. Geisler, W. W. Haythorn, and W. A. Steger, *Naval Res. Logis. Quart.*, **10**, 1:25–54 (1963).

4. Planet: Part 1—Availability and Base Cadre Simulator, B. J. Voosen and D. Goldman, RAND Corporation, RM 4659 PR, 1967.

5. A GPSS Simulation of Carrier Operations and Avionic Maintenance, G. A. Walz and G. Damiani, *Proc. IEEE Auto. Support Sys. Symp.*, St. Louis, Mo., 1969.

6. Aerospace Applications of GPSS, J. L. Hooley, *Proc. IBM Seminar Oper. Res. Aeros. Ind.*, 1968.

11

Illustrative Example IV— Performance of a Computer System

The technique of a simulation is challenged when the analyst attempts to apply modeling to the design of computer systems. This chapter considers the problems of using simulation for this purpose. The application of a computer to message switching is a limited example. In the model the analyst considers the implications of direct access storage on the message switching time.

When a computer system is simulated, many individual computer instructions are required to simulate a simple function. If a simulation were formulated to the level of detail of individual computer instructions, the simulation could run 1000 times slower than the actual instruction. Therefore simulation to the individual instruction level of detail is a tool for special subsystem analysis. A typical example might be the influence of component tolerances on the performance of proposed logic circuits.[1-3]

Frequently the focus of interest lies at the other end of the spectrum. For example predict the capability of a computer system to process a typical day's batch of computer tasks or for a more complex system determine the amount of housekeeping and useful CPU time for a multiprocessing system. To find out how the proposed system would perform, the analyst need only run his jobs on such a computer system,[4]—if he can find one! When this can be done the procedure is quick and relatively inexpensive. Difficulties arise under various sets of circumstances:

- The computer system is to be reconfigured either with different hardware or with changed operating system software.
- There are numerous significantly different "typical" days.
- A nonexisting system is to be analyzed.

So far we have considered the situation in which simulation may be too expensive and that in which simulation is not needed because the actual computer operation can be directly studied. There is a middle ground between these two techniques where simulation can be useful for predicting computer system performance. It is characterized by coarsening the level of detail to consider functions implemented by hundreds or thousands of instructions and by considering new or complex computer systems rather than existing ones which may be tested in the field. There is, however, an inconsistency, because the "difficult" computer systems frequently require attention almost to the level of detail characterized by the individual instruction. Generally "difficult" computer systems are those with interactions of several programs—one answering real-time inquiries, another processing lower priority tasks, and a third processing data input and output. The designers of these systems could use computer simulations to trade off among numerous hardware possibilities. These include size of "scratch pad" storage, large-scale slower access speed and resident and removable packs, core storage, drum size and access time, disk file organization for tape file speed. On the system level—we can use additional hardware such as an input/output processor, directly couple a small computer to the large machine, or change the quantity and speed of the channels between the CPU and the rest of the system. The software possibilities are even more numerous: fixed or dynamic buffer allocation, buffer numbers and sizes, priorities between tasks, extent of anticipatory processing prior to actual task processing, and, most complex, evaluation of partitioning schemes—fixed, dynamic, priority, timed, and interactive.

These factors can be evaluated using simulation, *but only for a specific software system.* The software influence in complex computer systems may be equal to or greater than that of the hardware. In addition the intricate and combinatorial nature of the controlling program may make it necessary to simulate software behavior at a very detailed level. This, in turn, is complicated by the difficulty of knowing what the program will do under "typical" conditions. In "difficult" computer systems the interaction between the hardware and software varies depending on the system load, its characteristics, the serialization of tasks, and the competition between high and low priority tasks. To further complicate the difficulty as loads increase, the per unit task capability of the system decreases since individual tasks take longer and more items must be searched. The total throughput increases, but at slower and slower rates until the system may actually come to a halt once all its available storage is filled. While the example in this chapter is not at this level of complexity, it does provide insight into how the computer-oriented system designer may use simulation.[5-14]

Simulation of Communication-Computer System

The example[11] selected considers the use of simulation to compare the performance of different systems under increasing loads. A single computer is used for communication network message switching. These messages originate at remote locations and are destined for other locations in the system. The computer controls the communication network, switches the messages, and checks for message loss or duplication.

Messages are received by the central site, stored, and forwarded to their destination. Messages destined for terminals which are closed for the night or temporarily out of service are retained at the central site for later delivery. Statistics are maintained about the performance of both individual lines and the message switching functions. Historical records are maintained for message retrieval at a later time. The application of the computer to perform this type of store-and-forward message switching is well established.[12] The role of simulation is to develop a model to aid in determining the characteristics of the computer system. Several restrictive choices have been made to limit the number of alternatives and to reduce the simulation running time.

- The characteristics of the computer system are treated from the functional point of view. This reduced the level of detail to a macro rather than a micro level. Individual instructions are ignored and only the times and characteristics of overall functions are included.
- The tasks for the computer system are restricted to those defined, with the frequency and characteristics of each task available from input data. The software performance is defined for each function.
- The functions simulated assume full time availability of the system. No background tasks are considered.
- Specific hardware is described through a set of equipment characteristics. The definitions are gross, such as the time required to seek one record in a file. When different equipments are substituted for one another the same characteristics must be defined.

This model was developed using GPSS. The limited objective was to predict possible behavior of the direct access storage unit in the computer system part of a message-switching system under a number of different conditions. The storage device was used to store messages until they could be processed. The remainder of the communication-computer system was considered constant.

Message-Switching System

Messages originate at terminals on a number of communication lines leading into the switching center where they are stored and then sent to their destinations as shown in Figure 11.1. The messages—in the form of segments— are processed by the central processor of the computer and stored on the direct access storage device or disk. The incoming messages are assembled character by character in core storage at line speed in small "primary" buffers, which are refilled many times during the message arrival. The "segment," the portion of the message held in a primary buffer, is considered the basic quantity of data the computer analyzes and transfers to the disk storage device.

During the period when the segment is waiting to be transferred to the disk, an even smaller secondary buffer takes over the function of assembling the incoming message. The primary buffer must be unloaded before the secondary buffer is full or part of the incoming message will be lost—an intolerable situation. The normal sequence of operations, shown in Figure 11.2, fills the primary buffer with a segment, and then the secondary buffer takes over while the primary buffer is transferred to disk. The secondary buffer is then transferred to the primary buffer, with the remainder of the segment bypassing the secondary buffer to go directly to the primary buffer. The process is repeated until the full message is received. Obviously one critical system design considera-

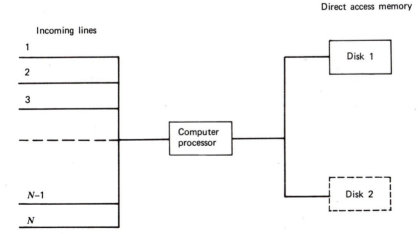

Figure 11.1 Overview of incoming lines and message processing.

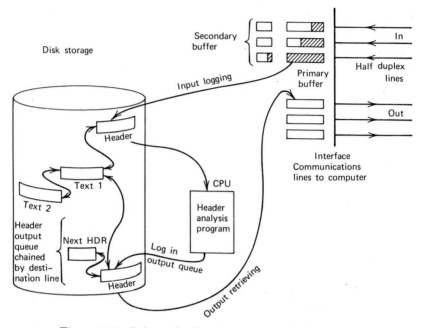

Figure 11.2 Information flow for message switching system.

tion is to assure that the secondary buffer subsystem is adequate. Figure 11.3 indicates the timing considerations involved both when the access times are adequately short and when they are too long and some data are lost.

The first segment of the message is the header identifying origin, destination, length, and the time when the message entered the system. The first segment is stored in a special location; the remaining segments are transferred to the first available disk location. They are stored sequentially; that is, segments from all lines are intermixed to minimize disk seek-time. Many segments are stored on each disk track, and only after all the positions available have been filled will the disk access arm move to the next disk area.

Control of where the message segments were stored is maintained through a set of forward and backward pointers. The first segment stored on the disk reserves the position where the next segment will be recorded and retains this address. The next segment stored contains pointers both to the next segment and to the previous segment. The last segment contains only a backward pointer.

Forwarding of the message to its destination by the switching system may wait until the entire message has been received. Another procedure allows the forwarding to begin after an arbitrary number of segments have been received. The number of segments received before forwarding starts is given the term "cut through" value. Either method may be used; so both must be considered in the model design. Computer processing for either is almost identical. First, the computer determines that the segment is a message header and stores it on the disk in a location reserved for headers. These headers form a queue of messages waiting for output communication lines. Forward and backward pointers structure the queue according to the output line and in a first-in, first-out order. When the output line is free, the first waiting message header is transferred from the disk to the primary buffer for the output line, as shown in Figure 11.4. Retransmission can start on the output line either as soon as possible or after the entire message has been stored. Retrieval of the next segment of the message from the disk is overlapped with the transmission of the previous segment.

The priority structure favors the incoming message over the outgoing one because loss of an incoming character represents an actual failure of the system. The failure to transfer data to the outgoing lines means

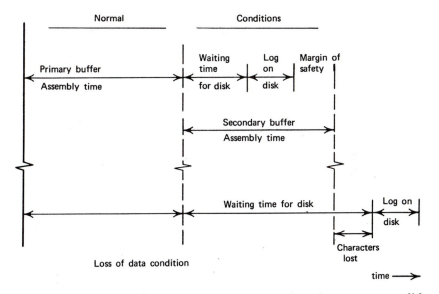

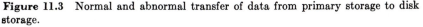

Figure 11.3 Normal and abnormal transfer of data from primary storage to disk storage.

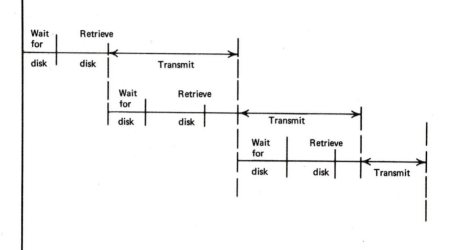

Figure 11.4   Timing considerations of outgoing message formation.

only some idle line time. Within the computer system this dictates that the movement of message segments to the disk and the analysis of message headers will have higher priority than the processing of segments from the disk to the line buffers.

The model is designed to evaluate the performance of the disk subsystem, therefore the details of the data processing are unnecessary, provided the computer is not loaded with other processing. The scope of the programs the data processor must be prepared to handle are quite extensive as shown in Figure 11.5. Only four subroutines are needed for the functions in the model: line control, header control, input/output interface, and disk input/output. In a full treatment of the total problem many more interactions with other subroutines would be involved.

Goals of Simulation

The structure of this model is general. Therefore it must be tailored for the particular system being studied. The critical variables that must be held constant are the levels of message traffic and message length distribution. The variables in the design of the computer system are primarily related to a different configurations of disk storage units. The system design is the result of trade-offs among the following factors:

- Number of disks required.
- Desirable segment size.
- Service provided from initiation of message transmission to receipt at destination.

Analytical solutions could be devised for determining the influence of the computer storage subsystem on performance of the message switching system with competition, interference, and congestion arising from simultaneously servicing numerous lines. However, the anticipated minimum size of a message-switching system requiring dedicated computer servicing of incoming and outgoing communication lines is about 50 lines. This produces numerous conflict situations that are best resolved through simulation. In addition embellishing the system to make it approximate the characteristics of a complete message switching system with multidestination messages and internal priorities cannot be handled analytically. Finally there is the opportunity to attempt to reduce the system resources committed by dynamic allocation of segment buffers and using excess computer capability for other types of programs. The influence of these factors cannot be determined analytically.

Simulation Results

The usefulness of the simulation for analysis of the message-switching system can best be seen by going through the evaluation of a sample system. The system chosen is characteristic of a moderate sized company with remote locations for sales offices, plants, and warehouses tied to and through a home office. Internal communications are so numerous that fast, efficient, and economical written records are a necessity. These are obtained either through public or private wire communications networks. The traffic is composed of an almost continuous flow of inquiries, responses, reports, directives, and sales orders, with strong requirements for reliability and speed.

The minimum system to be evaluated was based on a computer central processing unit and one disk storage unit. The traffic during the peak period required 100 lines at line speeds of 10 characters per second, standard 100-word per minute teletype circuits. The line utilization was limited to 70 percent of line capacity. The average message length was 550 characters, of which 100 were used for the header. Variations in message length were considered for both the linear and exponential

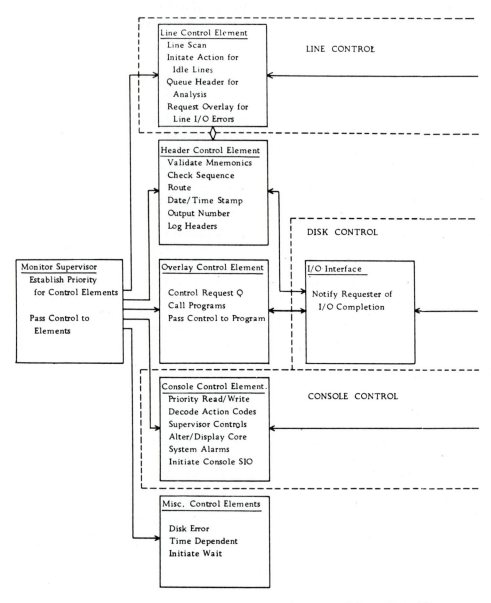

Figure 11.5 Message

284

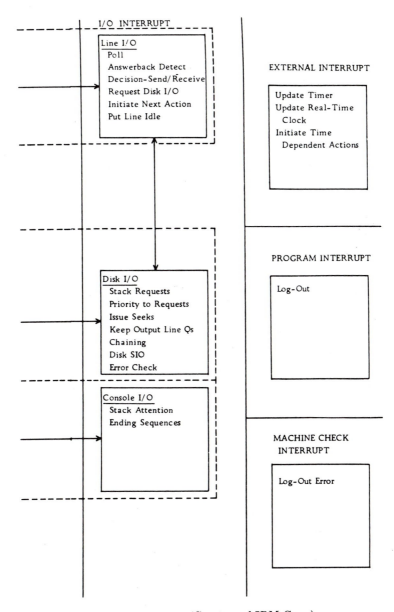

I/O INTERRUPT

Line I/O
Poll
Answerback Detect
Decision-Send/Receive
Request Disk I/O
Initiate Next Action
Put Line Idle

EXTERNAL INTERRUPT

Update Timer
Update Real-Time
  Clock
Initiate Time
  Dependent Actions

PROGRAM INTERRUPT

Log-Out

Disk I/O
Stack Requests
Priority to Requests
Issue Seeks
Keep Output Line Qs
Chaining
Disk SIO
Error Check

Console I/O
Stack Attention
Ending Sequences

MACHINE CHECK
INTERRUPT

Log-Out Error

switching system programs.   (Courtesy of IBM Corp.)

variations about the mean. The initial values for the items characteristic of the message-switching system were: 100-character primary buffer, 200 segments stored on the disk before a disk seek was required, and one disk unit.

The importance of understanding the input data can best be illustrated by the variation in system performance resulting only from the choice of message length distribution. In both cases the mean message length is 550 characters. The comparison between exponential and linear message length distributions indicates the very different lengths of time that messages spend in the system. The exponential distribution can frequently be handled more quickly than the linear, but it has the property of occasionally taking a long time to clear the system. The result is to tie up storage and to extend the mean switching time as shown in Figure 11.6, to 164 seconds from 142 seconds for the linear case. The linear case is more predictable, having bounded limits on either side of the mean value. Its extreme value is about twice the mean, about 300 seconds. For the exponential distribution, however, at 300 seconds almost one message in 10 is still in the system. Even more noticeable is the change in times for the delays before completion of disk seeks, as shown in Figure 11.7. The average delay in seek time is increased to 29.3 seconds from 26.0 for the linear distribution.

The other aspect of input data, message traffic volume, also must be considered in predicting system performance. Since the lines are

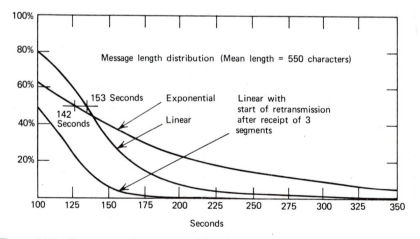

Figure 11.6 Percentage of messages being switched in times higher than x. (Courtesy of IBM Corp.)

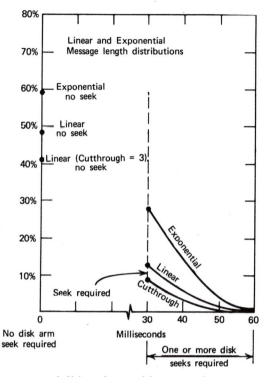

Figure 11.7 Percentage of disk seeks requiring more than *x* msec. (Courtesy of IBM Corp.)

assumed to be 70% loaded, the changes in this factor are significant in terms of disk utilization. The three values used—50, 80, and 100 lines—show that disk usage rises at a considerably more rapid rate than the increase in the number of lines shown in Table 11.1. However, switching systems performance was essentially the same for this comparison, which used linear message length distribution. There was a building up of potential input delays, as is shown in Figure 11.8. The 50-line case has much more of a chance of completing a message before the next one comes in. The segment gets recorded on the disk more readily and, likewise, is more quickly retrieved for the outgoing lines. The output delays are very similar to the input delays in distribution, as shown in Figure 11.9, but the average delay is greater.

Since the disk is the sole element of the system that appears to be a potential bottleneck, the logical exploration would be to determine

Table 11.1. *System Performance Statistics for Linear Distribution*

Number of lines	50	80	100
Messages/system/hour	1120	1800	2240
Mean waiting time to log on disk (m seconds)	41.9	57.3	68.9
Mean waiting time to retrieve from disk (m seconds)	57.9	94.1	142.6
Disk utilization	19.9%	38.7%	56.1%
Mean seek time (m seconds)	15.9	21.5	26.0
Mean switching time (seconds)	141.7	141.6	142.4
Maximum number of characters of buffer storage in use at any one time	4.5K	6.8K	9.2K
Average number of buffer characters required	3.0K	5.2K	7.0K
CPU utilization	4.9%	8.9%	12.1%

the effect of adding a second disk to the system. In this case the messages would be assigned alternately to one disk and then the other. The disk utilization drops off faster than would be the case in a linear relationship—23% utilization for each disk as contrasted with 56% for only one disk. Figures 11.10 and 11.11 show the marked improvement gained

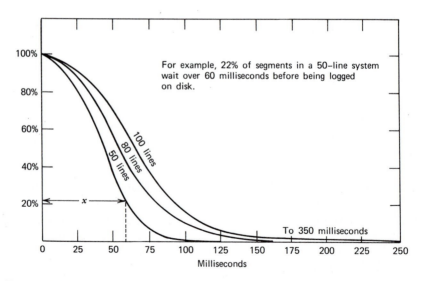

Figure 11.8 Percentage of segments waiting longer than x before being logged on disk. (Courtesy of IBM Corp.)

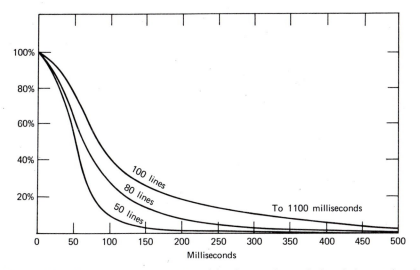

Figure 11.9 Percentage of segments waiting longer than x before being retrieved from disk. (Courtesy of IBM Corp.)

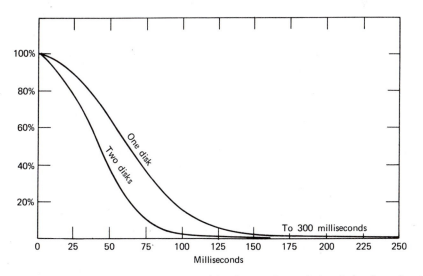

Figure 11.10 Percentage of segments taking longer than x before being logged on disk.   One disk versus two disks; 100 lines, priority of output over input.   (Courtesy of IBM Corp.)

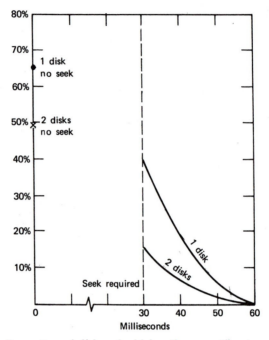

Figure 11.11   Percentage of disk seeks higher than *x*.   (Courtesy of IBM Corp.)

both from the time spent in the system and the possibility of being serviced without a disk seek by adding a second disk. This factor might also be used in extending the model to include the reliability of the system. In this case two disks provide a possibility of continued operation in spite of the failure of one unit.

The other main question to be resolved in the design is the amount of core storage used for the segment buffers. If the computer would also be used for other purposes, the amount of core available would be an important consideration, since for 100 lines, 20,000 characters of storage are required for incoming and outgoing 100-character buffers. Three different segment sizes were considered: 50, 75, and 100 characters. Table 11.2 shows that disk utilization goes up more than linearly, which would also mean greater time used by the computer system to process the small segments. Figure 11.12 compares the time required to store incoming segments of different lengths. The value of this potential system design would depend on cost factors that were not included in the model.

The last item in this example considers a change in system procedures

Table 11.2.

Buffer size in characters	50 line 100 characters	50 line 75 characters	50 line 50 characters
Mean waiting time to log on disk (*m* seconds)	41.9	44.6	51.9
Disk utilization	10%	27%	42%
Maximum number of buffer characters in use at a time	4.5K	3.5K	2.4K
Average number of buffer characters in use at a time	3.0K	2.4K	1.8K
CPU utilization	5.0%	6.7%	10.6%

rather than hardware. Instead of storing the entire message on a disk prior to forwarding it to the outgoing buffers and lines, only a limited number of segments would be stored before the beginning of the outgoing transmission. Meanwhile the remainder of the transmission continues to be received. For example if the cut-through value were three, these

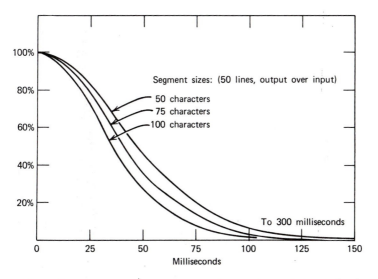

Figure 11.12 Percentage of segments waiting longer than *x* before being logged. (Courtesy of IBM Corp.)

segments would be recorded on the disk before there is an outgoing transmission. The mean time spent in the switching system is reduced to 105 seconds from 142 for the linear distribution of message lengths. The reduction in seek-times for the disks reflects the reduced number of stored messages waiting for outgoing buffers. The total amount of stored data is the same. The change in procedure merely allows the outgoing transmission to start before the disk access arm has moved to the next area.

Model Structure

Because this model was organized for ease of use by users not familiar with the workings of the model or GPSS, a series of initial cards was used to set all the values used in the particular simulation. The user

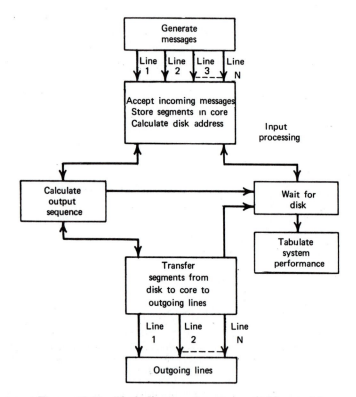

Figure 11.13 Block diagram of message switching model.

had only to set a very limited number of items on a few cards to specify a particular system:

- Number of characters per segment.
- Total number of lines.
- Planned communication line utilization.
- Mean message length.
- Type of distribution.
- Number of segments accessible without a disk seek.
- Number of disks.
- Line speed.
- Control over whether the entire message must be recorded before transmission.

The running time on an IBM 7094 was 8–10 minutes for the simulation of a 10-minute initialization of the system by filling it with messages followed by 30 minutes of simulated operation.

The model uses a total of 150 GPSS blocks. Figure 11.13 shows the overall block diagram of the model. The main subfunctions are the creation of messages for all remote stations, processing through the computer to determine where to store the segments, processing the headers into a sequence for output, competition for disk access and tabulation of seek times, and output processing of segments to outgoing lines.

The model is basically a straight-line sequence of actions, with competition for the disk being interspersed at points along the way. The main burdens are accounting the space allocation on the disk and the formation of queues before each access to the disk. The model can handle a number of lines, but the indirect addressing of the various parts of the model permits a single set of blocks to control the full series of actions. This results in an overall model of 150 blocks, a relatively small model for this amount of system representation.

Summary

The modeling of the communications network and the controlling of the store and forward functions of the computer are simulated on an IBM 7090 in about one-third of real-time. To accomplish this, the model treats the problem in gross functional terms. There is no detail of the program level to mirror the way the actual program performs. Rather the smallest element of concern is at the level of the function being performed. For example where is the disk arm and how far will it have to move to record or retrieve the segments?

This model gives quantitative insight into the use of disk storage in a message-switching system. This insight could not be obtained by analytical means. But the disk is a small part of the computer system. If the model were also to detail the core storage function, it would grow several times and be correspondingly slower.

The two most critical extensions for the model would be:

1 To study the system performance when, instead of being the only task, it becomes one of many possible tasks operating concurrently;

2 To find out how would other direct-action tasks, such as inquiry-response, fit into the system.

To accomplish these extensions would require full understanding of the controlling software, which determines what will be done, when where, and what resources are needed, and how conflicts will be resolved. It would require a far more detailed model in regard to basic controlling software, operating system, monitor, or supervisor, and it would, of course, be much slower. Thus we arrive at a dilemma, should one develop and use a model superficially detailed to save time and money and possibly obtain misleading data; or should one develop the detailed model, and be finished after the system has been built, although probably not programmed?

Computer simulation modeling cannot remain static! There are extreme needs, and money is being spent to develop a computer simulation tool. This step requires a special simulation language geared to computer hardware and software. The general purpose simulation languages require too much effort and money to solve even present-day computer system problems.[13-16] This chapter has merely introduced the subject. Although there has been considerable effort expended in the area of general purpose simulation languages, thus far the tool to solve the intermixed hardware and software simulation has not been developed.

PROBLEMS

1. The level of detail of a model depends on its potential application. In particular complex systems lend themselves to partitioning in order to control the number of variables in the simulation. Then, at a later time, the parts of the total system model may be combined for particular applications. Compare where and to what extent partitioning of models could be used in the following computer systems.

   (a) Batch processing of the job stream in full sequential operation.

   (b) Batch processing of job stream execution with overlapped input and output.

    (c)   (b) above with a separate computer to preprocess both input and output.

    (d)   Industrial process control application where a small computer is used to control only one application.

    (e)   Time-shared system including the computer complex of (c) above, remote terminals, and the communications lines.

    (f)   Software control for selection of next job, determination of priority, control of length of job execution, and selection of sequence of files to be used.

2.   The performance of a communications and computer complex consisting of central computer and 100 active remote data terminals may be predicted by simulation. Compare the pros and cons of the following approaches:

    (a)   Collect historical data and use an existing system, either in a laboratory or under field conditions.

    (b)   Predict performance based on statistics gathered from a sequence of 100 runs.

    (c)   Use a deterministic model with a standard data terminal job mix.

3.   How would the above be used to predict system performance for the following changes:

    (a)   Job mixes.

    (b)   Computer hardware configuration.

    (c)   Communications services.

4.   Instead of a selection based on average performance, the systems are to be compared on the basis of worst case operation—compare alternate means of obtaining insight to system performance.

5.   Two alternatives for the control of communications lines are polling of terminals and responding to requests for service. Each places a different task on the central computer. Contrast the simulation approaches to model each alternative. In real life the models should also include:

    (a)   priority selection of terminals.

    (b)   repetition of garbled transmission.

    (c)   great variation in message length.

Indicate to what extent these factors complicate the simulation approach.

6.   Real-time computer systems have the property of providing slower service due to more processing per transaction as the work load increases. Indicate specific ways the following nonlinear relationships might be simulated:

    (a)   Search of longer files.

    (b)   Movement of the disk arm versus no movement and movement over greater distances.

    (c)   Recourse to overflow records.

    (d)   Interruption of processing to handle a higher priority input/output command with additional record keeping.

BIBLIOGRAPHY

1. Computer-Aided Design: Simulation of Digital Design Logic, G. G. Hays, *IEEE Trans. Comput.*, **C-18**, 1:1–10 (1969).
2. Exclusive Simulation of Activity in Digital Networks, E. G. Ulrich, Commun. *Assoc. Comput. Mach.*, **12**, 2:102–110 (1969).
3. A Three-Value Computer Design Verification System, J. S. Jephson, R. P. McQuarrie, and R. E. Vogelsberg, *IBM Sys. J.*, **8**, 3:178–188 (1969).
4. The Role of Simulation in Computing, T. A. Humphrey, *IEEE Trans. Sys. Sci. Cyber.*, **SSC-4**, 4:393–395 (1968).
5. A Programming Language for Simulating Digital Systems, R. M. McClure, *J. Assoc. Comput. Mach.*, **12**, 1:83–89 (1965).
6. Simulation of a Multiprocessor Computer System, J. H. Katz, *AFIPS SJCC*, 127–139 (1966).
7. An Experimental Model of System/360, J. H. Katz, *Commun. Assoc. Comput. Mach.*, **10**, 11:694–702 (1967).
8. Evaluating Computer Systems Through Simulation, L. R. Huesmann and R. P. Goldberg, *Comput. J.*, **10**, 2 (1967).
9. The Simulation of Time Sharing Systems, N. R. Nielson, *Commun. Assoc. Comput. Mach.*, **10**, 7:397–412 (1967).
10. Computer Simulation of Computer Performance, N. R. Nielson, *Proc. 22nd Assoc. Computer Machines Conf.*, Thompson, Washington, D. C., 581–590, 1967.
11. Private correspondence, J. R. Maneschi.
12. On Teleprocessing System Design: The Role of Digital Simulation, P. H. Seaman, *IBM Sys. J.*, **5**, 3:175–189 (1966).
13. Simulation of a Multiprocessing System Using GPSS, F. C. Holland and R. A. Merakallio, *IEEE Trans. Sys. Sci. Cyber.*, **4**, 4:395–400 (1968).
14. The Effects of Input/Output Activity on the Average Instruction Execution Time of a Real-Time Computer System, S. G. Catania, *Proc. 3rd Conf. Appl. Simul.*, 105–113, Los Angeles, 1969.
15. Calibrating the Simulation Model of the IBM System/360 Time Sharing System, P. E. Barker, *ibid.*, 130–141.
16. Simulation of the Time-Varying Load on Future Remote-Access Immediate-Response Computer Systems, H. A. Anderson, Jr., *ibid.*, 142–163.
17. Jobstream Simulation Using a Channel Multiprogramming Feature, S. E. McAulay, *4th Conf. Appl. Simulation*, New York, 190–194, 1970.
18. A Simulation Study of Cost of Delays in Computer Systems, S. R. Clark and T. A. Bourke, *ibid.*, 195–200.
19. A Modular Simulation of TSS/360, J. W. McCredie and S. J. Schlesinger, *ibid.*, 201–206.
20. Simulation of the ILLIAC IV—B6500 Real-Time Computing System, H. R. Downs, N. R. Nielson, and E. T. Watanabe, *ibid.*, 207–212.

12

Illustrative Example V—
Auto Traffic Flow Through
a Series of Intersections

One significant use of simulation is to evaluate the behavior of existing or planned vehicle traffic networks. Simulation is a practical tool for the traffic engineer to use in a variety of problem areas, ranging from the strategic "What facilities are required?" to the tactical "How can these facilities be improved?"

A Vehicle Traffic Simulator (VTS)[1-3] has been developed as a tactical tool for traffic system engineers. VTS is an activity-oriented simulation program written in GPSS II and assembly language. This early program has the ambitious goal of enabling the designer himself to use the program. Typical examples of the uses of VTS would be to study the effects of:

- Different traffic control signal setting.
- Eliminating left and right turns.
- Changing lane directions.
- Increasing facilities and providing special left-turn lanes.

The objective of this chapter is to provide a detailed example of the use of simulation by making the reader familiar with VTS. While this chapter will not enable the reader to use VTS, he will be able to see how a simulation model is organized, implemented, verified, and structured to satisfy the user's requirements. A user who is not familiar with simulation languages will therefore see a simulation program tailored to meet his needs with a minimum of learning effort. VTS utilizes the traffic engineer's jargon, provides a format for data entry, simulates the problem, and delivers output in a meaningful manner. The traffic engineer can concentrate his effort to feed input data and evaluate results.

Simulation provides the systems designer with a tool to compare alter-

natives and to develop a series of specific criteria for traffic control
in particular systems. This is achieved by following each vehicle *individually* and gathering statistics for the performance of each *alternative*
system. Some of the possible criteria available from simulation are:

- Vehicle throughput.
- Total vehicle delay.
- Average speed.
- Distribution and number of vehicle stops.
- Distribution of vehicle delays.

From these output data, the traffic engineer can evaluate methods
of traffic control and the effectiveness of proposed hardware.

VTS uses GPSS II and relies on GPSS II for its structure. However,
a number of special assembly language subroutines have been added
to organize data structures, improve execution speed, provide data packing, and present special graphic input data and output formats. To
employ VTS the user supplies traffic network geometry, input data containing traffic-flow rates, left- and right-turn percentages, and signal
light settings.

Network Geometry

A VTS traffic network is a collection of entities called *intersection
modules,* which describe the geometry of each intersection. Each module
consists of junctions, junction cells, lanes, signal light facilities, and
left-turn zones. The number of lanes and the dimensions of the lanes
are unlimited for each intersection configuration.

Junction. A junction is the area of the road intersection where vehicles compete with one another. In Figure 12.1 the area identified
by the heavy border encloses the junction part of the intersection module.

Junction Cell. Each junction is divided into a checkerboard arrangement of junction cells of arbitrary size. The boundaries of these cells
are determined by pairs of intersecting lanes in the junction. The cell
is the minimum area that may be occupied by competing vehicles when
attempting to cross the intersection. Larger-size vehicles may occupy
more than a single cell at one time. In Figure 12.1 each cell is numbered
in a designated sequence, shown as $C_1, C_2, \ldots C_n$.

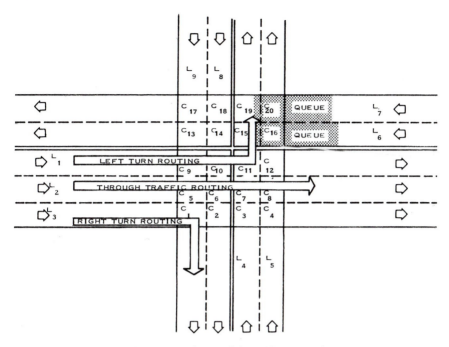

Figure 12.1 An Intersection module. (Courtesy of IBM Corp.)

Lane. A lane is the portion of a road connecting the preceding junction to the next one.

Signal Light Facility. The portion of the signal cycle allotted to traffic streams having the right-of-way constitutes a single phase of the signal light. Thus the green and amber time for the traffic streams in lanes L_1, L_2, L_3, L_4, and L_5 are one phase of traffic. A signal light facility is related to each phase of traffic. This is a GPSS facility with the property of allowing only one transaction to control the signal light. The intersection unit shown in Figure 12.1 has two traffic phases, one phase for each pair of directions shown.

Left-Turn Zone. A left-turn zone is the region of the junction that must be free of all vehicles before a left turn may be attempted. Such a zone consists of all opposing junction cells and oncoming queues. The cross-hatched area in Figure 12.1 is the left-turn zone for traffic moving in an easterly direction.

The physical composition of most intersections can be described by

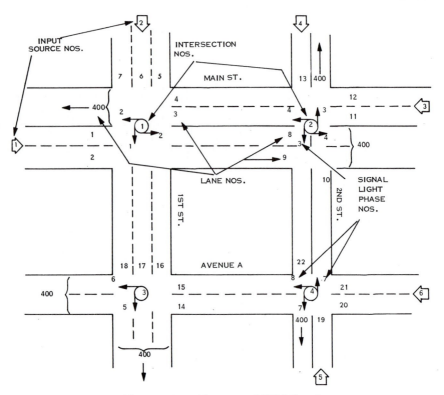

Figure 12.2   (Courtesy of IBM Corp.)

modules combining the above elements. A set of rules translate the
network geometry into the specific structure established for VTS.
Figure 12.2 illustrates the conventions governing the four elements.

   (a)   *Intersections*—numbered consecutively left to right and row by
         row beginning with the number 1. The intersections in Figure
         12.2 go from 1 to 4.
   (b)   *Lanes*—all lanes approaching an intersection are numbered in a
         counterclockwise rotation, beginning with 1 at the top lane of
         the left arm of the intersection. The numbers must be consec-
         utive through successive intersections.
   (c)   *Input sources.* Points in the network where vehicles originate
         are numbered in the same way as lanes, counterclockwise and
         consecutively through successive intersections. The input

sources of Figure 12.2 are enclosed in the arrowheads and are numbered 1 through 6.

(d) *Signal light phases.* Associated with each intersection are signal light phases, for each direction. The phases are numbered consecutively through successively numbered intersections. Thus assuming only two phases per intersection, intersection 1 has signal phases 1 and 2; intersection 2 has phases 3 and 4 and so forth. Figure 12.2 signal phases include 1 through 8. Left turn only phases would increase the number if they were used.

Geometric Input Format and Data

Data entry describing intersections for VTS is accomplished by a set of seven cards for each pair of intersections. The format is fixed and a special form is used to relate to the network geometry shown in Figure 12.2 with the input data. The left side of the form represents one intersection, the right side another, with each intersection (shown in Figure 12.3a and b) appearing as it would to an observer directly overhead. Thus, the left part contains the intersection identification and six items of data describing traffic lanes. The four arms of the intersection can be related to any desired compass orientation. Each arm of an intersection contains the following information:

(a) Number of approaching lanes.
(b) Length of these lanes in feet.
(c) Percentage of total traffic making right turns.
(d) Percentage of total traffic making left turns.
(e) Adjacent intersection number—the number of the connecting intersection for this arm. If the adjacent intersection is an input or output, this number is 0.
(f) Merge indicator—a number in this field indicates that turning vehicles are permitted if there is an acceptable gap to merge into moving traffic streams even if the signal is red as in the case of a right turn on red signal. Left turn merging is permitted for one-way traffic streams only. The numbering convention is:

(1) Merging is permitted from the left lane only.
(2) Merging is permitted from both left and right lanes.
(3) Merging is permitted from the right lane only.

A zero or blank means that no merging is permitted from any lane.

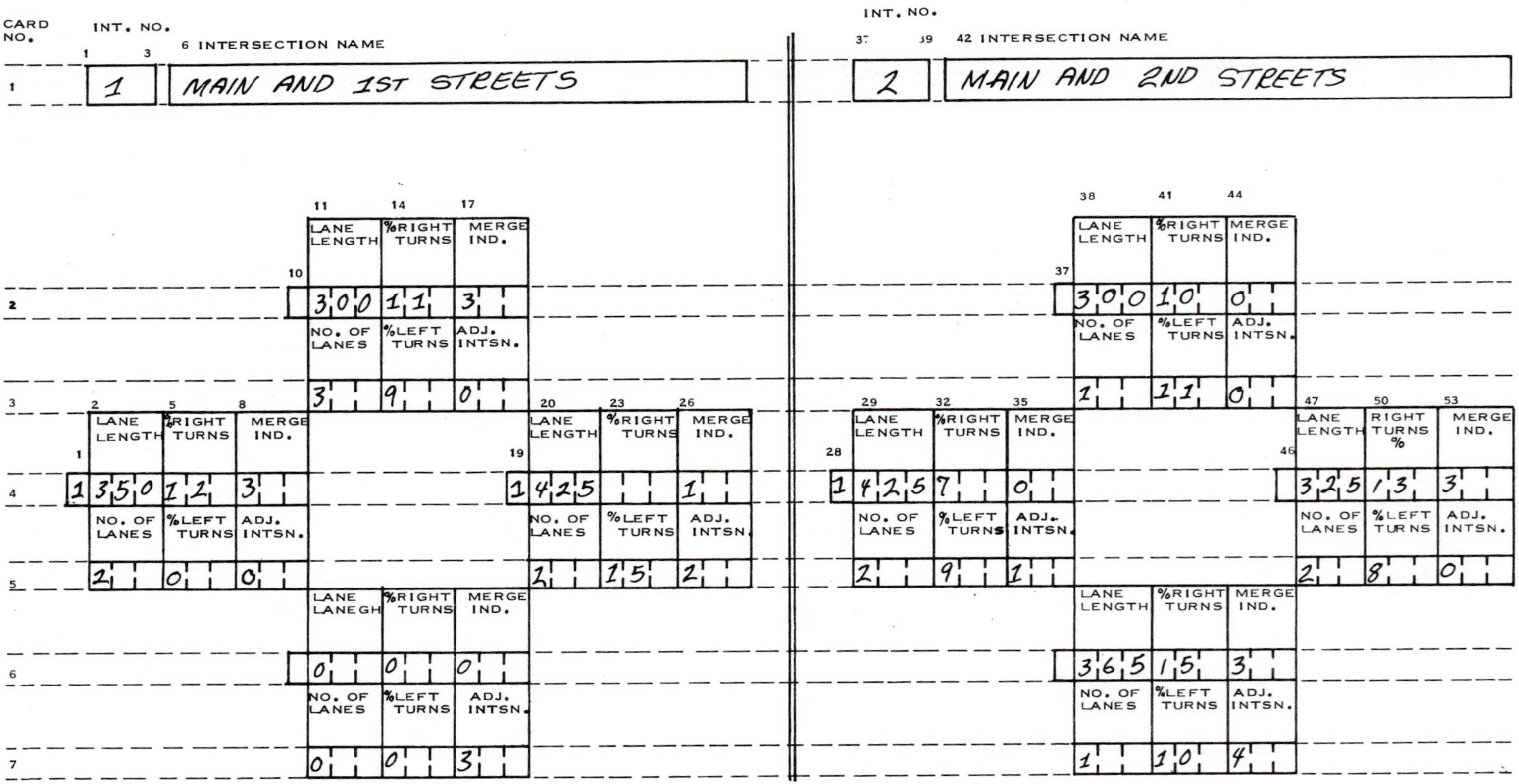

Figure 12.3a (Courtesy of IBM Corp.)

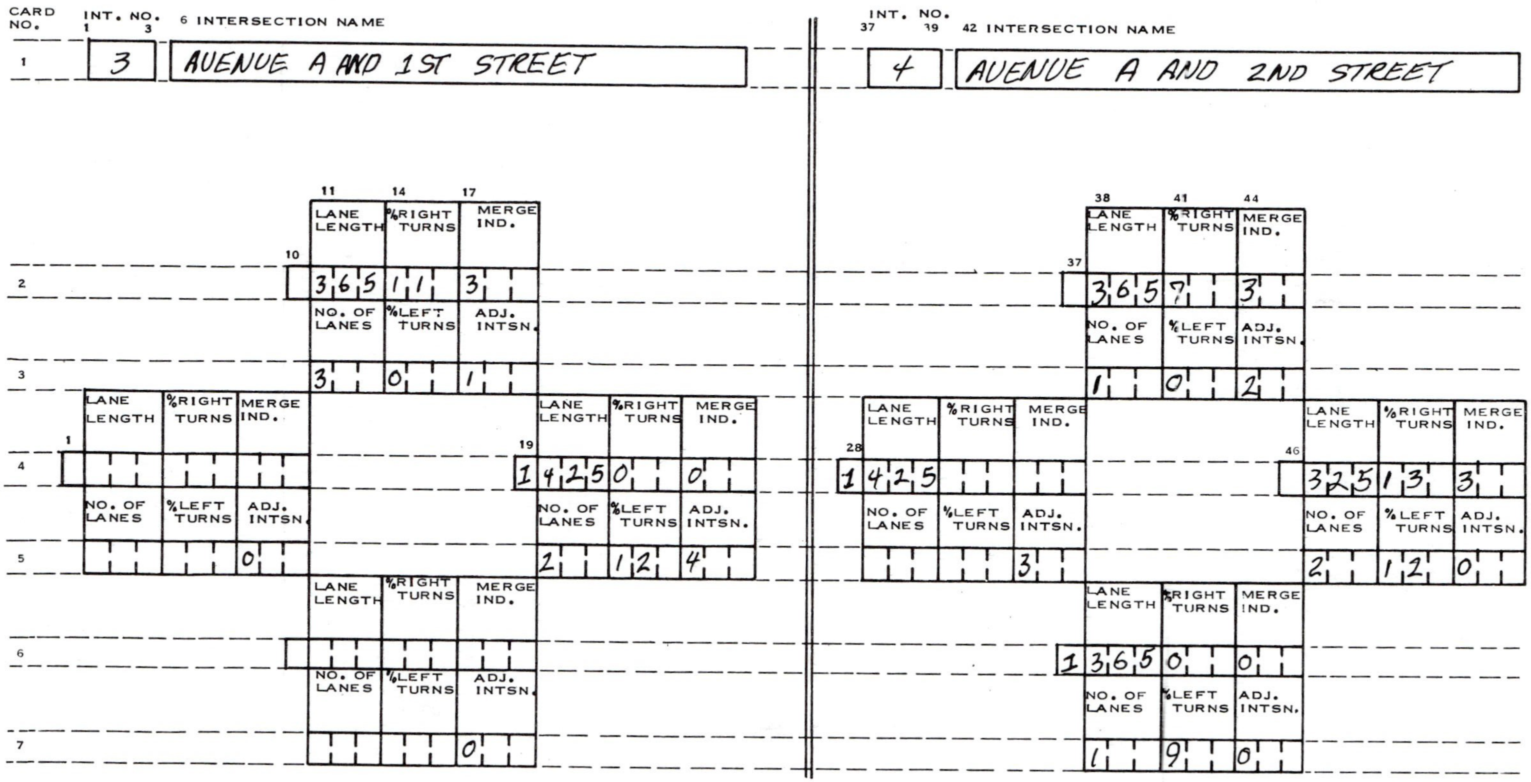

Figure 12.3b (Courtesy of IBM Corp.)

303

Blanks appearing in any of the data fields are interpreted as zeros. If a blank or zero appears either in the lane-length field or in the number-of-lanes field, VTS indicates an error.

Additional VTS input data consist of GPSS FUNCTIONS. These data are used by the program to describe:

- Mean time between vehicles at network input sources.
- Speed distributions.
- Vehicle length distribution.
- Cycle lengths for signal settings.
- Offset and split percentages for signal settings.

Simulation of Vehicle Flow

The geometry of the traffic network is structured by mapping the intersection modules representing each area and then linking them together intersection by intersection. Each intersection module is simulated in a single subroutine that processes vehicles for the entire network. Input data FUNCTIONS provide particular vehicle and network properties as vehicle velocity, vehicle length (in feet), signal light number, junction number, routing, queue number, and lane number.. For illustration purposes, the use of FUNCTION capability of GPSS to enter vehicle velocity will be covered in depth. The actual velocity for each vehicle is calculated from its lane position and a set of assumptions governing the behavior of vehicles. For the following discussion, the basic assumption is that vehicle velocities are normally distributed in contrast to other distributions or historical data. This implies knowledge of the mean velocities and their standard deviation. As a result of using the normal distribution, 68.3% of all velocities will be between +1 and −1 standard deviations. Figures 12.4a and 4b show the normal distribution and the cumulative normal distribution. From these, it may be seen that for a mean of 30 mph and for a standard deviation of 4 mph, 68.3% of all velocities will be between 26 and 34 mph. Within two standard deviations, 95.45% of the velocities will be in a range of 22 to 38 mph and so on for three or more standard deviations. However, the velocity selected will be modified so that no vehicle will be assigned a speed of less than five mph. Other values for the standard deviation would result in different ranges of velocities. A standard deviation of zero will set all velocities to the same value.

The actual calculation of velocity is performed using the VARI-

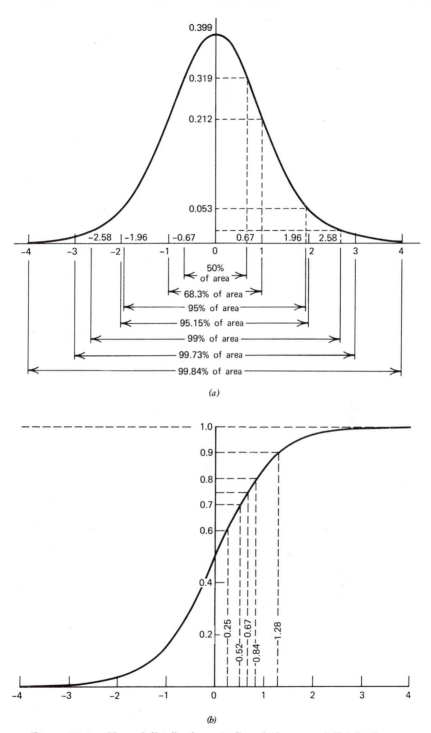

(a)

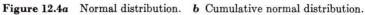

(b)

Figure 12.4a Normal distribution. *b* Cumulative normal distribution.

ABLE = FN9 + FN16*FN5 statement. FUNCTION 9 provides for
the situation shown in figure 12.6 the mean velocity for each lane,
namely, lanes 1–4, 8–13, 19, and 22, 27.9 mph; lanes 5–7, 16–18, 30 mph;
and lanes 14–15, 20–21, 25 mph. Likewise, the standard deviations for
the corresponding lanes will be provided through FUNCTION 16 and
have the values of 6.28, 8, and 4 mph respectively. Function 5 provide
the modifiers for selecting the specific value of standard deviation to
be added or subtracted from the mean velocity. Figures 12.5a and b
show the card formats for the FUNCTIONS representing mean velocities, value of standard deviations, and specific standard deviation
modifier.

A simplified diagram indicating the overall flow of vehicles within

Function   Title

9	Mean velocities

Purpose

Supplies mean velocity for each lane in a network.

Xn

Lane number.

Yn

Mean velocity in miles per hour.

Card Format

1	2	7	13	19	25	31	37	43	49	55	61	67
	9	Function		P3	D8		Mean velocity for each lane					
4		27.9	7	30.0	13	27.9	15	25.0	18	30.0	19	27.9
2	1	25.0	22	27.9								

Figure 12.5a   Composite of data from 7 data entry cards.

Function

Title

| 16 | Standard deviations for mean velocities of function 9 |

Purpose

| Supplies standard deviation for each lane in a network. |

Xn

| Lane number. |

Yn

| Standard deviation in miles per hour. |

Card Format

1	2	7	13	19	25	31	37	43	49	55	61	67
	16	Function	P3	D8		Standard deviation for each lane						
4		6.28	7	8.0	13	6.28	15	4.0	18	8.0	19	6.28
2	1	4.0	22	6.28								

Figure 12.5b. Composite of data from 7 data entry cards.

intersection modules is shown in Figure 12.6. As vehicles move through the intersection module, attributes are assigned by VTS, specifying street, lane, and junction characteristics. Each vehicle is located in a lane and queue, or in a junction region. VTS guides vehicles on pre-determined paths through the network to simulate established traffic-flow patterns. Routing FUNCTIONS specify the paths to be taken by the vehicles across the junction. A path is a list of junction cell numbers occupied by the vehicle while crossing the intersection. The last number of the list refers to the next lane segment number, which is used to direct the vehicle to the next intersection or final exit.

Table 12.1 gives the data for the distribution of turns for the street segment shown pictorially and in graphic form in figures 12.7 and 12.8.

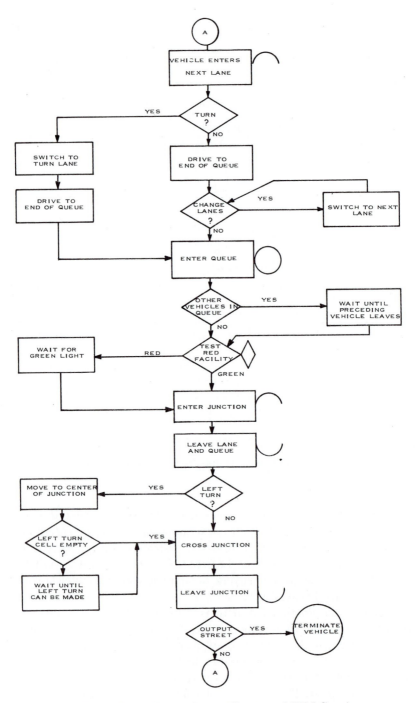

Figure 12.6 Vehicle flow. (Courtesy of IBM Corp.)

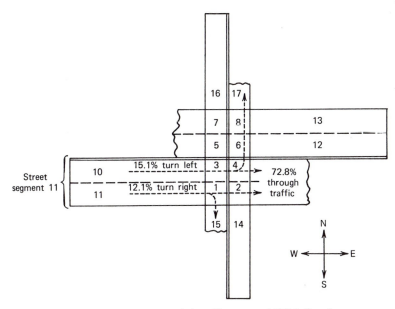

Figure 12.7 Intersection. (Courtesy of IBM Corp.)

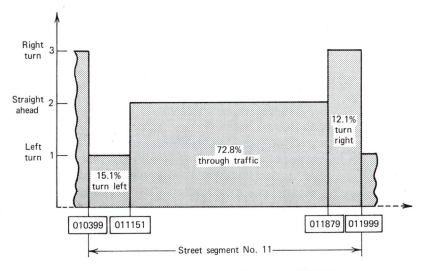

Figure 12.8 Traffic movement. (Courtesy of IBM Corp.)

The abscissa is a 6-digit packed number composed of the low-order three digits denoting the cumulative percentage from Table 12.1 multiplied by ten, and the three higher-order digits representing the street segment, 11. The ordinate specifies the routing FUNCTION number

Table 12.1　*Turn distribution—east traffic*

Turn	Percentage	Cumulative Percentage
1 Left	15.1	15.1
2 Through	72.8	87.9
3 Right	12.1	100.0

for the turn. For the case shown in Fig. 12.8, each vehicle obtains a pseudorandom number and adds this to the intersection number multiplied by one thousand. For this intersection, 0 to 150 results in a left turn, 151 to 878 routes the vehicle straight through the intersection, and 879 through 999 causes a right turn.

Figure 12.9 shows the routing FUNCTION associated with left turns. The abscissa, a 5-digit packed number, is composed of an indexing parameter that is incremented by 1 every time a vehicle enters a new

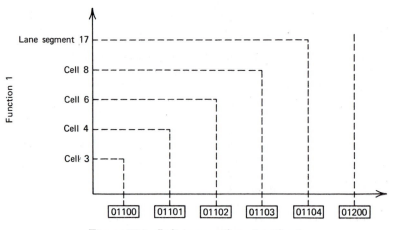

Figure 12.9　Left turn routing—function 1.

junction cell or lane segment in the two lower-order digits. The three higher-order digits correspond to the street segment number. Hence all left-turning vehicles are routed first to junction cell 3,01100, and then through cells 4, 6, and 8 and land segment 17 as the index is incremented to 01104. The contour of the path is established beforehand according to the traffic laws and existing practices. All left turns can be controlled by a single FUNCTION through the use of the three higher-order numbers to identify the intersection. Similarly, any other vehicle movement may be represented by using FUNCTIONS to guide vehicles on predetermined paths through the network, making it possible to simulate established traffic flow patterns, emergency vehicles, public conveyances, and any specially designated traffic.

Simulation of Network Control

The key elements in traffic control are the number, location, and control schemes for the signal lights. Provision for fixed variation in signal settings includes:

1. Signal cycle length—total time for a single control sequence of red, amber, and green lights.
2. Splits—percentage of signal cycle length for the red and green periods.
3. Offsets—percentage of cycle length for initial synchronization of consecutive traffic signals to maintain an uninterrupted flow.

A control loop shown in Figure 12.10 provides an independent subroutine control loop to generate a signal "regulator" for each traffic phase at every intersection. These transactions circulate only in the control loop, turning the signals on and off at fixed intervals. Vehicles arriving or waiting at a junction in the intersection module test a red signal associated with the phase; the junction is not entered or crossed until the signal has turned green. Amber signal time is included in the total green time. The control loop can operate in either a fixed-signal or a real-time adaptive mode.

Fixed-mode signal settings are specified by two FUNCTIONS: one for signal-cycle length and the other for split and offset percentages. The cycle-length FUNCTION lists each traffic phase with corresponding cycle-length duration. The split FUNCTION associates percentages of green and red cycle length times with each phase. The cycle-length FUNCTION provides the time in seconds for each phase. The offset

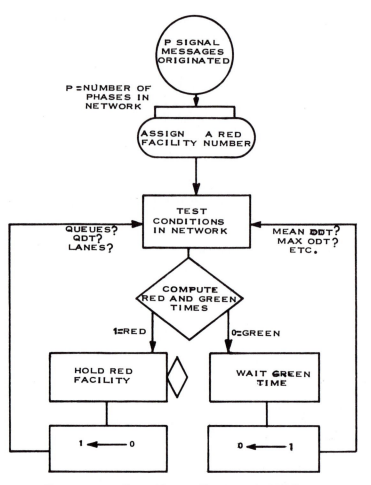

Figure 12.10   Control loop.   (Courtesy of IBM Corp.)

time obtained from the split FUNCTION is used to initiate all signals
from the same reference time.   At the outset of a simulation run the
offset time is calculated as a percentage of the times associated with
the green or red split.   The regulation of the signal light is dependent
on the time the facility is in the red condition.   After the green time
has elapsed, the associated red signal facility is held for the remainder
of the cycle.   This operation continues either for the entire run or until
signal settings are changed by the insertion of a new set of offset and
split FUNCTION data cards.   The procedure described permits inclu-

sion of fixed sequences of settings in the control mechanism of a model, which may be changed by the user after any time period.

The limitations of the fixed mode control of signal setting is not inherent in the model structure. An adaptive mode based on the number of vehicles in the intersection queue could be used for some or all intersections. The adaptive mode would use a FUNCTION to select a VARIABLE statement which relates the duration of the signal setting with the queue length. The complexity of the VARIABLE statement would determine whether only one or several intersections are controlled as an integrated system. Reference may be readily made to current values of conditions within the system—queue lengths, signal settings, and volume of traffic. If planned, recent history could be introduced in terms of maximum delay already encountered, size of queue lengths previously endured or whether a specific delay has been within the 90th percentile. These values may be used to provide a comparison for dynamic reference by the real-time adaptive control algorithms. For example, when a lane queue exceeds a percentage of its historical mean, the green setting at the previous intersection may be changed until the queue is reduced. This could be used to prevent traffic from blocking an intersection as traffic builds up to the congestion point.

Simulator Outputs

Outputs from the simulator provide standard traffic measures such as the following:

- Distribution of queue delay times.
- Distribution of queue lengths.
- Number of vehicles waiting in each queue after each signal change.
- Average utilization of lane capacities.
- Distribution of total travel times for all vehicles.
- Distribution of total travel times for specially routed vehicles.
- Total number of vehicles passing through each intersection.
- Distribution of average speeds.
- Vehicle throughput.
- Network status at specified time intervals.

Two types of output are available—graphic displays of network status with histograms of accumulated statistics, and detailed tabular listings of other data. Most traffic simulations will require only the graphic displays and histograms; therefore, there is an option to suppress the tabular listings.

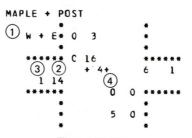

Figure 12.11a

① Green phase is shown in upper left-hand corner of intersection.
  S + N indicates green phase for south-north directions.
  W
  + indicates green for west-east direction.
  E

  R
  R + R indicates all red cycle.
  R

② Numbers at junction indicate number of vehicles in queue.
③ Numbers behind these indicate number of vehicles moving in lane.
④ Numbers in junction indicate number of vehicles inside intersection.

Graphic Presentation—Networks

The networks subroutine provides an instantaneous "snapshot" of the network or of selected intersections in the network at time intervals specified by the user. It consists, as shown in Figure 12.11a, of the following for each intersection:

- Intersection name.
- Current green phase, that is, the direction in which traffic is now moving.
- Number of vehicles currently waiting in queues at the intersection.
- Number of vehicles currently moving in the lanes behind the queues.
- Number of vehicles in the intersection.

The overview of the geometry of the numerous intersection of the network is shown in Figure 12.11b.

- Current simulation clock time.
- Reset clock time (the time since all statistics were set to 0).
- Compass orientation of the network.

Graphic Presentation—Histograms

The histogram subroutine provides plots of the frequency distributions and/or the cumulative frequency distributions as shown in Figure 12.12a

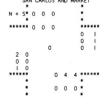

Figure 12.11*b*

CURRENT CLOCK TIME = 7 MINS. 30.0 SECS.   RESET CLOCK TIME = 5 MINS. 0.0 SECS.

******************AVERAGE SPEEDS - ALL VEHICLES******************
FREQUENCY DISTRIBUTION
ENTRIES IN DISTRIBUTION   MEAN AVER. SPEED (MPH)   TABLE NUMBER
466   13.2   72

```
30 + 139 ----------------------------------------------------------------------

PER      . ACT-
CENT     . UAL
         .
20 + 93 ----------------------------------------------------------------------
         .                ******
FREQU    .                *     **********
ENCY     .                *     *        **
10 + 46 --------********--******---*-------*-------------------------------------
         .       *      *     *        *
         .       *      *     *        *            ******
         .       *      *     *        *            *     **********
         .       *      *     *        *     ******-*--------------*----------+
         .       *      *     *        *     *              *           ******
0   . ******     *      *     *        *     *              *     *          ******
     .--------+-------+-------+-------+-------+-------+-------+-------+-------
        .0     5.0    10.0    15.0    20.0    25.0    30.0    35.0    40.0
                                                              SPEED (MPH)
```

Figure 12.12a   (Courtesy of IBM Corp.)

and *b* for the tabulated data in the simulation. Histograms are available for the following distributions:

(a) Queue delay times for specific queues.
(b) Queue lengths for specific queues.
(c) Queue profile with respect to time (size of queue at user-specified time intervals).
(d) Cumulative queue delay times—all queues.
(f) Cumulative trip travel times—all vehicles.
(g) Average speeds—all vehicles.

Detailed Tabular Listings

The standard GPSS II tabular listing provide detailed information in the normal formats. These include the block counts and savevalues which relate primarily to the operation of the model. The tabular statistical outputs provide detailed data for the utilization of facilities and storages. Individual tables accumulate the number of times a tabulated value fall within a specific range for delay times, queue length, and queue profile. Overall, the standard listings provide information on each of the following:

(a) Current (absolute) and reset (relative) clock time.
(b) Current number of vehicles waiting in queues.
(c) Total number of vehicles currently in lanes.
(d) Length, in feet, of current queues.
(e) Speed of a platoon of moving vehicles.
(f) Average time signal light was red and green and number of times signal changed.
(g) Average occupancy of each lane.
(h) Average occupancy of each intersection.
(i) Number of vehicles to have entered each intersection during the run.
(j) Average time spent in each intersection.
(k) Current number of vehicles in intersections.
(l) Detailed tables from which histograms are prepared.

Computer Program

The VTS program is written in GPSS II and FAP for the IBM 7090. Running time for the simulator varies with the size of the network. The ratios of simulation running time to real or actual time

CURRENT CLOCK TIME = 7 MINS. 30.0 SECS.   RESET CLOCK TIME = 5 MINS. 0.0 SECS.

***************AVERAGE SPEECS - ALL VEHICLES***********************
CUMULATIVE FREQUENCY DISTRIBUTION
MEAN AVER. SPEED (MPH)
13.2

ENTRIES IN DISTRIBUTION
466

TABLE NUMBER
72

```
100 +  466 -----------------------------------------------------
     .  ACT-
     .  UAL
 90 +  419 -----------------------------------------------------
FREQU . FREQU
ENCY  . ENCY
 80 +  372 -----------------------------------------------------

 70 +  326 -----------------------------------------------------

 60 +  279 -----------------------------------------------------
     .
     .
```

PER CENT

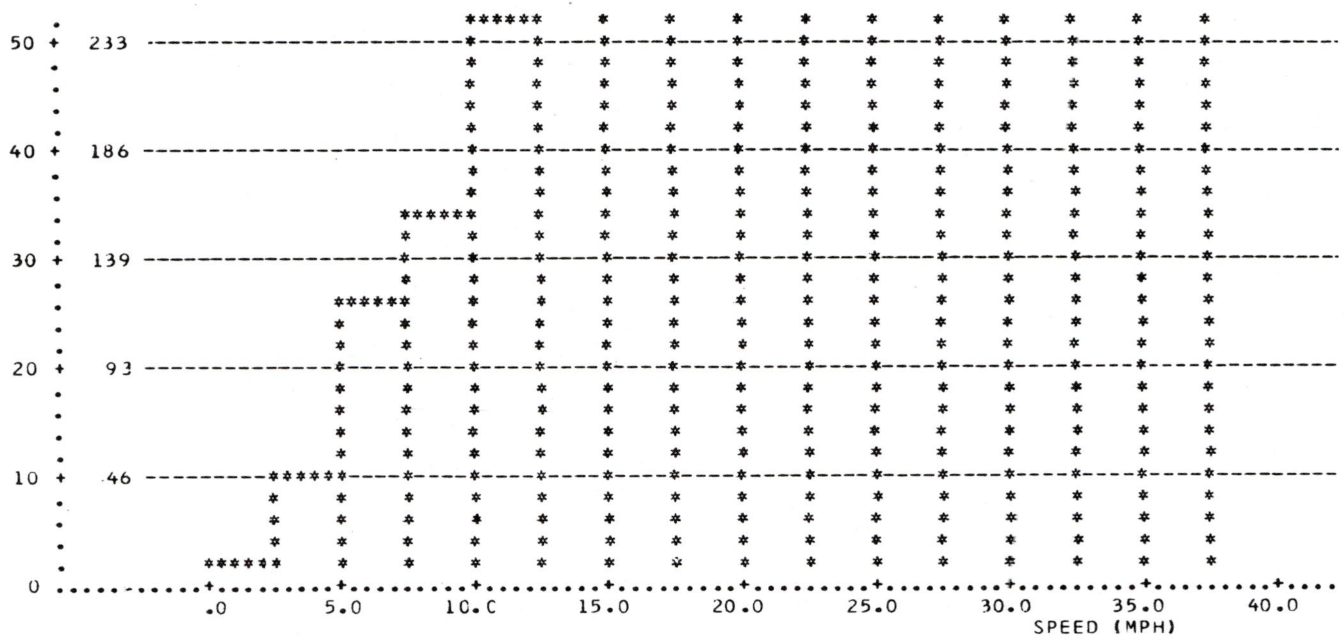

Figure 12.12*b* (Courtesy of IBM Corp.)

319

were found to range from 0.25:1 for single intersections, to 3.5:1 for large networks of 25 or more intersections. For most applications, the size of the program is limited by the number of vehicles that can be in the network concurrently (average of 8 core locations required for each vehicle).

The total number of concurrently active vehicles in the network can be approximated by the following formula:

$$V \doteq \frac{1000000 - 700L}{3V_q + 14V_e}$$

where V = total number of concurrent vehicles.

V_q = expected maximum percentage of vehicles in queues.

V_c = expected maximum percentage of vehicles moving in lanes.

L = total number of lanes in network.

Thus for a network of 150 lanes and an expected 60:40 ratio of vehicles in queues and lanes, the total number of vehicles is shown by:

$$V \doteq \frac{1000000 - 700 * 150}{3 * 60 + 14 * 40} = \frac{895000}{740} \doteq 1209$$

This formula has been derived for obtaining an estimate of vehicle capacity, not a precise quantity. Many other variables could affect this value (number of intersections, number of turns, number of vehicles in intersections, etc.); only the most significant were made precise in the formula. It should be noted that a greater percentage of vehicles in queues results in a greater capacity.

GPSS Model Dimensions

	400 Blocks	
	50 Functions	Vehicle Routing, exponential and normal distributions, signal settings, interarrival intervals.
	125 Variables	
(Up to)	75 Tables	(User specifies number of tables).
	1200 Logic switches	Junction cells, queue retainers.
	1200 Savevalues	Queues, speed of platoons, left-turn zones.
(Up to)	325 Storages	Lanes, intersections.
	75 Facilities	Signals.
	0 Queues	(Used savevalues instead).

11 Help routines coded in assembly language

    1—reads graphical input deck and converts it to GPSS functions (about 30).

    1—converts user-supplied functions (from seconds to 0.01 seconds) + housekeeping duties.

    1—Prints out network maps.

1—Prints out histograms.

2—Used to amalgamate and separate cars into 1 queue at intersection (requires 3 cars/vehicle + 6 + 2 words).

5—used to decrease running time and save core storage (combines
different block type capabilities into single help routines—saves
on overhead, housekeeping, and error checking).

Validation

How closely does the simulation approximate reality? To gain confidence that the model adequately reflects real life, some field measurements were made and compared with the simulation results. Figure
12.13 shows a graph of queue length measured in the field and queue
profile with respect to time as a result of the output data. The queue
lengths are at 80-second intervals and were measured between 5:00 and
5:15 p.m. during the *peak 15-minute period of the day*. The eastbound
volume is 300 vehicles per *15 minutes* over two through lanes and one

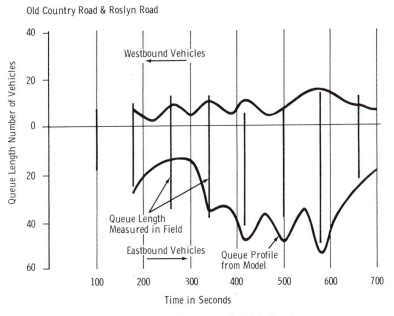

Figure 12.13 (Courtesy of IBM Corp.)

left-turn lane. The westbound volume is 220 vehicles per 15 minutes. This intersection was one in a network of five signalized intersections in front of the Nassau County governmental buildings along Old Country Road in Long Island, New York. The average speeds measured along this strip for east and westbound vehicles respectively *were 7.3 and 10.2 mph.* The model output showed 6.3 and 9.0 mph, which was considered a good match. With these comparisons it was felt that the model parameters were representative of the study area.

Results from Simulations

Results provide some feeling for the usefulness and validity of the simulation. The study for Nassau County consisted of a string of *five* intersections along Post Avenue in Westbury, just north of Old Country Road (as shown in Figure 12.14).

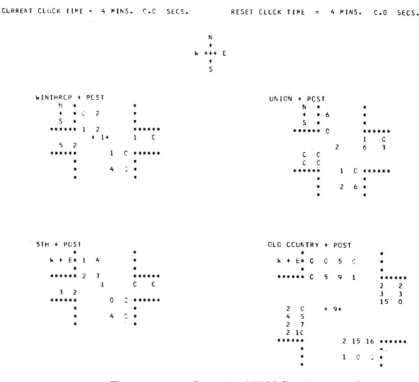

Figure 12.14   (Courtesy of IBM Corp.)

All of the intersections along Post Avenue carry heavy north-south traffic but relatively light side traffic. All the signals are *two-phase*. Old Country Road and Post Avenue intersection has heavy traffic volumes on all approaches with heavy turning movements. This intersection has four phases. Three signal settings were tested:

(a) The existing settings of 60-second cycles, with the setting for the intersection of Post Avenue and Old Country Road being 80 seconds.

(b) The first four intersections with a 60-second cycle, with an average progression set up for the traffic along Post Avenue for both north and southbound traffic, and with Old Country and Post again at 80 seconds.

(c) all five signals at 80 seconds with an average progression for both directions.

From results presented in Figures 12.15 through 12.19, it was concluded that the four intersections at 60 seconds and the one at 80 seconds were the best two-phase solution, and this was recommended to the traffic division of the Nassau County Department of Public Works.

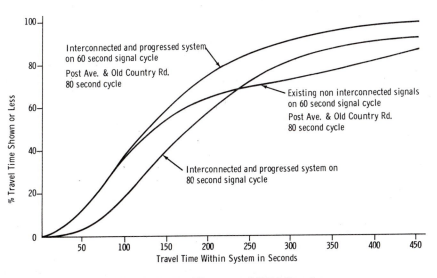

Figure 12.15 (Courtesy of IBM Corp.)

CUMULATIVE QUEUE DELAY TIMES

Post Avenue - Winthrop Avenue to Old Country Road

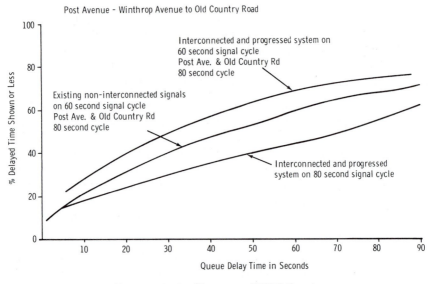

Figure 12.16   (Courtesy of IBM Corp.)

CUMULATIVE SPEED DISTRIBUTIONS
Post Avenue - Winthrop Avenue to Old Country Road

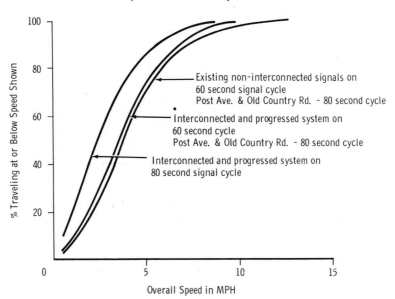

Figure 12.17   (Courtesy of IBM Corp.)

CUMULATIVE QUEUE DELAY TIMES

Through Lane, South Approach   Post Avenue and Winthrop A···

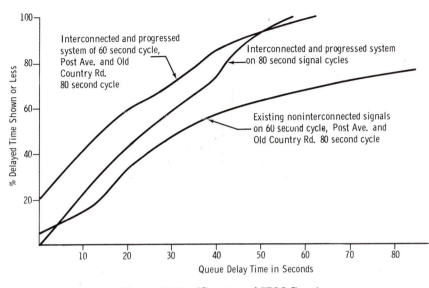

Figure 12.18   (Courtesy of IBM Corp.)

CUMULATIVE QUEUE DELAY TIMES

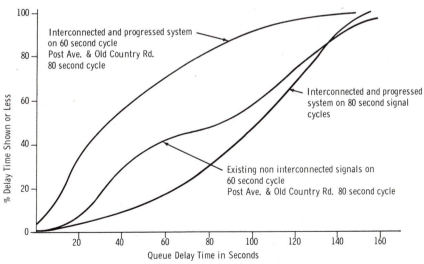

Figure 12.19   (Courtesy of IBM Corp.)

325

CURRENT CLOCK TIME = 4 MINS.   0.0 SECS.
RESET CLOCK TIME = 1 MINS.   0.0 SECS.

```
              N
              +
        W + + + E
              +
              S
```

```
    GLEN COVE AND OLD COUNTRY
            *                    *
    W + E *  1  1   1            *
    * * * * * *  1  1   4        * * * * *
                               1    7
                               0    4
                               0    1
                  + 4+         3    1
       0    6
       0    2
       1    1
       0    4
    * * * * * *              0  0  0 * * * * * *
            *                           *
            *                0  0  1 *
            *                           *
```

Figure 12.20 Intersection used to study signal light phases.   (Courtesy of IBM Corp.)

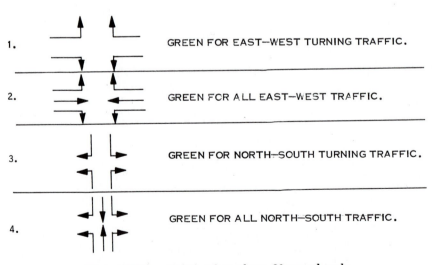

1. GREEN FOR EAST—WEST TURNING TRAFFIC.

2. GREEN FOR ALL EAST—WEST TRAFFIC.

3. GREEN FOR NORTH—SOUTH TURNING TRAFFIC.

4. GREEN FOR ALL NORTH—SOUTH TRAFFIC.

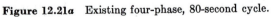

Figure 12.21a Existing four-phase, 80-second cycle.

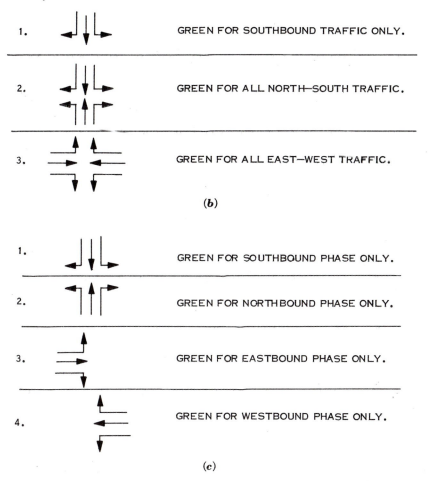

1.   GREEN FOR SOUTHBOUND TRAFFIC ONLY.

2.   GREEN FOR ALL NORTH–SOUTH TRAFFIC.

3.   GREEN FOR ALL EAST–WEST TRAFFIC.

(b)

1.   GREEN FOR SOUTHBOUND PHASE ONLY.

2.   GREEN FOR NORTHBOUND PHASE ONLY.

3.   GREEN FOR EASTBOUND PHASE ONLY.

4.   GREEN FOR WESTBOUND PHASE ONLY.

(c)

Figure 12.21b Southbound lead three-phase, 80-second cycle. *c* Four phases, one for each direction—80-second cycle.

Another effective use to which the program was applied was in the phasing or sequencing of the green signal. The standard method by which traffic is controlled is by variation of the length of the total signal cycle time as well as the red and green periods. Another approach is to alter the "phasing" of the signals, that is, to establish optimum sequencing of opposing green phases.

Most signal settings provide just one green phase for each perpendicular direction at an intersection. Others have separate green phases for

Glen Cove Road and Old Country Road

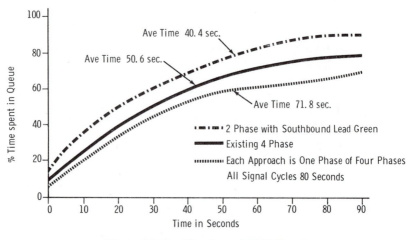

Figure 12.22 (Courtesy of IBM Corp.)

CUMULATIVE SPEED DISTRIBUTIONS

Glen Cove Road and Old Country Road

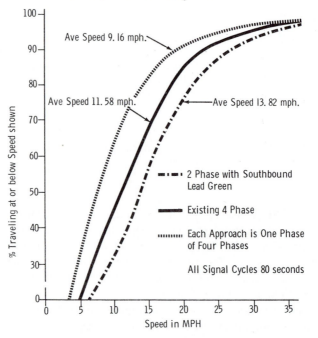

Figure 12.23 (Courtesy of IBM Corp.)

QUEUE LENGTH COMPARISON

Glen Cove Road and Old Country Road

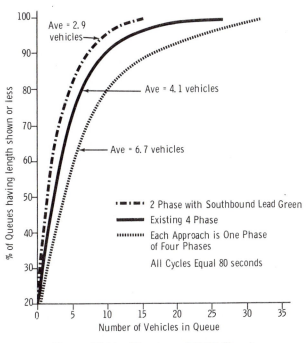

Figure 12.24 (Courtesy of IBM Corp.)

turning vehicles left, right, or both. Traffic engineers are not convinced that some other method of phasing would not offer substantial delay reductions and shorter queue lengths. Simulation affords an opportunity to test diverse phasing schemes and to measure their performance.

A simulation experiment was performed with VTS on the intersection shown in Figure 12.20. The following green signal sequences were tested for the location under the same traffic loads: Figure 12.21a, b, and c.

Three criteria were used in measuring the traffic flow at this intersection:

(a) Queue delay time.
(b) Queue lengths.
(c) Cumulative speed distribution.

Figures 12.22 through 12.24 consistently demonstrate the superiority of the southbound lead phasing.

Summary

Insight to traffic control situations can be obtained and verified through simulation.[4-6] Complex control systems can be modeled and the resulting computer time cost is relatively insignificant. All of .these capabilities are available to the traffic engineer, in a format consistent with his background and experience. However, the input data require effort to be processed in the correct form. The outputs are plotted to give the engineer a direct overview of the situation being modeled.

From the viewpoint of modeling techniques, VTS is an excellent example of what can be achieved with GPSS II, a basic simulation language. Conversion of this model to GPSS V would improve its flexibility and simplify some of the complex addressing schemes. The matrix savevalue has a structure which more readily fits the need of simulating an intersection or a network of intersections. More significantly this effort stressed the need to orient the simulation system to the user not the model developer. The equivalent effort to provide the HELP routines would still be needed to manipulate input data, conserve core storage, present special printouts, and improve running time. The major limitation is the maximum number of vehicles simultaneously in the model. To enlarge the number of vehicles considerably would require some significant changes to GPSS in order to move transactions out of core storage and onto peripheral storage devices. One approach is to move the tail of the future events chain onto a storage device and bring transactions into core as space becomes available.

VTS provides excellent insight into the effort required to provide a generalized simulation tool for a specific application. A tool that may be used by the trained specialist in the field rather than be restricted to being used by the modeler. This goal is achieved because the model is specialized to a particular purpose with a well-defined problem. The man-machine interaction is the basis for successful use of the simulation approach.

PROBLEMS

1. Data entry represents one of the more troublesome aspects of simulation. Instead of using seven tab cards to enter intersection data as shown in Figure 12.3, show how a CRT display containing only 12 lines of 80 characters each could be used.

2. The control of turning traffic uses a function as shown in Figure 12.9. This technique is required for GPSS II. However, for

GPSS/360 a variety of different techniques could be used. Suggest and compare alternatives.

3. Indicate how alternative presentations of the output data could make a simpler presentation in contrast to the network snapshot of Figure 12.11*b* and histograms of Figures 12.12*a* and *b*.

4. The simulation of large numbers of vehicles can require extensive amounts of computer time. Suggest approaches to reduce the amount of computer running time and what the implications of these changes would be.

5. Assume the capability for man-machine interaction to change specific settings while the simulation is running. Indicate those data needed to enable human intervention. Show how these data should be presented for quick and accurate human reaction. What are the advantages and disadvantages of human intervention versus automatic program.

6. If the simulation were to become the actual traffic control system, specify those data which must be part of the routine system operation and for unusual situations those data required for human intervention. Would different types of traffic sensors influence the control system design?

BIBLIOGRAPHY

1. A Vehicle Traffic Simulator, A. M. Blum, *IBM Sys. J.,* **3,** 1:41–50 (1964).
2. A General Purpose Digital Traffic Simulator, A. M. Blum, in Proceedings of the Analysis and Control of Traffic Flow Symposium, *SAE Automot. Eng. Conf.,* **680167,** 10–25, Detroit, Mich., 1968.
3. A General Purpose Digital Traffic Simulator, A. M. Blum, *Trans. Soc. Auto. Eng., Proc.,* P22:10–25, 1968; and *Simulation,* **14,** 1:9–25 (1970).
4. Simulation of a Traffic Network, J. H. Katz, *Commun. Assoc. Comput. Mach.,* **6,** 8:480–486 (1963).
5. Analysis of On-Ramp Capacities by Monte Carlo Simulation, R. F. Dawson and H. L. Michael, *Highway Res. Rec.,* **118**:1–20 (1966).
6. Optimizing a Local Street System by Simulation, N. Voskoglou and R. J. Wheeler, *Traffic Quart.,* **23,** 2:179–196 (1969).

Part III

COMMENTARY

The five dissimilar systems that provided our simulation examples are a minute representation of the systems which could potentially benefit from simulation. Rather than continue to explore one specific system after another, let us determine instead the general characteristics of system problems which benefit from simulation. Furthermore to gain perspective, we must categorize whether the benefits to be obtained are from greater understanding, quantitative comparisons, verification of intuition, equivalent operational experience, or combinations characteristic of the problem.

With the understanding gained from the five examples, we must generalize and establish the limits of the value of this approach. By making comparisons among the examples for underlying similarities and basic dissimilarities, we shall be in a position to estimate whether we can extrapolate to other areas and with what degree of confidence. Our knowledge of the limits of simulation technology should encourage us with those problems that are suitable and help us avoid those that are not.

We must obtain a reasonable perspective of the value of the results. Were they valid? Did simulation adequately represent the system? What is our degree of confidence in these results? How effectively have we expended effort to model the system and collect data? Were these efforts similar for various examples or are we limited in the conclusions we can draw for these or other applications?

Simulation is a rapidly changing technology. We must anticipate those problems which, while beyond our present ability to solve, will be within our ability tomorrow. Our look into the future must differentiate between what is needed to simulate systems and what is needed to reduce the costs of model development so that it becomes more feasible.   With these factors established, we should be able to foresee where our technology will take us in the future and what systems will not only be simulated but also controlled by the all seeing look-ahead ability of simulation.

13

What Could Be Learned From The Examples—My Problem Is Not Like Any of Those

Problem Relationships

The examples used in the previous chapters illustrate a few problems that system designers may approach with discrete event simulation. The variety of system problems ranges broadly: information flow, hardware characteristics, minute details, and abstract concepts. In each case the problem is based on the familiar elements of timing relationships and resource management. Time appears in a variety of forms, such as when to begin, how long to run, why stop, and how well utilized. Resources appear as facilities, equipments, personnel, and stock levels. The interactions among these elements in a simulation as in the real-world produces the quantitative data the system designer needs: schedules, queues, service times, reorder levels, resource utilization, and success and failure probabilities. Ideally, the system designer can combine these critical elements into a single figure of merit.

The adequacy of the interrelationship of the critical elements is the key to the usefulness of the figure of merit. To attain this objective, the system designer finds it necessary to structure the problem, limit the study area, understand the logical relationships, and compare and predict system behavior. For our five illustrative examples these tasks are contrasted with regard to approaches, measurement techniques, parameter variation, verification methods, model detail, and simulation techniques.

Problem Similarities

The illustrative examples seem different and unrelated because of emphasis on the application area rather than the system elements. Now, instead of concentrating on the applications, consider the elements common to the different examples.

- *Measurement criteria.* The figures of merit used to measure each system appear different but fundamentally are related to performance and cost.
- *Measurement intervals.* Timeliness of the selected data varies—sometimes during, and at other times at the end of, the simulation. Frequently the individual item is followed to find the prediction of absolute performance.

The broad features of the various examples are compared in Table 13.1. The criteria of measurement at the detail level is the one area that varies most from problem to problem. This is to be expected, since these are the application criteria.

Costs are an inherent reason for undertaking a study. The system designer must be able to predict costs with confidence. The cost data may not be complete, as was the case in both the auto traffic study and the railroad performance evaluation. In the remaining studies, the cost data are clearly present as the utilization of resources and the quality of service. Cost data requires finding the total number of rail road passenger cars required for different alternatives, solid-state output from the production line, utilization of computer system resources to service communications lines, and effort needed to eliminate targets. In each case, the designer has been able to translate information from widely differing areas into common denominators of cost and comparative performance. Cost has the singular advantage of being as close to a universal criterion as exists.

Measurement Approaches

A problem may permit measurements only at the end of a long period. The modeler will have built up to this point with a series of intermediate results to gain confidence in the model, and then run the entire model for a single result. Intermediate results do not determine the railroad passenger fleet size. The key item is the maximum number of cars *ever* needed on this schedule.

Likewise, the auto traffic simulation could be evaluated on the basis of final results. This may require the system designer to include gross accumulations of service criteria into a single figure of merit to compare performance. An example would be to sample the queue size at selected

Table 13.1

	Measurement	When Measurements Are Taken	Factor Varied	Verification	Detail Level
1. Auto traffic	Service provided	At specific intervals	Traffic signal settings Lane geometry	Compare to actual data	Each car Each lane Each area for turn
2. Railroad	Passenger service Fleet inventory	For the entire interval (day)	Schedule Fleet size	Primarily for relative comparison	Group of passengers Individual cars
3. Job shop	Product output Utilization of facilities and personnel Cost	Daily Weekly	Mix of products Facilities Personnel Strategy	Absolute—resolve the exceptions	Individual orders, machines, people
4. Weapon system	Targets eliminated Costs	Long duration	Targets Weapons Loss ratios Support system	Compare anticipated relative performance Spot check data to observe instance	Individual aircraft, people
5. Computer system application	Resources required	For specific intervals	Computer Core size Number of lines Storage devices	Relative comparison	Gross Individual message

times. However, if verification were attempted, then intermediate data must be available for comparison with field data.

In the job shop the measurement intervals are rigidly set to consider the shop performance over a minimum interval of a single shift and to accumulate data over a number of shifts. Both are needed, because there is the risk in cumulative data of overlooking a critical condition. The statistics gathered from the simulation must provide the overview, equipment utilization, people needed, indication of system balance, and the critical case of the longest queue.

Factors Varied

Each problem has characteristic factors which are varied by the system designer during the process of system analysis. Since these factors are application oriented, it is necessary to understand the fundamental implications of the significant factors. This is achieved by using different input data and changing model logic to determine sensitivity.

- *Use different input data.* This is the most obvious and first choice. The structure of the model is not varied, only the input data. Traffic signals can have different timing phases, new machines can complete the work faster, the number of communication lines can be greater, and the accuracy of a weapons system can be improved.

  In the cases where a direct comparison is desired, the change in the input data must be controlled. Changing the schedule of the railroad can, for the same fleet of passenger cars and the identical passenger demand, significantly affect the system performance. To control this type of reaction the scenario must keep constant as many factors as possible. Likewise, in the weapons system, the weather, initial target requests, and number of urgent requests never change. In the production model the form of input data changes. Originally the input data are based on probabilities; then, as time passes and actual data become available, there is a replacement of the original data with actual data. The model shifts from being probabilistic into one which is deterministic.

- *Modify the logical structure.* For the same input data the system behavior changes. While the schedule remains the same, the number of cars in a train may be made up of cars from two storage yards instead of one. This change modifies the logical structure. Another case could be when either of two machines with different productivity could be used for the same process, but the choice depends on the queues awaiting processing. The computer message switching can be used to compare the operational procedures of full or partial storage before forwarding.

• *Vary both data and logic at the same time.* Have the model determine the passenger car fleet needed, with different storage yard locations and with different passenger-demand data. Likewise, the strategy of machine selection may vary according to queue size and arbitrary assignment of certain orders to one type of machine.

Verification

The best examples of verification of simulation predictions with actual experience should come from the computer system designers. They have well documented existing computer systems, input data, and the advantage of having the computer. In spite of these advantages, the verification of simulations of complete computer systems is quite rare. Obviously, the weapons system evaluation is a simulation which cannot be verified. The emphasis must be on relative comparisons. The verification of automobile traffic simulations can be expected only for the existing cases, in contrast to numerous traffic studies predicting the relative performance of different systems. The situation in each of these areas is dynamic and the role of simulation is expanding with the increased use of system analyses in the design process.

Scheduling simulations can be readily verified. This is the case for the job shop, in which the resource allocation can be made from the simulation and then compared with actual production. When the present schedule is projected into the future, nondeterministic situations arise which are less subject to verification. In these instances the value of simulation is in its ability to offer a prediction of the possible consequences of certain courses of action. The uncertainties may be greater when the prediction period is far into the future, but these should be reduced as the data base broadens. Meanwhile continued changes to the model help to make it conform to additional experience.

Degree of Detail

Some simulations seem to automatically indicate the level of detail. Each individual car, job, message, or mission seems to readily fit the desired detail for tactical types of simulations. Such automatic indication would not occur if the number of missions were measured from a more strategic viewpoint. Then an aggregated measure might be desired, such as the summary cost of all squadron operations versus total benefit.

The level of detail is significant, because the simulation cost for computer time can vary greatly. The prime example is in the simulation of the computer system in which, if each instruction were simulated in detail, the running time is prohibitive. A decision must be made either to spend large amounts of money for computer time or to consider less detail. Usually, the choice is in favor of greater coarseness. There is another alternative. The detailed model can be used to find out the properties of a typical and heavily used subroutine. Then the coarse model uses the results from the previous subroutine simulation as input data and other subroutines can be compared on a relative basis. However, there is less confidence in the results and some interrelationships may be lost.

One example of collecting detail into larger units for later use occurs in the railroad example. The demand model is developed to handle individual passengers and is run only once. Its results after aggregation become the input to the schedule model. As a result, groups of passengers are processed as an entity rather than individuals.

Constraints

The examples indicate large differences in the portion of the problem modeled. The job shop, with cost data, does represent a significant portion of the problem. The inventory is controlled, the work is scheduled, new equipments are forecast, manpower needs are anticipated, and the details of day-to-day operation are included. The grosser coverage of interactions with sales effort and feedback from actual field performance are ignored, but even these areas could readily be included.

The railroad scheduling model covers less peripheral areas. The interactions with pricing of trips and frequency of service would be significant broadening of the model. Compared to the overall problem of passenger transportation, the rail service model is limited to a portion of the problem. This is consistent with the tactical rather than strategic nature of the model. The interactions between rail transportation and the private car or air transportation can be considered to occur only at specific constrained interfaces.

The auto traffic congestion model could be considered either well bounded or wide open, depending on how far the degrees of interaction are carried. A few intersections may be analyzed for their peak conditions without considering the remaining road and traffic structure. When it comes to predicting the all-weather worst-case conditions for the road design, the model would have to be extended to interact with considerably more factors and be far more complicated. Such analyses are sometimes useful, as the case of a snowstorm which ties up auto traffic. The situation of interest then could be the case where the clogged roads prevent the snowplows from moving and thereby hasten the traffic tie-up. The model in this case might also be used to determine the number and staging locations for the snowplows.

One danger in selecting constraints is the setting of limits which prevent meaningful interactions from arising. Computer throughput is not a constant, but is related to internal conditions. Under normal conditions, internal queues are small and times to find records are short; but under increasingly heavy loads, the time increases more rapidly than the useful output. This occurs because the files are growing and more internal processing is required to find each record. At some point the system rapidly changes from being still productive to being completely overloaded; and then when there is no more working file area, the system can no longer recover and becomes inoperable.

The constraints involved in the determination of the boundaries of the weapons system are purely arbitrary. Each system is subject to its own criteria of effectiveness and its relationship to the other military functions. Therefore, the extent of each model is different. In one case it may apply to an individual ship; in another to a single aircraft; and more frequently to groups of military units, aircraft, and ships. The full degree of interaction occurs when the model is designed to indicate the performance of a complete military activity. These models, in general, have not been attempted. When the degree of complexity becomes so involved, strategic war-gaming models have been developed rather than a detailed tactical model.

Nesting of Models

The models used for illustration represent a broad cross section. In each case, the model is only part of the problem. The entire problem is never represented. However, in the process of reaching toward greater reality and closer representation of the real world, the scope of the models has grown. This enlargement of the treatment of the problem area can make the development of the model into a greater undertaking than is desirable, funded, or representative of available data. To overcome this difficulty, the overall model may be developed as a series of smaller, more attainable interlocking submodels.

The case of auto traffic breaks down into the model to develop the inputs of competing traffic and the model to follow the individual car. This nesting of two models has the advantage of running each variation in competing systems against the same traffic scenario with a reduction in computer running time, and the disadvantage of the elimination of interactions between the two models. The competing traffic is generated only once and then used as needed rather than requiring computer time to generate it for every run. This separation of input data from the model is characteristic of several of the other models. The passenger demand in the railroad model is developed and stored through a separate demand model. The mix of jobs to be run through the computer system is standardized and then used as a reference to judge the system performance. The job shop, since it is a deterministic model, only uses the actual data when handling historical and current data. Only when the model is projected into the future does the need for possible future work loads require an equivalent to the demand submodel. In this example, the scenario has the added requirement of always requiring the same work loads for the projected time. Then as the calendar advances values replace the projections that were used up to this point. In this manner, each month's anticipated orders are replaced during the month by the actual order.

The weapons system model illustrates the hierarchy of models which cover an entire system in a tactical model. First there are the demands that the system is required to attempt to satisfy. These are the mission requirements—the scenario of specific targets to be attacked with their characteristics. This part of the model is analogous to the demand submodels. The tactical mission performance can be considered as the main model. Here the prediction of the possible performance of the weapons system is treated in an overall fashion, a percentage of targets identified accurately or hit. The planes respond to the mission request takeoff, attack, and return. Within this model there are submodels or extensive subroutines to cover the mission phase under analysis. If the attack phase were studied to determine changes in overall results due to greater precision in the method of target identification, it would result in one elaborate submodel in the total set of submodels.

The proof of any projected system can not lie only in the mission phase. The best attack system, after all, is of no use if it is unmaintainable. Therefore, the system study areas must continue to include the behavior of the support system. The support model is also a combination of two areas, namely, the requirements to support the system under analysis and additional support activities for other systems. There is further extension to the area of logistic support. The complexity of the system does not permit complete repair in the field. As a result, some of the complex and vital units must be returned to a depot through a logistic pipeline. Again, the pipeline could be a full-scale model, but the results from typical pipeline studies might provide sufficiently accurate data to predict that aspect of system performance. Therefore, the modeled system consists of a number of detailed models, each developed for analysis of specific areas. Instead of using these models to simulate the entire system, an additional number of simplified models are developed which use results obtained from corresponding detailed models while preserving general interactions. Then the overall system simulation consists of taking results from each detailed area and substituting the results from detailed models in simplified ones. The risks in this procedure lie in sacrificing detailed interactions. However, since every system is a subsystem of a large system, there is a point at which the amount of interaction must be constrained.

Similarities among the Models

One advantage of the systems approach to simulation is that concepts developed for one application can be used again in other models. This advantage can exist in spite of great differences in the level of detail of the models, the specific functions being modeled, and the uses to which the model is to be put.

The most prominent similarity among the examples is in the approach toward problem decomposition. The demand submodel of the railroad reappears as competing traffic, sales orders, and mission requests. Learning how to attack the problem is the major difficulty. It is helpful to have similar experiences for guidance when approaching a new problem. This does not mean that one should use old solutions, but rather that one has a greater degree of confidence when applying previous experience to the initial contacts of a new application area. Of course, prior experience is very helpful when similar situations reoccur. One such situation is competition for resources. The job shop may be more elaborate in setting up, controlling, and retaining statistics of the problem. But the same problem exists in the servicing of the aircraft in the weapons model and, to a similar degree, in the utilization of resources, core storage, peripheral equipment, mass storage, buffers, and communications lines in the computer model.

Measuring system performance is another area in which experience from one simulation may apply to others. The actual computer output when it is in text and reduced to a few numbers is unique. But the processing and collection of data from which final results are obtained must be handled in similar ways. The utilization of a machine is obtained from the same method of data collection used to measure the utilization of a communication line, aircraft, highway, railbed, or, for that matter, personnel. Nor are only simple results required, statistically processed results means, standard deviations, regression analysis results and curve fitting may contribute to the figure of merit.

Another characteristic of these examples is the use of list processing techniques. The answers sought are not only average values, but worst cases: step-by-step histories of paths orders followed through the shop, check-points of flights over enemy territory, or competing traffic levels our auto encountered at each intersection. In each case a trail must be kept of *some* individual transactions. Once this routine is developed, it is used over and over with little tailoring for the particular case.

Type of Results Sought

The emphasis in all of these examples is on the comparison of results from different runs against either each other or an ideal. This control over the degrees of freedom of the simulation ties in with the use of the scenario rather than statistical inference. If the study of competing auto traffic were reduced to a problem that could be represented statistically, results could be obtained more easily through analytical methods. Probabilistic relationships can be solved in closed form much more readily than by simulation. However, interactions, timing of lights relative to each other and to traffic, preclude the use of simple techniques but are easily handled by simulation. The selection of one set of signal light setting over another is clear when the traffic can be controlled to occur in the same manner each time and the results compared. When this approach is not sufficient for the prediction of the actual performance of a series of intersections, the problem must be simulated many times and statistical inference gained from a number of separate runs. Here simulation is at the dual disadvantage of long computer runs and the statistical behavior of the pseudorandom number generator over the range of numbers used.

The example of computer message switching demonstrates one of the problems of determining the trade-off between the level of detail and the statistical validity of the results. When the model is being developed, the uncertainty about the critical elements causes the designer to include details which later appear to be adequately represented by a simpler model. The results from one stage can be used to speed up the operation of the model under rules used in the previous model. But are these results indicative of the expected results for a wide variety of programs? This is where the design of the detailed model and the analysis of the results from it have to use a number of statistical tools. The number of trials must be related to the results of previous studies of the pseudorandom number generator. Then, within the region of numbers that conform to the required criteria, the number of trials can be anticipated.

Applications of These Simulations

Simulation was used differently in each example. Even in similar application areas the purposes were different. The auto traffic study had a relatively limited goal of how to time the signals at a number of interrelated intersections. The sensitivity analysis was performed prior to and independent of the simulation. Verification served to confirm that this was the sensitive area. The problem was well defined before and did not change during the simulation. The inherent flexibility permitted wide variation in the types of traffic, the densities and the geography; but this was a model for the prediction of the intersections controlled by signal lights, not a limited access highway.

The railroad model was not as well defined and restricted. There were two major areas of investigation: the characteristics of the passenger vehicle and the schedule used. Here the model could be used independently for either use. The sensitivity to the speed of the vehicles could be a special study in its own right. Like the auto model, the number of interrelated factors was so great as to make either the scheduling or sensitivity analysis dependent on the use of the computer. Since the model was less limited in its use and objectives, it was made more flexible. The model could be concentrated in a few areas or enlarged to cover the entire range of ground transportation problems. The outputs were like the auto model, limited to exactly what was being requested at the time. The level of detail was established by the desire to obtain specific outputs, but the model structure was suitable for modifications to include greater detail. For example, acceleration characteristics of the train could be included. The level of detail could also be reduced by making simplifying assumptions relative to passenger demand. Validation of this model was rather simple whether it was used for development of new schedules or for analysis of vehicle speed, since the model was deterministic.

The job shop model has a general structure. There is little restriction on conflicts for resources. Since the system covers initial sales order, utilization of equipment and personnel resources, control of in-process inventory and scheduling of both immediate and distant-future items, the model emphasis must be on a flexible modular structure. The general nature of this model means that variations of it could be used in numerous other ways with little additional effort to define the model further and a new set of input data.

The weapons analysis model is representative of a relatively abstract area of simulation. In this application the system is, by its nature, not well enough defined. Therefore, the first aim of the simulation is to aid in problem definition. The initial coarse overall model is further developed and expanded with the aid of the understanding gained from the model. The significant areas of uncertainty can be isolated, using a sensitivity analysis. Once the areas are known, the model is expanded to better represent the application. This further step imposes the requirement of salvaging some of the effort to date as part of the new model. The weapons system model is initially least defined and most subject to growth as new areas are modeled. Thus the requirements for growth and nesting capability in the model are accentuated. Not only must the model interface with a number of other models, but it must also be useful without hard data and aid in the search for needed data. As this type of model evolves with new inputs and greater understanding, it loses some of its abstract nature and becomes the current representation of a specific problem.

The simulation of computer systems can be used to acquire knowledge about existing and potential systems. Computer throughput is sensitive to the conflicts restricting internal data flow, thus imposing a need for representation of the multiplicity of hardware and software function within the computer. This problem is beyond the simple representation of the system components. There is a need to have in varying detail the times required and properties of the potential system components under various operational regimes. It is not a small undertaking to have computer system structure and data available—even more so since there is no certainty that these will be the needed regimes.

Output Characteristics

One common characteristic of all these models is the effort to provide the user with an output tailored to his needs. In each case, effort was expended on the problem of how the decision maker will see results and how readily he will be able to understand them. The greatest effort in this direction was in the auto traffic model. Here there was a large effort to simplify the man-machine interface. The other models also used specially tailored output. The railroad model was coupled with a postprocessor model to take some of its output and plot it on a graphic plotter. The remaining output was in clear language for the convenience of the model user. Similarly, the job shop model recognizes the needs of different people using the model and provides special inputs and outputs for each user. These vary from prepunched cards to provide input data for each stage to very detailed tabular and graphic outputs. The user's needs are considered. Some results are a detailed picture covering great detail, while others are only management by exception.

As a brief glimpse into the future, one might expect that all the examples would have output variables plotted; these could be compositely displayed as a number of points in a plane. Then the user could use visual pattern recognition to see whether the desired set of properties had been explored and if the results look promising. These would provide the insight to use new input data and new criteria for system evaluation. Once these factors became well enough understood, the human contribution could be reduced or even eliminated; and a self-organizing program would design the system.

Other Problem Areas

These examples present some of the numerous current uses of simulation. Other examples could be found to show how the system designer has used simulation. Analyses of distribution and transportation systems would not introduce simulation concepts different from those used in the auto traffic, railroad, job shop, and computer system models, whether the applications were to ports, airports, terminals, or warehouses. The degree of detail required for all these system designs is equivalent. The outputs would serve different purposes, but the input data would be quite similar: demand, schedules, and resources.

Design of production systems for mass-produced items would be larger than the job shop model. An additional concept needed is the merging of two subassembly lines to form a single product line. This concept can be implemented within the level of detail characteristic of these models. The fundamental problem is still the simulation of multiple competitions for limited resources.

Queuing models frequently appear in analyses of limited resources. There are numerous types of queuing studies. Simulation is restricted to uses within the storage size, and speed restriction of available computers. However, when the problem is larger than the simulation can handle, it may no longer be necessary to follow individual transactions. Therefore, analytical techniques may provide a sufficient solution.[1] This is especially valid for the typical queuing problems of telephone exchanges, telephone operator requirements, and communication line capacities.[2-5]

A model of the business environment with emphasis on inventory levels, reorder points, production cycle times, rate of new product development and costs is basic to full analysis of industrial production. The industrial dynamics concepts of DYNAMO[6,7] has been suggested for this area. However, it would appear that the flexibility and logical capability of simulation languages provide more typical characteristics of the business environment than does the approximation to a feedback system. The basis for management decision may be the expected mean value, but frequently the confidence level is of greater importance to the decision maker. The increased penalties associated with a worst case may override the solution indicated by the mean value.

The weapons system model includes the concepts of predicting failures, determining repair times and considering the logistics support system. These concepts could be extended to include other aspects of reliability/maintainability/availability system design.[8-10] One example of the difference between the analytic and simulation approaches is when predicting the incidence of failure. In the analytic approach a constant failure rate (FR) is used. The probability of failure during a portion of the mission is determined from total mission duration risk based on a failure rate which is the risk per unit interval.

```
Mission risk = FR*TIME
```

The simulation approach need not be quite so simple. Instead of a constant failure rate, a varying failure rate could be used. One that varies with the duration of the mission or, more likely, with the previous history of the equipment. One such varying failure rate is shown in Figure 13.1, for a "bath tub" curve. Here the failure rate drops rapidly as the period of "infant mortality" is exceeded, remains constant for some time, and then increases as parts begin to wear out. To determine when a failure occurs in the simulation, a pseudorandom number is selected which is less than the normalized mission risk. For example, if the failure rate were normalized to 2000 chances per million and the mission duration were 10 hours, a draw of a random number of less than 20 when the maximum value is 1000 indicates a failure.

To use a varying failure rate the number of operating hours must

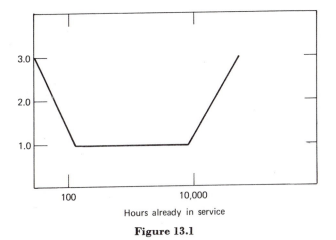

Figure 13.1

be retained for each individual equipment. The failure rate for the mission is then selected according to the service history.

Another aspect of real life which sometimes should be considered is the maintenance-induced failure. This is simulated by an evaluation at the end of the repair cycle of whether or not a maintenance-induced failure occurred; and if it did happen when was it discovered. A variation of the maintenance-induced failure is the misdiagnosed repair action. Here either the wrong part was replaced and the equipment returned to service still in an unsatisfactory condition or multiple failures are found, when in fact there is only a single failure.

The duration of a maintenance action is another area for simulation, since it could depend on a nominal expectation modified by experience, training, environment, and test equipment. A complex situation frequently arises in the calculation of the repair time, namely, a bimodel characteristic. In Figure 13.2 curve I applies when the repair man recognizes the problem; curve II, when he doesn't.

The human element appears in the weapons system and job shop models, but neither model has as its goal the prediction of how human performance affects system performance. This area has been simulated in great detail.[11-17] To model the human performance, extreme detail is required, with ample opportunity for logical selection, multiple interaction, and conditioned response. One difficulty in modeling human performance has been to obtain input data prior to the construction and operation of a system. The difficulty of obtaining input data is further aggravated by the widely different estimates the experts provide.[18] One

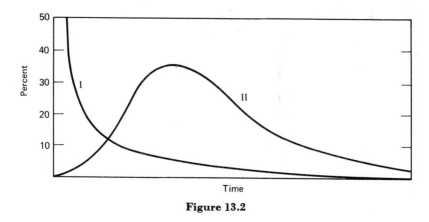

Figure 13.2

technique to minimize this problem is to substitute for the question "How long on the average will this task take?" a series of questions:

•   What is the longest time?
•   What is the shortest time?
•   Between what limits should the time be?

Then these times can be related, as shown in Figure 13.3.

The weighting for the various areas is subject to the individual problem. But the significant factor is that while various experts provide wildly different average values, their estimates are much closer when they are constrained to optimistic, pessimistic, and intent of design replies.   The average values obtained from various experts using this technique are much closer together than the original estimates.

Activity-oriented languages may be better suited than general simulation languages for some types of applications.   For example, when simulating very rare occurrences, as in the probabilities of neutrons hitting

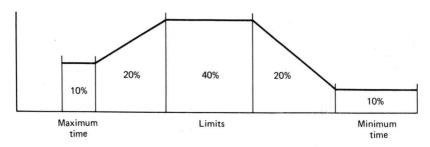

Figure 13.3

nuclei, the available random number generators and the running times would become very awkward. This problem has been handled by special programs.

Another area where an activity-oriented language might be preferred is the simulation of computer systems. Complex computer systems with multiple programs, remote terminals, and extensive peripheral equipment require simulation to predict their performance. When the system is very complex, especially because of the interactions between hardware and software, a special simulation language might be preferred. This would be the case with extensive software in the system controlling the internal data flow. Executives, monitors, and operating systems are some of the basic software packages that must be simulated. In an activity-oriented language it would be possible for the software representation to be included as a part of the basic language. The result would be to reduce the amount of effort required to model each computer system. A second feature of such an activity-language approach would be to contain the library of hardware and software characteristics. This library would enable the system designer to specify the hardware by name and have its characteristics inserted in the model. Likewise, the software characteristics would be used as needed.

Business and war games may be considered as examples of simulations. However, they are not usually in the province of the system designer nor have they usually been implemented using simulation languages, FORTRAN has been the common vehicle. But since these games are closely related to the manager's "What if?" of the job shop example, the use of simulation languages will increase. This becomes even more apparent as interplay between opposing sides becomes an integral part of the resolution of management choices.[19-35]

There is one more area where simulation techniques are useful, econometric models. There are two types of these models. One type is the global model, which does not follow individual items through the system. Instead, the various factors are evaluated on a probabilistic basis. This approach can be solved analytically although a sampling with random numbers may be needed to reduce the amount of calculation. The other type of model is more consistent with the use of simulation languages. Individual items are followed through the model to represent various aspects of the economy or the local ecology. For example, a company's gross sales may be a function of the GNP, its advertising budget, and acceptance of new products. These three factors might be interrelated. A simulation model would use random numbers to determine the probability of the GNP being at one level, the related advertising budget, and the emphasis of that budget on new products. The characteristic of this model is the following of individual items through the model to establish and maintain interrelationships. In a similar manner the influence of air pollution on the plant life and in turn ecology of a local area would be analyzed.[36-41]

It is hoped that this chapter has served to bring into focus the similarities among problems that lend themselves to the simulation approach. The variety of problems is great. But the techniques are flexible and dynamic. There is sufficient reason for optimism to use simulation since there are many difficult problems around.

BIBLIOGRAPHY

1. *System Analysis for Effective Planning*, B. H. Rudwick, Wiley, New York, 1969.
2. *Elements of Queueing Theory*, T. L. Saaty, McGraw-Hill, New York, 1961.
3. *Introduction to the Theory of Queues*, L. Takács, Oxford, New York, 1962.
4. Seven More Years of Queues, T. L. Saaty, *Naval Res. Logis. Quart.* **13,** 4:447–476 (1966).
5. Stochastic Properties of Peak Short-Time Traffic Counts, G. F. Newell, *Trans. Sci.*, **1,** 3:167–183 (1967).
6. Industrial Dynamics—After the First Decade, J. W. Forrester, An appreciation of Industrial Dynamics, H. I. Ansoff and D. P. Slevin, *Manag. Sci.*, **14,** 7:383–415 (1968).
7. Industrial Dynamics—A Reply to Ansoff and Slevin, J. W. Forrester, *Manag. Sci.* **14,** 9:601–618 (1968).
8. Quantifying Human Performance for Reliability Analysis, W. B. Asken and T. L. Regulinski, *Human Factors,* **11,** 4:393–396 (1969).
9. Prediction of a Naval Vessel's Performance, R. C. Baxter, *IEEE Trans. Sys. Sci. Cyber.*, **SSC-4,** 4:382–387 (1968).
10. Equipment Maintenance Studies Using a Combination of Discrete Event and Continuous System Simulation, H. G. Hixson, *Proc. 3rd Conf. Appl. Simulation,* 76–85, Los Angeles, 1969.
11. Computer Simulation of Consumer Behavior, W. D. Wells, *Harvard Bus. Rev.*, **41,** 3:93–98 (1963).
12. A Computer Experiment in Elementary Social Behavior, J. T. Gullahorn and J. E. Gullahorn, *IEEE Trans. Sys. Sci. Cyber.*, **SSC-1,** 1:45–51 (1965).
13. Crisiscom: A Computer Simulation of Human Information Processing During a Crisis, A. R. Kessler and I. de Sola Pool *IEEE Trans. Sys. Sci. Cyber.*, **SSC-1,** 1:45–51 (1965).
14. Information Processing Behaviour in a Crisis Situation—A Simulation Study, P. N. Rastogi, *Simul.*, **13,** 4:177–186 (1969).
15. Analysis of Sourcing and Disaster Planning, G. P. Carlson, *4th Conf. Appl. Simul.*, 178–181, New York, 1970.
16. Estimation of Manpower Forecast Variation by GPSS Simulation, P. Hicks, *Digest 2nd Conf. Appl. Simul.*, New York, 1968.
17. Computer Techniques for Analyzing the Microstructure of Serial Action Work in Industry, J. W. Rigney and D. L. Towne, *Human Factors,* **11,** 2:113–121 (1969).

18. Modeling as Applied to the Evaluation of Alternative Systems, J. Reitman, R. Malmgren, L. N. Miller, and R. Kelly, *Rec. Sys. Sci. Cybern. Conf. IEEE,* 1964.

19. Computers in Top Level Decision Making, R. H. Brady, *Harvard Bus. Rev.,* **45,** 4:67–76 (1967).

20. Simulation in a Decentralized Planning Environment, P. H. S. Redwood and J. J. Schengili, *4th Conf. Appl. Simul.,* 8–11, New York, 1970.

21. Simulating with SIMSCRIPT, H. M. Markowitz, *Manag. Sci.* **12,** 10:B396–B405 (1966).

22. Airline System Simulator, W. A. Gunn, *Oper. Res.,* **12,** 2:206–227 (1964).

23. Airline Simulation for Analysis of Commercial Airplane Markets, L. R. Howard and D. O. Eberhadt, *Trans. Sci.,* **1,** 3:131–157 (1967).

24. Company Model for Income Prediction, M. Araten and B. G. Price, *IEEE Trans. Sys. Sci. Cyber.,* **SSC-4,** 4:379–381 (1968).

25. A Simulation Study of Routing and Control in Communications Networks, J. H. Weber, *Bell Sys. Tech. J.,* **43,** 6:2639–2676 (1964).

26. Planning a Multilevel Car Park, E. Broad Jr., *Indus. Eng.,* **1,** 9:17–23 (1969).

27. Simulation of Elevator System for World's Tallest Building, J. J. Browne and J. J. Kelly, *Trans. Sci.,* **2,** 1:35–56 (1968).

28. Simulation and Cost-Effectiveness Analysis of New York's Emergency Ambulance Service, E. S. Savas, *Manag. Sci.* **15,** 12:B608–B627, (1969).

29. A Simulation Model of Fire Department Operations: Design and Preliminary Results, G. M. Carter and E. J. Ignall, *IEEE Trans. Sys. Sci. Cyber.,* **SSC-6,** 4:282–293 (1970).

30. The Simulation of Hospital Systems, R. B. Fetter and J. D. Thompson, *Oper. Res.,* **13,** 5:689–711 (1965).

31. A Simulation of Hospital Admission Policy, W. G. Smith and M. B. Solomon, Jr., *Comm. Assoc. Comput. Mach.* **9,** 5:329–339 (1966).

32. Simulation Study of a Hospital Emergency Command System, S. A. Levine, *3rd, Conf. Appl. Simul.,* 248–267, Los Angeles, 1969.

33. A Dynamic Model for Planning Patient Care in Hospitals, A. K. C. Wong and T. Au, *4th Conf. Appl. Simul.,* 45–51, New York, 1970.

34. An Antisubmarine Warfare Model for Convoy Protection, W. W. Fenn, *IEEE Trans. Sys. Sci. Cyber.,* **SSC-4,** 4:413–418 (1968).

35. A Simulated Port Facility in a Theatre of Operations, R. E. Davis, Jr., R. W. Faulkender, and W. W. Hines, *Naval Res. Logis. Quart.,* **16,** 2:259–269 (1969).

36. *Ecology and Resource Management,* K. E. F. Watt, McGraw-Hill, New York, 1968.

37. An Air Pollution Model of Connecticut, G. R. Hilst, *Proc. IBM Sci. Computing Symp. Water and Air Resource Manag.,* Form 320-1953:251–274, 1968.

38. *Systems Simulation for Regional Analysis,* H. R. Hamilton, S. E. Goldstone, J. W. Milliman, A. L. Pugh, III, E. R. Roberts, and A. Zellner, M.I.T. Press, Cambridge, Mass., 1969.

39. *Urban Dynamics,* J. W. Forrester, MIT Press, Cambridge Mass. 1970.

40. *Simulation Techniques for Design of Water Resources Systems,* M. M. Hufschmidt and M. B. Fiering, Harvard University Press, Cambridge, Mass., 1966.

41. An Energy Storage System, B. H. Easter, *Simul.* **13,** 3:155–164 (1969).

14

Pitfalls and Problems—If
I Knew Then What
I Know Now . . .

From the system designer's viewpoint, nothing is ever perfect. Simulation, like any new technique, can be expected to have drawbacks as well as advantages. So far the advantages have been stressed. The disadvantages, while not minor, tend either to be related to the complexities of system design with the attendant properties of complex problems, or to the inadequacies of the computer. When these problems are brought under control for a specific application simulation becomes a most useful tool. Furthermore, other than for intuition it is the cheapest form of insurance.[1-3]

There are three major classes of difficulty with the use of simulation:

- Customer understanding simulation results.
- Fitting the model to the problem.
- Avoiding simulation language constraints.

The first is the most serious and obvious. In the design of an incompletely understood complex system, the customer frequently asks to have a set of functions modeled when in fact he defines a different set or omits critical elements. Because this problem comes up in many forms, problem definition is the first area which must be clarified. After there is adequate problem definition, there must be recognition of both the limitations of simulation results—whether of a statistical nature, a relative comparison, or a deterministic schedule—and in the choice of the level of detail. The latter must be controlled to avoid unnecessary detail which will take too long to model, execute, and deplete the available funding. The final class of problems exists because the simulation languages do something different from what is expected. This may result from language limitations, inadequate documentation, and modeler inexperience. In reality the simulation languages mirror the complexity of the problem. Therefore there is no guarantee that a complex application can be resolved by the use of a model. For example, there is still no claim to an adequate model of the socio-economic interactions produced by racial tensions in a community.

Customer Understanding of the Model

If the customer does not develop the model himself, then the modeler may find it difficult to comprehend to what degree simulation should be applied. Unfortunately, this is frequently what happens as there are two separate groups who must collaborate—one problem oriented and the other simulation oriented. The customer's intent may be made clear through specifications, conference notes, commentaries on earlier versions of the model, simulation results from the different versions, and joint use of the model. These practices will bridge the gap in proportion to the closeness of the working relationships. The difficulties increase when the customer leaves the model and the modeler alone for long periods. The simulation language itself helps to bridge the gap through its assist in problem definition. Lists of assumptions must be stated, reviewed, and updated. The simulation should be part of the sequence of information describing the system development rather than occasional instances when copious results are delivered. The interim results must be reviewed with the customer to re-estimate the model's veracity and the pertinence of its abstractions. Since the model should be flexible, the customer's suggestions can be incorporated; and this better be done before the modeler considers the model finished. Following this procedure also enables the modeler and customer to obtain similar perspectives, namely, an understanding of the values of input data, their range of validity, model rules, and system constraints. Only when a close relationship between customer and system design team is possible, will the problem definition pitfall be avoided.

Sometimes the close relationship between customer and modeler is not possible, as in the instance in which the customer does not adequately understand his problem. Under these circumstances the desired practice is to use the model to document the system definition step by step as its form and status change. Then the ability is not lost to revert to an earlier stage of system design and proceed along a different track. The effort to clearly document the model and retain each significant version is not unreasonable. When there is doubt as to what is significant, it is better to store more rather than less. To retain versions of the model, card decks may be duplicated and stored. A file of separate printouts is useful. For really lengthy models, magnetic tapes may provide the historical continuity. The important factor to consider is that modeling a complex system is a dynamic iterative process. It is the use of higher-order simulation languages which prevent this process from becoming exhorbitant. Since good system design practice demands problem definition, the model will add only a fraction to the cost of problem definition.

Pitfalls in Pitting the Model to the Problem

There are areas which can turn the development of a model into an ocean of trouble. The rule seems to be small models, small troubles; large models, much more than the proportionate amount of trouble. In general the degree of difficulty with the model development depends on computer capability required, time needed for each run, and quantity and format of input data.

Frequently the prime modeling consideration is how to restrict the size of the model to the available computer core storage. *The rule is to keep the model as coarse as possible until forced to make it more detailed.* The amount of computer necessary for a coarse model of a large problem is the same as that required for a detailed model of a small portion of the application. Therefore, do not start with too much detail. Add the detail later as the problem becomes better defined. While this appears obvious, it means considerable anticipation of what might happen to the model and an emphasis on an open-ended design. The concern with the available computer core storage also relates to the degree of machine independence for the model. The early coarse versions may be debugged and tested on small computers, but the final versions may need larger and faster machines for economic production runs. Models of complex systems can exceed computer capability. Careful consideration of the model purpose can help to restrict model size. Then, if the model proves the insight correct, there are savings in time and cost. If the intuition is wrong, however, it is helpful to have found it out at least expense. Experience so far does not indicate what is the maximum model size. Numerous models requiring over a million bytes of core storage have been developed. Some larger models have been partitioned to require only that amount of core storage at any one time.

Another method of restricting model size considers the set of constraints selected. The model should be able to interact with either more or less peripheral influences, depending on the current problem definition.

Artificial constraints may at one stage of development impose one set of boundaries. Later these boundaries will be changed as new submodels are added. The modeler has the choice of whether to build a single coarse overview model and detail parts of it to greater depth as needed later, or he may decide to partition the problem into areas that are relatively independent and construct a series of interrelated nested submodels. The degree of success with nested models depends on each submodel being tested separately, with adequate confidence in its validity. In actuality, this goal is rarely achieved for highly detailed models, because they aggregate into an unwieldy model and there is the unintentional assumption of independence among factors in different areas which in fact may not be independent.

Intimately related to the decision to control the size of the model is the need to consider the running time of the model. The cost of running the model must remain small relative to the gain from the simulation. The running time influences costs in two ways: (1) the actual cost of computer time, which is usually considerably less than the manpower cost, and (2) long running times may lead to long turnaround times. The difficulty here is that turnaround time influences the productivity associated with model development. The solution to the running time problem goes again back to the adequacy of the problem definition effort, which controls the fineness of detail. Costs of both development and production running times have to be anticipated and the model detail and structure must be modified accordingly. The approach of using a coarse model initially for sensitivity analysis has the virtue of indicating how the best use must be made of available computer resources. The problem is first investigated for sensitivity to conditions in particular areas; then if necessary, the model is expanded. The additional computer time is now used to solve these sensitive problem areas either with more detailed models or with the new development of submodels.

This discussion is not pertinent to all system-design problems. The typical example of where it does apply is in the design of computer systems (specifically, in trade-off analyses concerning computer system hardware and software). In this case, the fully detailed model down to each machine instruction could take 1000 times the normal computer running time to execute. The alternative is to reduce the degree of detail to the macro instruction, subroutine, or functional model level.

The other simulation examples do not require as extensive amounts of computer time compared to real time. The simulation of a railroad, for example, could be done in one thousandth of the normal operating time. The optimistic aspect of the concern with the level of detail is that as the computer speeds have increased, the cost of computer time has gone down sharply. This trend will continue into the future as machines become both bigger and faster at less incremental cost.

There is an alternative approach to constructing a model. Instead of starting with an overall model, only a part of the system is represented in a detailed submodel. Then the next part of the system is represented and added to the original. This process continues until the problem is completely modeled. Since this does not allow for the overview of the model, the level of detail is more difficult to establish at the start of the effort. It is also necessary to review the completed model to remove unnecessary detail. This can become a time-consuming process, because the model might even need to be redone. This approach should be predicated on the availability of a clearly defined statement of the entire problem to be modeled. Ideally this should be available before the modeling effort begins.

The occasion to use this approach is when the modeling tasks can be divided among several team members. While this may appear to be efficient, the need to communicate among team members slows up the process. The extent of this reduction of useful effort by the individual is uncertain. However, an estimate based on experience would indicate reduction of overall productivity by a large factor, such as two or three. The exception would be when the system naturally decomposes into elements which stand alone. An example of such a decomposition would be a demand submodel presenting some of the system requirements and a scenario, a series of relatively independent functional sub-models, a follow-up model to collect and process some data for the user and, where possible, separate housekeeping and data manipulation routines. The interface definitions are then set up as the first definition of the model by the original modeling team. Implicit in this approach is an adequate effort to define the system, consider the various aspects of the modeling problem, and establish the interfaces. One example of this type of system is the separation of an overall weapons system effectiveness application into tactical, base, logistics and production aspects.

Some Debugging Tricks

There are as many ways to debug models as there are system designers to create them. However, some universal approaches can underlie the approach for a particular problem. These approaches are independent of the simulation language used and fairly obvious.

1. Follow each class of item through the logic of the model.

2. Force rare events to occur for the purpose of testing significant paths.

3. Stop the simulation at a significant time and try to determine why each item is where it is.

4. Gather statistics, either for the full model or more likely from a simplified version, and compare the results with some hand calculations.

5. Record in full the history of a best and worst case and then rationalize these histories.

6. Use a different random number sequence to see if there are very different results; and if there are, try to find out why.

7. Devise a means of following the sequence of actions which are scheduled to occur at the same instant. Then see if the intended logic and priority regimes are actually followed.

These debugging approaches are not independent of the task of model development. It is necessary to consider how and where to introduce trails in the model. Special areas must be set up to save transient indicators. Generally, these items will be removed from the final model. But during the development process they are needed, increase the development effort and cost, and, it is hoped, will in the long run prove their worth with a smooth-running model at lower overall cost.

A successful debugging technique with GPSS is to sort the model deck according to block type and in the order in which the blocks appear. This produces a sequence of all blocks that could produce an action. This technique provides a catalog of actions in the model grouped by action type. The use of this debugging technique is illustrated by the grouping of ASSIGN blocks. When all parameters are directly assigned, it is easy to see where each parameter is assigned and how the parameters are used. This also serves to separate the directly and indirectly addressed parameters. The latter need more effort to determine their content, but even so, this technique helps. Not only ASSIGN blocks are grouped but also blocks concerning savevalues, facilities, storages, queues, logic switches and tables. This technique is a considerable aid to keep a concise record of the logic of a very lengthy model.

Data Pitfalls

A characteristic of the application of simulation to complex problems is the need for an adequate and frequently large data base. This data base is comprised of the unique input data for this application and the general reference data.[4-6]   The input data—good data, bad data, too much data, and too little data—sooner or later become the critical element in the simulation. A sensitivity analysis may aid in determining what data are necessary and to what accuracy. Sooner or later the simulation must handle some fórm of input data. The good data case may almost be dismissed as trivial. It means that the needed data are available, in the needed amount, in a format compatible with the model, and certified by the responsible authority as representative, accurate, and pertaining to this problem.

The situation the modeler usually faces requires searching for data. This can be a time-consuming operation. When no source of data can be found, the modeler must fall back on the use of pseudodata. This requires effort to find useful probabilistic relationships.

In today's computer age, the problem frequently is too much data. The computer has collected tape after tape, in a format for some other purpose, and now the problem is to extract only what is needed. This task is usually a considerable one. The original files were recorded with different formats, on separate tapes with incompatible equipment, and with copious extraneous material. But if the data are available, the budget must include the cost of processing them into useful form.

The fact that there are magnetic tape files does not guarantee the accuracy of the data. An investigation is needed into how the data were obtained and recorded, and for what purpose. To find out whether the available data are ùseful for new purposes, the system designer may have to process the data, to inspect large samples in great detail, and even investigate the original unprocessed data. Which of the methods to use depends on the importance of the input data, the costs of additional processing, and the possibilities for choice. Frequently all the techniques are partially applied to get some gauge of the value of the available data. There are advantages for the system designer in examining some of the raw data in detail while learning about the problem. Inconsistencies may stand out conspicuously. Statistical techniques to smooth the data can obscure these anomalies.

It is desirable to start searching for possible sources of reference data as early as possible. Once the data are found the problem of both getting them and converting them into a useful data bank can take considerable time. The individual records must be consistent with the computer system, the simulation language, and the access method. Therefore, it is useful to consider, almost from the beginning of the system design effort, the setting up of data banks or the decision to use existing data banks established for a different purpose. When new data must be collected and processed this can go on parallel with the development of the model. Data banks set up on random access files should be structured for use by many different applications. In the past a major limitation on the size of the data bank was the capacity of available random access devices, but with the development of large demountable disks the available capacity exceeds any possible simulation application.

Design of Simulation Runs

There are numerous statistical techniques for determining the degree of confidence in the results based on the number of trials, the number of events, and relationships among the significant factors. These techniques are based on some assumptions underlying the data and the character of the problem. Simulation, however, is a tool used after analytical approaches have been abandoned.[7-9] Therefore, statistical predictions which would have been valid for more conventional problems cannot be automatically transferred into the area of complex system design. The factors contributing to simulation results usually are not mutually independent, normally distributed, or found to possess negative exponential holding times. Under these circumstances statistical inference may be useful, but it can also be harmful when applied conventionally. A trivial example is the expected increase in confidence with more trials. This is true when the frequency distribution of pseudorandom numbers conforms to its expected distribution. As shown in Chapter 4, however, the mean values for the first few thousand numbers are frequently outside the expected values. The output may be a set of biased results. If the simulation were geared to a full cycle of pseudorandom numbers, this contradiction would not exist, since numbers are checked for their popularity. But when only a thousand numbers are used, the results after twice the number of trials may not be more accurate. Likewise, there is a difficulty in expressing absolute confidence in the result when there are several interlocking but weak relationships. If these are potentially significant, the number of trials must be very large to allow these weak interrelationships to have their proper representation in the results.

In addition to the internal statistical difficulties provided by the simulation language, there are numerous other difficulties caused by the structures of problems suitable for simulation. There are many interrelated factors often not linearly related but still dependent. For example, the time it takes a repairman to fix an automobile depends upon some of these factors:

- Is the trouble diagnosed?
- Would it be diagnosed more quickly by a more experienced or better-trained mechanic?
- Has only a symptom been eliminated, not the fundamental trouble?
- Is this trouble the result of a previous maintenance action?
- Will this repair cause another fault?

Obviously, in cases like these, simulation can be used; but statistical relationships could be considered with only very simplified versions of the problem. There are numerous references in the literature to the use of statistical tools with simulation. Under these circumstances, the general rule for the designers of complex systems is to rely heavily on comparative results between alternative designs rather than on absolute results. This does not preclude using optimizing techniques. System designers must seek the best solution on a relative basis from the alternatives considered.

Certain factors emerge from the use of simulation. The system designer has been forced to spend considerable time with the problem and in return has a precise definition. The various runs have given him the equivalent of years of experience with the system. The search for worst cases has uncovered unusual and strange sequences of events. These in turn cause new experiments which explore special situations which sometimes lead to new alternatives. Standard statistical tests can be applied, but their meanings are uncertain. There are no invariably useful statistical tests. It appears better to err on the side of less statistical support for conclusions than to claim statistical inference which cannot be supported.[10-21]

Simulation Language Constraints and Subtleties

In addition to the host of potential difficulties provided by the problem definition input data and the output results, there are quirks in the languages. These become evident through the simple and straightforward, but yet often frustrating and time-consuming, process of *using* a simulation language. These difficulties are basically caused by using a general language developed by someone else. Each language is aimed at solving some classes of problems. From the evolutionary viewpoint, these problems were the initial goals of the language designer. Now other features have been added; the language has been broadened, made more complicated, capable, and unfortunately obscure. Moreover, to further complicate the situation there is the trend toward separation between the groups that develop the languages and the users.

One fundamental problem the simulation language designer must resolve is the handling of events which in reality are concurrent. These events must be treated realistically by the language which uses a digital computer that operates internally in sequential fashion. For example, a train terminates at midnight, its passengers are discharged, and the statistics for *this* train are included in the previous day's activity. When hourly statistics are to be gathered, the people waiting to board the train on the hour may be included either in the waiting queues or among the passengers on board. This simple problem requires anticipating the problem, planning how to deal with it, making sure consistent policies are followed, and finally checking if the language will follow the solution. The structure of the language will, in some cases, permit solutions similar to real life; while in others, artificial solutions must be used. For this example there are several possibilities.

1. A rigid priority system is established which will have as its lowest item the collecting of statistics. In this case, all other possible activity must have been completed before reaching the lowest priority.

2. An artificial time increment is used to separate the collection of data from the passengers getting off and on the trains.

3. Before the statistics are gathered, a check is intentionally caused to determine that no other possible events could occur at this event time.

Allocation of resources in the presence of competition is another difficult problem. The job shop has several different tasks waiting to be accomplished. These require teams of different size and skill with particular equipment. Selection of the next task to be performed is difficult because of the many forms of competition. The previous task may have released three people and one multipurpose equipment; the highest waiting priority task requires four people. The simulation can skip over to the next highest priority task, wait, search for a fourth man, or interrupt another task under way to obtain the man. To get the simulation to skip over to the next highest priority task or to wait is straightforward. To interrupt some other task and then keep track of where and when to restart it is more difficult. Searching for a fourth by scheduling overtime is closer to the real world but, like the real world, can also be a problem.

Since these are typical types of problems for system designers, the approach to understanding what the simulation is doing must be based on proceeding cautiously step by step. At each step be sure that the simulation is doing what is intended through the use of special historical records to show that the desired logic has indeed been used. While this procedure is time consuming, it provides confidence in the model and the results.

As the languages are more widely used, other users will find some of the problems. The newer versions of the language will have a better means to express the relationships that have so far been difficult. Some of these problems will be used as examples in the instruction manuals. The level of training required to *fully* learn the simulation language requires hard work. The introduction of programmed instruction texts for specific languages will ease the learning process.

Acceptance of Results

The problems of the system designer getting the simulation to use the available data, live within the idiosyncrasies of the simulation language, and produce results are merely the beginning. The real task is to get the customer to have confidence in the simulation. This means confidence in the model, the data, the rules, and the results. This major area must not be left for last. The customer thinks he knows the problem. Therefore, he has to be satisfied throughout the system design process. Simulation has advantages in this process. The steps can be made meaningful to the customer during the design process. The model can document the problem and be checked by the customer during its development. This procedure will keep the customer informed; but more important, it will provide feedback to the system designer that the problem being solved is the problem that *should* be solved. This opportunity for close understanding between the man with the long-term problem (the customer) and the short-term assistant (the system designer) is one of the strong advantages of simulation. The value of simulation is in building a bridge between the customer and the system design teams. Simulation provides feedback when it is most needed, during the early stages of problem definition and alternative system evaluation. The customer can be appraised of the current state of the system design; and he can make a contribution when the design is still fluid, before the specifications are firm. These advantages do not automatically follow from the use of simulation; nor do they result from periodic meetings with the customer. If they are to occur, it must be because of deliberate attempts to connect the customer with the simulation results. For the system designer this implies that the model will have more text, less missing documentation, a better description of the input data, and a clear presentation of results. The interpretation of results by the customer himself, is the goal. Trade-off curves showing different effects have more meaning to the experienced customer than the explanations of the systems designer. The customer can more readily accept results from the simulation when he has been a part of the progress, and has been asked to interpret results rather than approve some "final" results. The satisfying part of the state of the art is that this close association can be achieved.

BIBLIOGRAPHY

1. Objectives of Simulation, A. K. Swanson, *IEEE Trans. Sys. Sci. Cyber.*, **SSC-4**, 4:370–373 (1968).
2. Simulation: A Management Planning Tool, E. P. King, *ibid.*, 373–376.
3. Some Tactical Problems in Digital Simulation, R. W. Conway, *Manag. Sci.*, **10**, 1:47–61 (1963).
4. System Simulation Data Collection, R. Lessing, *IEEE Trans. Sys. Sci. Cyber.*, **SSC-4**, 4:388–392 (1968).
5. Selective Sampling—A Technique for Reducing Sample Size in Simulations of Decision-Making Problems, M. E. Benner, *J. Ind. Eng.*, **14**, 6:291–296 (1963).
6. The Sizes of Simulation Samples Required to Compute Certain Inventory Characteristics with Stated Precision and Confidence, M. A. Geisler, *Manag. Sci.*, **10**, 2:261–286 (1964).
7. (a) The Setting of Maintenance Tolerance Limits, A. A. B. Pritsker, *J. Ind. Eng.*, **14**, 2:80–86 (1963). (b) The Monte Carlo Approach to Setting Maintenance Tolerance Limits, *J. Ind. Eng.*, **14**, 3:115–119 (1963).
8. (a) NEASIM: A General-Purpose Computer Simulation Program for Load-Loss Analysis of Multistage Central Office Switching Networks, R. F. Grantges and N. R. Sinowitz, *Bell Sys. Tech. J.*, **43**, 3:695–1004 (1964). (b) On Some Proposed Models for Traffic in Connecting Networks, V. E. Benes, *Bell Sys. Tech. J.*, **46**, 1:105–116 (1967).
9. Markov Chains and Simulations in an Inventory System, E. Naddor, *J. Ind. Eng.*, **14**, 2:91–98 (1963).
10. Problems in the Statistical Analysis of Simulation Experiments: The Comparison of Means and the Length of Sample Record, G. S. Fishman, *Commun. Assoc. Comput. Mach.*, **10**, 2:94–99 (1967).
11. The Allocation of Computer Time in Comparing Simulation Experiments, G. S. Fishman, *Oper. Res.*, **16**, 2:280–295 (1968).
12. Stopping Rules for Queuing Simulations, I. W. Kabak, *ibid.*, 431–437.
13. Verification of Computer Simulation Models, T. H. Naylor, J. M. Finger (Critiques, J. L. McKenney; W. E. Schrank and C. C. Holt), *Manag. Sci.*, **14**, 2:B92-B106 (1967).
14. Design of Computer Simulation Experiments for Industrial Systems, D. S. Burdick and T. H. Naylor, *Commun. Assoc. Comput. Mach.*, **9**, 5:329–339 (1966).
15. Methods for Analyzing Data from Computer Simulation Experiments, T. H.

Naylor, K. Wertz, and T. H. Wonnecott, *Commun. Assoc. Comput. Mach.*, **10**, 11:703–710 (1967).

16. On Monte Carlo Methods in Congestion Problems: I Searching for an Optimum in Discrete Situations; II Simulation of Queuing Systems, E. S. Page, *Oper. Res.*, **13**, 2:291–305 (1965).

17. Use of Linearized Nonlinear Regression for Simulations Involving Monte Carlo, J. E. Walsh, *Oper. Res.*, **10**, 3:228–235 (1963).

18. Simulations Involving Monte Carlo, J. E. Walsh, *Oper. Res.*, **12**, 2:206–229 (1964).

19. The Analysis of Simulation-Generated Time Series, G. S. Fishman and P. J. Kiviat, *Manag. Sci.*, **13**, No. 7:525–556 (1967). [Review, L. Kleinrock, *IEEE Transactions on Comput.*, **C-18**, 3:303 (1969)].

20. A Test of a Statistical Method for Computing Selected Inventory Model Characteristics by Simulation, M. A. Geisler, *Manag. Sci.*, **10**, 4:709–715 (1964).

21. The Direct Simulation Method—An Alternative to the Monte Carlo Method, D. Fischer, *4th Conf. Applic. Simul.*, 134–145, New York, 1970.

15

Man-Machine Interaction—
Communication is a
Two-Way Street

By this point it should be obvious that the members of the system design team should have confidence that simulation will help their effort. There may still be a lingering question, will the computer be an effective tool? This raises the final characteristic of an effective design tool. It must be responsive. The design team must have confidence that the answer to a question will be obtained and by a predicted time.[1] There are two aspects to this problem. The one faced first is to state the problem in a form for quick interpretation by the computer. Once this has been accomplished, the second is to put the computer output into a form for ready interpretation by the users. This latter area is where the most rapid progress is being made and concern of this chapter.

Insight into the performance of different system implementations can be presented through a variety of different computer output forms: tabular listings, graphs using the alphanumeric printer, graphs using incremental X-Y plotters, cathode ray displays using both alphanumeric and graphic presentations, and large-scale presentation devices. In turn these outputs may be available either concurrently with the running of the simulation program or at a later time.

Simulation—Part of an Iterative Design Process

The contribution of simulation to the system design process rests on whether the system designer can obtain useful insight into system characteristics before it is built, operated, or fully designed. This requires

367

the system designer to understand the system; and if his knowledge is faulty make modifications while there is still opportunity—a feedback process. The minimum requirements for the feedback process to work are:

- Get results before the need has disappeared.
- Show input data in meaningful form.
- Obtain results in a useful form.

To get the results quickly, the problem must be stated in a form for entry into the computer, input data must be obtained and entered, and results digested. The first two items have been covered through the discussion of criteria for the choice of simulation language and the structuring of the model and its data. Now let us explore the influence of the man-machine environment and interfaces from the viewpoints of our obtaining quick results and in what form.

Both the computer installation and the simulation language affect the extent of man-machine interaction and the strategy to be followed for a particular environment. There are inherent computer installation restrictions: turnaround time, extent of storage, speed, degree of time sharing, and display equipment. The wide variety of computer installations may be grouped into the following classes:

- Single job operation based on using card decks without external storage of data and sequential operation of each task in turn by the computer. The full computer system is occupied to read in cards, develop the model, run it and finally print results. Model size is restricted to available core storage less what is required for the operating system. This system lends itself to the use of console sense switches to control model paths and the printing of results during the model run. In general, this system has become obsolete for simulation, since the large amount of core storage required for simulation has forced a usage which makes more efficient use of a computer with this amount of core storage.
- More efficient machine utilization while still using batch operation is obtained when the reading in of input cards and printing of results are overlapped with the execution of the model. Now, the use of console sense switches becomes meaningless, since the printout is obtained after the model or a portion of it is finished. Some core storage is needed for simultaneous input and output; therefore, either a large computer is required or additional storage is available.

In the ultimate version of this system additional computers are used to convert the card decks to magnetic tape for input and store the results on magnetic tape prior to printing. In this process the time required from submittal of decks to obtaining results becomes progressively longer as the system becomes larger while the computer is considered more "efficient." Replacement of the magnetic tapes by disk files speeds up the process since the delay for printing is limited to the storage capacity of the disk used for this purpose.

- The addition of a monitoring cathode ray display to a system forces some changes in the operating procedures. If the display is to be useful, it must present results and await human reaction. Unfortunately, the cost of this large installation becomes even higher if a significant part of the time it is idle awaiting human reaction. Therefore, it has become advantageous to partition the core storage among different tasks and allow the computer to do one or another according to a priority scheme. Then when the system is awaiting human intervention other jobs not requiring display can proceed. But now this has become a large and most expensive installation. It is economical when we use only a small portion of the machine for a limited time.

- Under conditions in which the amount of data and the speed with which they must be presented is within the capabilities of a remote printer or display and its connecting communication lines, a time-sharing environment can be used. Unfortunately, the amount of data characteristics of simulation language and their models makes it difficult to construct and debug large models using remote time-sharing facilities. However, once a model is available and stored at the computer it may be run for slight changes in input data and presentation of limited results. Then output is sent back to the remote location for approval, interaction and change. This is an approach which has considerable merit when a model is to be used by a variety of users and at several locations.

Simulation has been used successfully under each of the above computer configurations. Since the state of the art is far from static the most complicated systems are becoming increasingly attractive with regard to both cost and to the speed of obtaining results. Eventually simulation languages will be geared to operation under the concept of the computer utility.[2-4] The other extreme, development of inexpensive computer configurations which are usable for simulation languages large models, and man-machine interaction, appears to be still in the future—unless we use them to solve only trivial cases.

A simulation program being executed within the main frame of the computer provides little opportunity for display of simulation progress unless special efforts are made to gear the simulation to real time. This may be actual real time as required in a training situation or a speeded up situation for a "what if" decision. Since it is readily possible to stop the program execution under predetermined conditions, all that is required is the addition of programs to present these intermediate results. The simulation program may then continue or wait for human reaction. If it were to continue, in the speeded up case, there may be little gained, because the next display may appear long before the design team has had the opportunity to understand the significance of the previous output. Use can be made of this approach when hard copy is produced for these outputs for the case where the design team can no longer follow and the simulation can be restarted later. If this is not the case it is necessary to slow down the presentation of results. In effect this requires the "tuning" of a model for responsive man-computer interaction.

In another approach each interim result is presented to the man for reaction before continuing the simulation. This has the effect of stopping the computer frequently and presenting partial results. This makes it impossible for the simulation to be geared to real time. In addition, frequent stopping of the simulation adds considerable overhead to the running time and can occupy a large portion of core storage for considerable periods. Until this approach is "tuned," there may be insufficient data available per interruption for a decision to be made.

Either of the above approaches can produce a satisfactory result. A practical technique to reach a "tuned" system is to interrupt the simulation at major times determined either through continued analysis of data or cyclically. At that moment a snap-shot of all data in the simulation is set up for later printout and partial results are displayed. This reduces the number of interruptions and permits the probing of results, since the full set of simulation statistics are available later. Both for the printouts and displays there are advantages to providing results in graphic rather than tabular form. Since printers are always a part of the computer system, this is easy to implement. Ideally, the man-machine interface should be able to present a limited amount of data while a lengthy simulation is in execution. These data should provide sufficient information only to halt or continue the simulation. However,

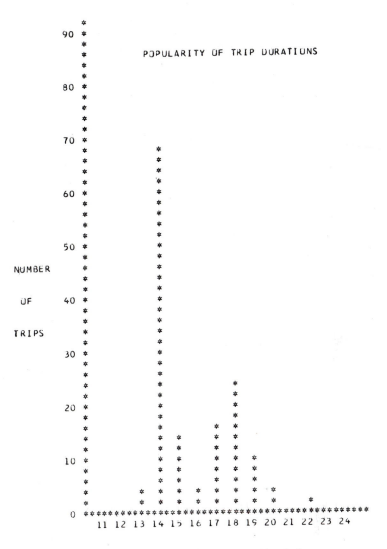

Figure 15.1a Histogram of labelled graphic printout generated by printing devices.

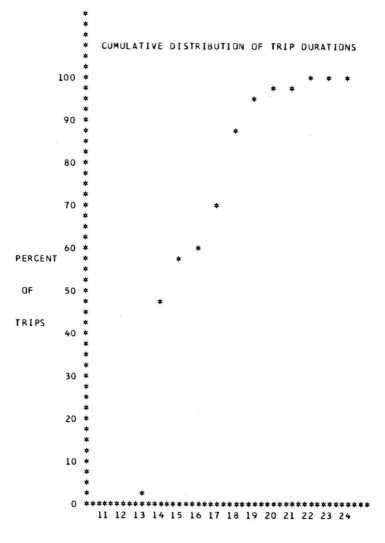

Figure 15.1*b* Plot generated by printing device.

the halt condition should allow options: terminate selective presentation of data, opportunity to change conditions within the simulation, restart, and continue.[5-6]

Data Presentation

Various types of printers can generate graphic data presentations either in the form of histogram or continuous data presentation. The advantages gained from using the printer for graphic output are that it is already there and can be readily intermixed with normal simulation output as shown in Figure 15.1a and b. The usual form of the printer is the high-speed line-at-a-time type which can print up to 3000 lines per minute or a page of output every second. The disadvantages are in the lack of resolution of the graph and poor labeling of the parts of

```
                        ASSUMPTIONS


     BASE OF INITIAL ASSUMPTIONS

     1.  HARDWARE MATURITY ASSUMED AT 65TH UNIT
     2.  IN-PLANT-RETURNS ASSUMED TO ADD 35% TO ESTIMATED MATURE ATP
         TEST TIMES
     3.  MONTHLY PRODUCTION TEST TIMES AVAILABLE
         2 SHIFTS = 16HRS/DAY - 80 HRS/WK
         4 1/3 WKS/MO
         15% SHOP LOST TIME + 10% T.E. DOWN TIME = 25%
         TST TIME AVAILABLE = (80*4 1/3)-25% = 347-87 = 260 HOURS
     4.  PROD TST TIMES CALCULATED ON 93% UNIT LEARNING CURVE
     5.  10 WEEK SETBACK CALCULATED FOR LRU LEVEL
     6.  O&R CALCULATED ON 60 HRS/WK

     GROUND RULE A

     1.  1 1/4 SHIFTS = 100% IST SHIFT + 25% 2ND SHIFT
     2.  4 WKS/ MO    20 DAYS/ MO
     3.  50% IN-PLANT-RETURNS
     4.  25% LOST TIME
     5.  10% OVERTIME 1ST SHIFT
     6.  93% LEARNING CURVE
     7.  O&R CALCULATED ON 8 HRS/DAY + 10% OT

     GROUND RULE B

     1.  1 1/2 SHIFTS = 100% 1ST SHIFT + 50% 2ND SHIFT
     2-7 SAME AS GROUND RULE A
```

Figure 15.2a

the graph. This technique has advantages compared to using long data listings, but except for histograms, it is not a particularly effective presentation method outside the system design team. It may be used to highlight assumptions and some input data as shown in Figure 15.2a and b.

A second approach is to use an off-line hard copy drafting or drawing device, the X-Y plotter. This device has the advantage of producing a highly accurate graph, with excellent resolution and convenient labeling of the desired curve. This form of plot, shown in Figure 15.3a and b, is most useful for showing the relationships among continuous functions and, to a lesser degree, among discrete valued data. Since large quantities of data may be plotted on one sheet, obscure patterns may be recognized in the data. When the X-Y plotter is an off-line device, the results are available after the simulation has been run. Therefore this device is most useful in obtaining high-quality graphs for inclusion in reports.

A third approach circumvents the problem of delay because of printer or plotter speed by using a cathode ray tube display. Some display units even have built in copying devices in order to provide hard copy. The actual cathode ray tube presentation can resemble either the printer or plotter, as shown in Figure 15.4. The potential resolution for the cathode ray display is better than the printer but worse than the plotter. However, there are other potential methods of presentation on the CRT using fainter lines, dotted lines, overlays, mixture of data with back-

MATRIX HALFWORD SAVEVALUE ACRFT

COL.	LANDG SPEED KNOTS	LANDG TIME MN*10	TKOFF TIME MN*10	TAXI SPEED FT/MN
ROW				
ACRFT TYPE 1	150	5	6	800
ACRFT TYPE 2	120	6	7	750
ACRFT TYPE CUB	55	10	8	200

Figure 15.2b Standard printer output tailored to present key information.

ground, and, most important, the creation of a dynamic display—one which can move, stretch, and be blown up.[7-9]

During system design there are different requirements for the presentation of results. A summary of anticipated costs and particular aspects of the system performance can be obtained from dynamic cathode ray tube presentations. At the stage of early investigation, when the system is being studied to determine which are the sensitive areas, the cathode ray tube has advantages, since it permits the system designers to quickly

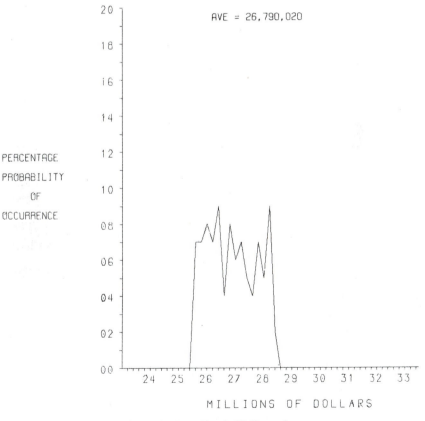

Figure 15.3a   Simple X-Y graph.

Random spikes for 10,000 trails of GPSS/360 RN1 seed set to 37

(b)

Figure 15.3*b* Complex X-Y graphic presentation.

obtain specific answers. Unfortunately, frequently there are other more difficult types of data presentation also required. Sometimes there are situations in which the many factors in the design of a complex system should be presented at the same time. This is a difficult task for a single data display device—the simultaneous display of a multitude of factors can be arranged only through extensive preparation and planning. The display then becomes the end rather than the means toward the end. This problem might be overcome with the use of very extensive displays. Instead of a single CRT several units are used, each displaying a particular aspect of the simulation. For these multiple displays to be dynamic, the use of the CRT would be most expensive. The solution to the problem of a multitude of simultaneous displays depends on solving the hardware problem of economically tying many displays to a single computer system.

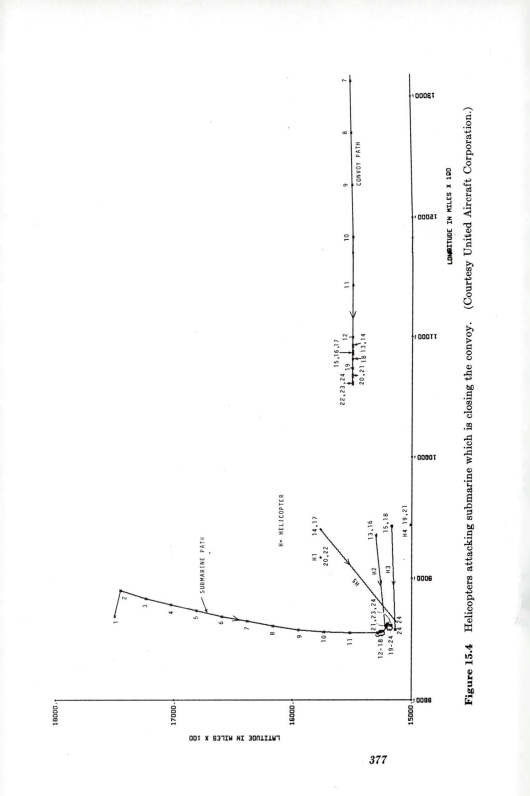

Figure 15.4 Helicopters attacking submarine which is closing the convoy. (Courtesy United Aircraft Corporation.)

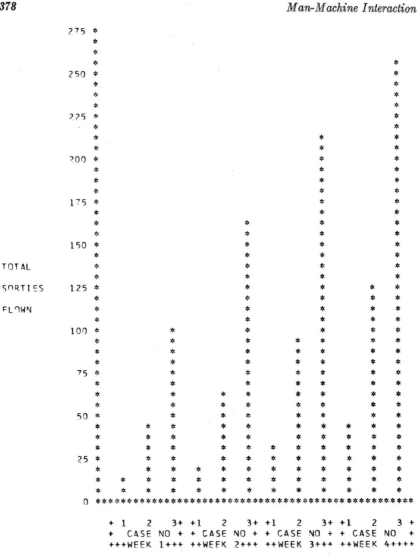

Figure 15.5

Display of Numerous Results

The use of the printer, while not dynamic or readily changeable, has advantages for multiple plots of being both the normal output and relatively easy for a number of plots to be viewed at the same time. This solution, shown in Figure 15.5, seems to be the best alternative for the everyday trade-off analysis. Here the data describing numerous interactions can be seen for one moment in time. The plotter approach circumvents the ease of getting too much data by using only what is needed. This process reduces the effort to find out what is needed, since the plots are limited to sensitive areas and the excess data are thrown away. Nevertheless, care must be exercised so that the back-up data does not drown the study team. It is useful to have a large wall area to mount the results for comparison. The advantages of this approach are its low cost and quickness but only after the procedure is established through initial results.

The previous discussion shows that, for simulation to become a fully useful tool for the design of complex systems, a variety of approaches are going to be required for the improved communication among the system design team and between the system designers and the customer. The less complex factors may be presented using CRT displays; the full problem, however, will have to rely on numerous printed or plotted displays at least until the CRT display hardware and software are greatly improved. The important system design team goal is to shift from the use of tabular to graphic data as soon as possible.

Display Usage by Modelers

Aside from the need to tie in with the system design team and the customer, the modeler has some overriding requirements. He must first construct the model before he can debug, document, present input and output data, or provide assurance that the model is doing what it is supposed to.

Model Building

The model is entered into the computer, either from cards prepared in advance or by interactive construction by the modeler using a time-sharing terminal or CRT display and working from coding sheets or directly from the flow charts.[10] The display can be used to make corrections, either for the entire model or for each statement as shown in Figure 15.6a–c. First the all too numerous trivial syntax and grammatical errors the simulation language does not accept are eliminated. In this figure the display has a separate line for each statement. The statements are stored as part of a direct access data set which permanently resides on a disk. Only a part is brought into core storage to obtain the initial errors. These are corrected using the keyboard

Figure 15.6a IBM 2250 display unit consisting of CRT display, function and alphameric keyboards, and light pen.

to update the data set. The process is iterative until the model has no input errors. Elimination of errors during execution of the model depends on their complexity. The trivial errors are pinpointed sufficiently to permit debugging from the terminal. The obscure ones unfortunately require printouts in order to compare parallel activities.

Once the model has been stored and debugged as a data set on the disk, the current version of the model becomes the only one stored. If all versions stored for historical perspective, it becomes utterly unwieldy to control. The best solution to this problem is to retain the current model in the computer, but to provide flow charts to act as the latest documentation of the model. Figure 15.7a and b shows a flow chart representation of one part of a model. These can be generated directly from the model statements using the printer. Another way of preserving the documentation is through a complete listing of the model. Hard copy versions of the model can be used to provide the documentation of the progress of the project, since they show the changes and new concepts. Documentation of input data is also a difficult process. The best approach is to keep the full data in the form of a permanent data bank. Since in complex systems there will be copious amounts of data sooner or later, it is well to set up a separate system for data control during the early stages of model development.

Presentation of Data

During the construction of the model, the design team has special requirements for the model results. Since the model is to help define the problem, the tentative solutions have to be available for the team to consider as a group. This will expose areas of differences of interpretation and inconsistency. To get the information to the team quickly and in a way in which they can make changes and see "What if" this or that is tried, the CRT display is most useful.[11,12] The CRT display establishes an environment much closer to the use than is characteristic of the simulation language. This condition is established through both

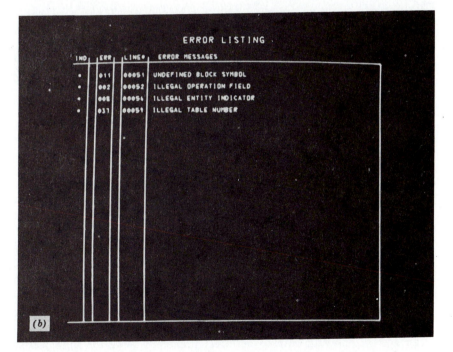

Figure 15.6b 2250 display of GPSS errors indicating line number for the error and the error type. Light pen detection of the line number automatically displays that line number in the center of the display as shown in (*a*). (Courtesy United Aircraft Corp.)

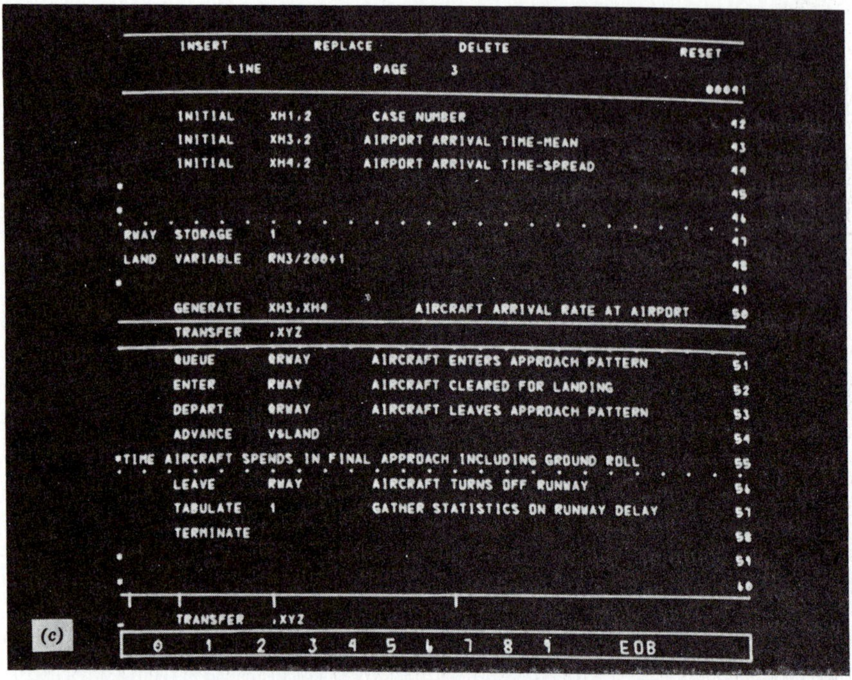

Figure 15.6c   2250 Display shows light pen selection menu at top, 20 model statements, and alphameric and numeric entry areas for statement entry and modification. (Courtesy United Aircraft Corp.)

the control of data and its graphic presentation. Some models have little input data and few results, merely a significant criterion such as the maximum queue length. Most models, however, require large amounts of input data, initial value, constraints, and even values for comparison. The entry of these data through the normal channels of coding sheets and verified card decks does not eliminate clerical errors. the means of data entry, although this is a timely, archaic, system. In contrast, the CRT allows direct entry of data to computer file. A two-man team can enter and check the data simultaneously. System design team effort is minimized using this approach. Figure 15.8 shows a display system allowing data to be directly entered using the light pen to select the location and the keyboard to enter the data. The guidance governing which data are entered where it is provided by row and column titles. A light pen selects the point in the array where a new datum is to be entered. Either a single point or the sequence of points belonging to a row or column may be entered. The skill level required to enter data is reduced and the frame of reference becomes consistent with the user's language. If large data banks already exist, these may be converted through special programs and only the data for special cases need be manually entered. Once the framework is established it need not be limited to the data entry. Columns may be reserved for results as well. This display system brings the simulation closer to the design team and in turn closer to the real world.[13]

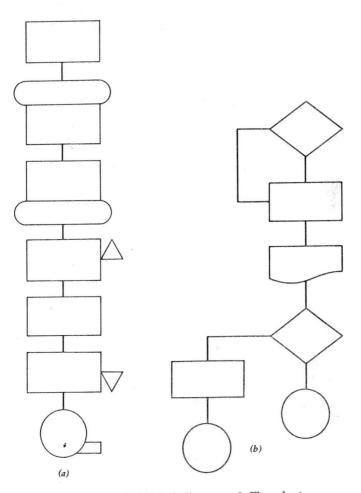

(a)   (b)

Figure 15.7a GPSS block diagram. *b* Flow chart.

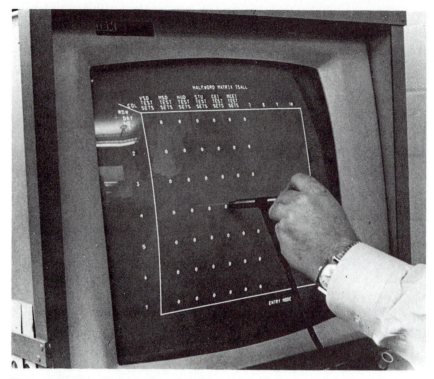

Figure 15.8 Selection of a location where data will be entered using the 2250 keyboard. (Courtesy United Aircraft Corp.)

Model Assurance

The models of complex systems need a special approach to determine that the model is indeed accurately portraying the problem. Primarily it is the need to be sure that the logic is consistent. Thus it is necessary to see how specific tasks are performed within the model. The need is for the model designer to follow specific factors through a running of the model one by one to see that the correct paths have been followed and that these logical paths through the model are the anticipated ones, or at least the rationalized ones. This result may be obtained with the cathode ray tube display, as shown in Figure 15.9. Here the movement of each transaction singled out for close inspection is observed every time it moves through a logical step. To understand the movement of the transaction through the model, it is necessary for the model

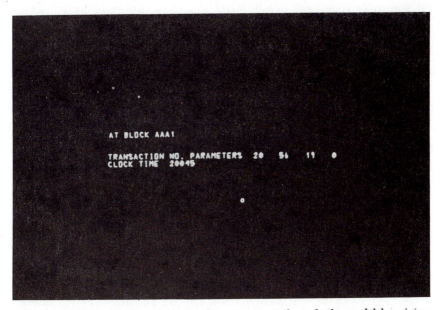

Figure 15.9 2250 display of transaction movement through the model by giving location, time, and transaction parameters.

to be stopped, evaluated, and continued, and restarted with changes or restarted at a previous time with changes. However, before changes are made it is desirable to be able to investigate a number of previous conditions to discover how the present situation came about. Since different items may be important during the process of assuring oneself that the model is representative of the system, different items may be singled out for observance. To single out a particular item may require a change in the model. This would be done after the model has been run using a separate run for the new item. Whether the tie-in with the modeler is through a display or a printout would greatly affect the speed with which the model is converted from an untried system to a final version of the model for use in trade-off studies.[14,15]

Dynamic Display

Model behavior during execution can be brought to the system design team through the display device. Since this may require additional effort to convert the model into one which provides a meaningful display, it is useful to consider what are the advantages.

- For fully debugged models being run with new data, there is the opportunity to observe that the system is being simulated according to plan. Frequently new data violate a hidden rule of the simulation with the fortunate result of a quick error and output. However, there is also the less fortunate case where an extremely long run produces no results since the new data established a loop.
- Interactions among related elements in the model seem to have been covered by the logic. Now when these elements are displayed together a series of violations of the rules becomes apparent. Obviously, the rules were not adequately defined. The display brought this to obvious attention long before the modeler could wade through all the printouts which contained the same information.
- Sometimes the knowledge of the problem does not inspire the confidence that the rules are available to the model builder. In this situation he can omit the decision logic from the model and rely on human intervention to make each decision as the situation arises. Moreover, the individual engaged to make the decision could be considerably more knowledgeable than the modeler. Obviously, this is not an easy technique, but it is valuable under special circumstances.

The display device has the capability of producing a dynamic display. The additional elements added to the model can be made to draw pictures and they can dynamically change the pictures as the simulation progresses. For this technique to be economical a series of generic subroutines to draw the picture have to be used. This reduces the amount of special effort required. To illustrate the use of this technique on a 2250 display unit, a series of pictures defining an airport and its airspace are shown in Figure 15.10. Figure 15.10*a* provides a mixture of static and dynamic information. The runway configuration with its descriptive phases and titles are static information. The runway lengths are to scale while the locations of the ground taxi queues and airborne hold area are not. However, the taxiways on the ground and the approach paths to the runways follow the usual routings. Dynamic information is displayed as the current quantities in the ground queues and hold area, simulated clock time, and the locations of all aircraft currently of interest, those taking off and on final approach. Four-digit numbers define the current aircraft location. The last three digits indicate tail number. While the first digit is used to indicate either arriving or departing traffic through 6 or 2, respectively. In Figure 15.10*a* at time 18 seconds into the simulation, Aircraft 6573 has left the hold area while 2575 is taking off. One minute later in Figure 15.10*b* 6573 has been joined by another, 6177, at least three miles behind it. A

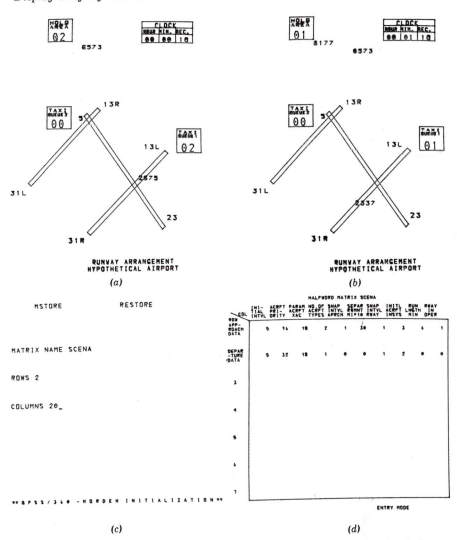

Figure 15.10 Composite of four 2250 presentations showing *a* airport at simulation start, *b* display one simulated minute later, *c* selection of model control matrix, SCENA, *d* data in control matrix. (Courtesy United Aircraft Corp.)

different aircraft 2337 is now taking off. The movements of the aircraft on the display are updated every simulated six seconds.

At this point it is desired to illustrate man-computer interaction by stopping the simulation and then changing the runway use. Instead of using two separate runways all new traffic will be directed to use runway 5. The actual stopping of the model is set up through an interrupt from the 2250 function keyboard. Then the driving scenario is requested, as shown in Figure 15.10c, while the actual matrix, which controls the model is shown in Figure 5.10d. The code indicating the runways in use appears as the last element of the first row 1.

This dynamic display enables problem-oriented individuals to run the airport without understanding the way the simulation was developed and structured. They can satisfy themselves that the aircraft are being properly handled in the simulation. Not only can windshifts be accommodated, but different types of aircraft can be entered. The goal is to have the simulation progress fast enough. This was true for this case using an IBM 360/50 which displayed the model twice as fast as real time.[16]

Time-Shared Environment

The widespread availability of time shared computer services suggests that simulation should also benefit from these capabilities. The most obvious benefit is to run a simulation from a remote location and enhance the portability of the model. To obtain the full interactive capabilities of the CRT terminal through a communication line is far from simple. Therefore, the technical question is how to make best use of a slow speed remote terminal. These are usually teletypes or electric typewriters. There are small screen CRT displays but usually their amount of information displayed at one time is limited.

The Computer system also presents difficulties in the time-shared environment since simulation applications require the equivalent of large amounts of core. For these reasons the progress in developing a full time-shared simulation capability has been slow.

The most advanced simulation capability in the time-shared environment is a system using the "virtual" machine approach. One such system, National CSS Inc. allows each user to have his own virtual machine with 256,000 bytes of core. Under these circumstances it has been possible to provide the following GPSS services as well as additional features.[17]

- Permanent Matrix Library on Disk. Basic data for many models can be stored and used as needed by each model. One model can retrieve data from the library, use it, put new data into the library, and then let the next model use those data.
- Data Entry and Manipulation. Enter, update, and modify data in the permanent library without using models. Instead direct access with titles for matrices enable manipulation of data without special skills.
- Interactive GPSS Model Development. Enable the development of

```
NGPSS RESET AT 10.09.11
***** ***** ** ********

REQUEST 1?--?

    Reply with:

        Parameters: NAME  HLP  MSR  RDS  TTL  JOB  OPTS

          Commands: ALT  RUN  RESET  EDIT  MINIT

           Replies: BACK  QRY  END  QUIT  FINISH


        For an explanation of any of the above, type:

                ?  abcd1  abcd2  abcd3...

        where "abcd1 abcd2 abcd3..." are one or more
        parameters, commands, and/or replies.

        To illustrate:

                REQUEST 4?--? rds msr edit end

REQUEST 1?--name elevates

REQUEST 2?--rds elevates

REQUEST 3?--msr demo

REQUEST 4?--ttl baxter

REQUEST 5?--run


ASSEMBLY PHASE BEGUN
INPUT PHASE BEGUN
 OPTIMIZED CORE USAGE

EXECUTION PHASE BEGUN

THIS MODEL IS SET UP TO SIMULATE FOR THIRTY MINUTES
ELEVATOR SERVICE IN BUILDINGS UP
TO TWENTY FLOORS AND TWENTY ELEVATORS
FOR YOUR BUILDING ENTER IN ORDER - NUMBER OF FLOORS AND ELEVATORS
AND ELEVATOR CAPACITY

15 8 20

    THE NUMBER OF PEOPLE WAITING IN THE LOBBY IS --65
THE MAXIMUM NO. WAITING WAS 109
DO YOU WANT TO CHANGE ELEV. CAPACITIES? (Y/N) y

ENTER  8 VALUES 22 22 22 22 22 22 22 22

OUTPUT PHASE BEGUN
NORMAL TERMINATION OF GPSS/NORDEN RUN
 RELATIVE CLOCK              101  ABSOLUTE CLOCK              101
DISPLAY MODULE ENTERED
```

Figure 15.11 Start and first 10 minutes of simulated elevator service shown on hardcopy remote printer.

the model to be debugged on-line with indication of Assembly and Input errors. Special treatment of Execution errors to truly reflect the interactive environment by allowing selective seaching of model statistics.

- Man-Machine Interaction. Provide a window into the model so that the model may be stopped, restarted after investigation of progress, moreover, in those case where progress can be related to the modeler's judgment, allow the modeler to change values in the model and then continue.

- Report Generation. Compile selectively reports using a free-format language which includes capabilities for iteration, arithmetic expressions, conditional phrases, display of numeric and alpha information, access to all GPSS entities, and selective compilation based on model logic.

```
display user
USER CHAIN      TOTAL       AVERAGE      CURRENT      AVERAGE      MAXIMUM
                ENTRIES     TIME/TRANS   CONTENTS     CONTENTS     CONTENTS
        1         40         28.324        10          11.217        20
        2         35         24.485         3           8.485        20
        3         35         13.885        18           4.811        20
        4         40         28.224        15          11.178        20
        5         43         17.418        20           7.415        20
        6         32         23.187         9           7.346        20
        7         35         19.114         9           6.623        20
        8         40         21.024        20           8.326        20
        9        365         11.380        65          41.128       109
```

```
-display blocks
BLOCK COUNTS
BLOCK CURRENT  TOTAL    BLOCK CURRENT  TOTAL    BLOCK CURRENT  TOTAL
    1      0      1        3      0      9        4      0      1
  FFH      0      8     FFFA      0      6      FFF      0     23
 FFFC      0     17       16      2     17       17      0     15
  FAB      0    172       20      2    172       21      0    170
   22      4    112      FAA      0    166       24      0      9
  FGA      0      2       28      0      1     FGA1      0      0
   32      0      1      AAB      0     10      ABB      0    515
   37      0    150       38      1     10       39      0      9
  AAA      0    365      BAA      0    300      CAA      0    196
   50      0      1     DAA1      0      6      DAA      0     90
   55      0      6       57      0      1      DBB      0      8
   61      0      1       82      0      0      GAA      0      1
   88      1      2      GAZ      0      1
```

```
-status current
    5 XACS ON CURRENT EVENTS CHAIN
```

```
-status future
    5 XACS ON FUTURE EVENTS CHAIN
```

```
-s block gaz
BLOCK      89
```

Figure 15.12 GPSS statistics for the first 10 minutes returned as a result of specific requests.

```
d mh elevs 1-15 1-8
MATRIX HALFWORD SAVEVALUE ELEVS
```

COL.	1	2	3	4	5	6	7	8
ROW								
1	22	22	22	22	22	22	22	22
2					1			2
3					1			1
4			4		3			1
5			1	4	3			2
6			1	1				2
7			2	1	4			1
8	2		3	1	1		1	2
9	2		1	1	4	2		
10	2		2	1	1	1	2	2
11	1		2		1	2	2	1
12				2		2		2
13			1	1		4	2	2
14	1	1		1				1
15	2	2	1	2	1			1

```
-return
OPTIMIZED CORE USAGE
EXECUTION PHASE BEGUN
!THE NUMBER OF PEOPLE WAITING IN THE LOBBY IS --93
THE MAXIMUM NO. WAITING WAS 126
DO YOU WANT TO CHANGE ELEV. CAPACITIES? (Y/N) n
OUTPUT PHASE BEGUN
NORMAL TERMINATION OF GPSS/NORDEN RUN
 RELATIVE CLOCK            201  ABSOLUTE CLOCK           201
DISPLAY MODULE ENTERED
```

Figure 15.13 Selective display of a portion of a 20 by 20 matrix and return to simulate the next 10 minutes.

```
d us
```

USER CHAIN	TOTAL ENTRIES	AVERAGE TIME/TRANS	CURRENT CONTENTS	AVERAGE CONTENTS	MAXIMUM CONTENTS
1	62	31.580		9.741	22
2	79	24.455	22	9.611	22
3	57	26.824	3	7.606	22
4	62	34.161	4	10.537	22
5	65	29.353	8	9.492	22
6	76	22.855	21	8.641	22
7	79	21.151	22	8.313	22
8	62	32.854	11	10.134	22
9	635	18.864	93	59.597	126

```
-d b gaz
BLOCK COUNTS
```

BLOCK	CURRENT	TOTAL
GAZ	0	2

```
-d mh fact1
MATRIX HALFWORD SAVEVALUE FACT1
```

COL.	TIME BETWN ARRVL	TIME BETWN FLOOR	EXIT TIME MEAN	EXIT TIME SPRD	CURNT QUEUE SIZE	RUN LNGTH MINTS	NMBR OF FLRS	NMBR OF ELVTS	ELVTR CAP-ACITY	MAX QUEUE SIZE
ROW										
UP	10	1	3	2	93	30	15	8	20	126
DOWN					1234					

```
-return


        OPTIMIZED CORE USAGE
        EXECUTION PHASE BEGUN
        !THE NUMBER OF PEOPLE WAITING IN THE LOBBY IS --37
        THE MAXIMUM NO. WAITING WAS 145
        OUTPUT PHASE BEGUN
        NORMAL TERMINATION OF GPSS/NORDEN RUN
         RELATIVE CLOCK          301  ABSOLUTE CLOCK          301
        DISPLAY MODULE ENTERED


        -rpg
        NRG OUTPUT PHASE BEGUN
        !
```

Figure 15.14   GPSS statistics for the second 10 minutes of simulation as a result of specific requests and the return to simulate the third 10 minutes.

PASSENGER DESTINATIONS FOR FIVE MINUTE

INTERVALS

FIVE MINUTE TIME INTERVALS

```
                FLOOR #15
    13    14    11    18    16     9

                FLOOR #14
     8     9     1    21    15    11

                FLOOR #13
    13    17     6    11    11     5

                FLOOR #12
     9    11     6     7     5     2

                FLOOR #11
    13    11     8    18    16    14

                FLOOR #10
    18    17     8    11     7     2

                FLOOR # 9
    18    17    13    22    16    11

                FLOOR # 8
    13    19    11    13    16    11

                FLOOR # 7
    18    21    14     8    16     5

                FLOOR # 6
    13    14     6    17    15     7

                FLOOR # 5
    19    22    12    18    14     7

                FLOOR # 4
    15    23    12     4     5     0

                FLOOR # 3
    21    12    10    18    13     7

                FLOOR # 2
    23    19    12    12    10     3

                FLOOR # 1
     0     0     0     0     0     0
```


THE MAXIMUM NUMBER IN THE LOBBY WAS 145
OPTIMIZED CORE USAGE
END OF GPSS EXECUTION

Figure 15.15 GPSS/360-NORDEN report generator output requested after 30 minutes of simulation.

```
REQUEST 1?--minit
 MIM: MSTORE LIBRARY NAME?--demo
 MIT: TITLES FILE NAME?--baxter
MINIT OUTPUT FILE IS DEMO LISTING
MINITIAL ENTERED


old fact1,2,10
DONE!
data
row(2),(1),1,2,3,4,5,6,7,8,9
printmat fact1
```

     MATRIX HALFWORD SAVEVALUE FACT1   FEBRUARY 1, 1971 AT 10.51.23 A.M.

COL.	TIME BETWN ARRVL	TIME BETWN FLOOR	EXIT TIME MEAN	EXIT TIME SPRD	CURNT QUEUE SIZE	RUN LNGTH MINTS	NMBR OF FLRS	NMBR OF ELVTS	ELVTR CAP-ACITY	MAX QUEUE SIZE
ROW										
UP	10	1	3	2	37	30	15	8	20	145
DOWN	1	2	3	4	5	6	7	8	.9	0

```
quit

REQUEST 1?--edit source
EDIT:
print  11
       REALLOCATE HMS,10,FMS,20,COM,22000,XAC,550,CHA,25
       SIMULATE                    RETURN OF MODEL FROM READ
       READ
       ABS
       START     1                  1    1
       START     1                  1    1
       START     1                  1    1
       MSTORE    MH2     FACT1
       MSTORE    MH3     ELEVS
       ENDABS
       END

 EDIT:
 q
 REQUEST 1?--edit elevator source
```

Figure 15.16 Independent access to the data bank through the data manipulation system and modification of the data in row 2. Printing of the READ/SAVE source file through the use of the interactive edit capability.

```
        REALLOCATE  HMS,10,FMS,20,COM,22000,XAC,550,CHA,25
        SIMULATE    1/12/71              ELEVATOR OPERATION UP ONLY
*
  PRMB1 EQU         7,M              THIS MODEL IS SET UP TO SIMULATE 30 MIN
  PRMB2 EQU         8,M              ELEVATOR SERVICE IN BUILDINGS
  PRMB3 EQU         9,M              TWENTY FLOORS AND TWENTY ELEVATORS
  PRMB4 EQU         10,M             FOR YOUR BUILDING ENTER THE NO FL&ELVS
  PRMB5 EQU         11,M             ELEVATOR CAPACITY
  BLANK EQU         12,M             BLANK LINE
*
  LINE1 EQU         1,M              THE NUMBER OF PEOPLE WAITING IN LOBBY
  LINE2 EQU         2,M              THE MAXIMUM NO. WAITING WAS --
  LINE3 EQU         3,M              DO YOU WANT TO CHANGE ELEVATOR CAPACITY
  LINE4 EQU         4,M              INPUT MATRIX FOR YES OR NO
  LINE5 EQU         5,M              ENTER NN VALUES
  YES   EQU         6,M              YES MATRIX
  NMBR  EQU         1,X              NUMBER OF ELEVATORS
  ELEVS EQU         3,Y              ELEVATOR CAPACITY AND USE
  FACT1 EQU         2,Y              SYSTEM PARAMETERS
*
  PRMB1 MATRIX      X,1,14
  PRMB2 MATRIX      X,1,10
  PRMB3 MATRIX      X,1,11
  PRMB4 MATRIX      X,1,17
  PRMB5 MATRIX      X,1,6
  BLANK MATRIX      X,1,1
*
  LINE1 MATRIX      X,1,12
  LINE2 MATRIX      X,1,8
  LINE3 MATRIX      X,1,12
  LINE4 MATRIX      X,1,1
  LINE5 MATRIX      X,1,4
  YES   MATRIX      X,1,1
*
*       INITIAL     MX$PRMB1(1,1),'THIS MODEL IS SET UP TO SIMULATE '
*       INITIAL     MX$PRMB1(1,34),'FOR THIRTY MINUTES '
*       INITIAL     MX$PRMB2(1,1),'ELEVATOR SERVICE IN BUILDINGS UP '
*       INITIAL     MX$PRMB3(1,1),'TO TWENTY FLOORS AND TWENTY ELEVATORS'
*       INITIAL     MX$PRMB4(1,1),'FOR YOUR BUILDING ENTER IN ORDER -'
*       INITIAL     MX$PRMB4(1,35),' NUMBER OF FLOORS AND ELEVATORS '
*       INITIAL     MX$PRMB5(1,1),'AND ELEVATOR CAPACITY'
*
*       INITIAL     MX$LINE1(1,1),'THE NUMBER OF PEOPLE'
*       INITIAL     MX$LINE1(1,21),' WAITING IN THE LOBBY IS --'
        INITIAL     MX$LINE2(1,1),'THE MAXIMUM NO. WAITING WAS --'
*       INITIAL     MX$LINE3(1,1),'DO YOU WANT TO CHANGE ELEV. '
*       INITIAL     MX$LINE3(1,29),'CAPACITIES? (Y/N)'
*       INITIAL     MX$LINE5(1,1),'ENTER NN VALUES'
*       INITIAL     MX$YES(1,1),'Y'
*       PAGE
```

Figure 15.17 Listing of model of elevator service used for illustration of terminal session.

```
        RESTORE     MX$PRMB1
        RESTORE     MX$PRMB2
        RESTORE     MX$PRMB3
        RESTORE     MX$PRMB4
        RESTORE     MX$PRMB5
*
        RESTORE     MX$LINE1
        RESTORE     MX$LINE3
        RESTORE     MX$LINE5
        RESTORE     MX$YES
*       VARIABLE STATEMENTS TO CONTROL THE TIMING AND FLOW OF PASSENGERS
*
UPSG    VARIABLE    BV$UPSG1*8+BV$UPSG2*P4*8/MH$ELEVS(1,P3)+1    LOBBY
UPSG1   BVARIABLE   Q$UPSG1'G'MH$ELEVS(1,P3)        MANY WAITING
UPSG2   BVARIABLE   Q$UPSG1'L'MH$ELEVS(1,P3)        SOME WAITING
ARIV1   VARIABLE    C1/50+1/2*1+1    TIME TO GET NEW DATA
UPSG1   VARIABLE    MH$BLDG1(P1,V$ARIV1)/5    NUMBER/MINUTE
LOBBY   VARIABLE    MH$FACT1(1,8)+1           LOBBY CHAIN NUMBER
*
BLDG1   MATRIX      H,20,20          PASSENGER DESTINATIONS
FACT1   MATRIX      H,2,10           SYSTEM PARAMETERS
ELEVS   MATRIX      H,20,20          ELEVATOR CAPACITY AND USE
*
        RESTORE     MH$FACT1
*
        SKIP        ,,I      DELETE INPUT SECOND LISTING
*       PAGE
        GENERATE    ,,1,1,64,4,H
        SAVEVALUE   LOBBY,V$LOBBY,H    LOBBY CHAIN NUMBER
        SPLIT       MH$FACT1(1,8),FFH,3   SET UP NUMBER OF ELEVATORS
        TERMINATE
FFH     ASSIGN      3-,1                NUMBER OF ELEVATOR
        TRANSFER    ,FFF
*
FFFA    ADVANCE     4                   TWENTY FOUR SECONDS BETWEEN STARTS
        LOGICR      ELVNO               LET NEXT TRANSACTION IN
        TEST G      Q$UPSG1,0,FGA       PASSENGERS ARE WAITING
FFF     GATE LR     ELVNO
        LOGICS      ELVNO               STOP TANSACTION
        UNLINK      XH$LOBBY,BAA,MH$ELEVS(1,P3),,,FFFA
FFFC    SAVEVALUE   ELVNO,P3,H          SELECT THE ELEVATOR TO USE
        ASSIGN      1,V$LOBBY           SELECT THE LOBBY CHAIN
        ASSIGN      4,CH*1              DETERMINE CONTENTS OF LOBBY CHAIN
        ADVANCE     V$UPSG              TIME TO LOAD PASSENGERS
        ASSIGN      2,MH$FACT1(1,7)     NUMBER OF POSSIBLE DESTINATIONS
        ASSIGN      1,1
FAB     ASSIGN      1+,1                PROVIDE ADDRESS OF NEXT FLOOR
        ADVANCE     MH$FACT1(1,2)       TIME TO MOVE BETWEEN FLOORS
        UNLINK      P3,CAA,ALL,1,P1,FAA    PASSENGERS EXIT AT FLOOR
```

```
        ADVANCE     MH$FACT1(1,3),MH$FACT1(1,4) TIME WITH SPREAD FOR EXIT
  FAA   LOOP        2,FAB
        ADVANCE     3              EIGHTEEN SECONDS TO RETURN TO LOBBY
        TRANSFER    ,FFF
*
  FGA   ADVANCE     10             WAIT FOR MORE PASSENGERS
        TEST E      Q$UPSG1,0,FFF  FOR MORE PASSENGERS RESTART
        ADVANCE     20             WAIT AGAIN
        TEST E      Q$UPSG1,0,FFF  FOR MORE PASSENGERS RESTART
  FGA1  TRANSFER    ,FFF           KEEP ALL ELEVATORS IN SERVICE
        TERMINATE
*
*
*       THIS ROUTINE GENERATES THE PASSENGERS
*
        GENERATE    ,,2,1,90,4,H
        ASSIGN      4,MH$FACT1(1,6) DURATION OF SIMULATED TIME
  AAB   ASSIGN      2,V$ARIV1       TIME SLOT WHEN PASSENGERS ARRIVE
*ARIV1  VARIABLE    C1/50+1/2*1+1   TIME TO GET NEW DATA
        ASSIGN      1,MH$FACT1(1,7) NUMBER OF FLOORS
  ABB   SPLIT       V$UPSG1,AAA     NUMBER GETTING OFF EACH FLOOR
        LOOP        1,ABB
        ADVANCE     MH$FACT1(1,1)   INTERVAL TILL NEXT TIME SLOT
        LOOP        4,AAB
*
  AAA   ADVANCE     1,1             RANDOMIZE ORDER OF ARRIVALS
        QUEUE       UPSG1           WAIT
        LINK        XH$LOBBY,FIFO
*
  BAA   DEPART      UPSG1           END OF WAITING
        LOGICR      ELVNO           ALLOW NEXT TRANSACTION IN
        MSAVEVALUE  ELEVS+,P1,XH$ELVNO,1,H      NUMBER TO EACH FLOOR
        ASSIGN      3,XH$ELVNO      NOTE ELEVATOR NUMBER
        LINK        XH$ELVNO,P1     PLACE PASSENGERS IN THE ELEVATOR
*
  CAA   MSAVEVALUE  ELEVS-,P1,P3,1,H    PASSENGERS LEAVE
        TERMINATE
*       PAGE
*INITIALIZATION PROGRAM ------------------------------------------------
*
  6     VARIABLE    RN2/60
  5     VARIABLE    (MH$BLDG1(P1,6)*100+RN2+200)/100
  4     VARIABLE    (MH$BLDG1(P1,6)*100+RN2+200)/100
  3     VARIABLE    RN2/60
  2     VARIABLE    (MH$BLDG1(P1,3)*100+RN2+200)/100
  1     VARIABLE    (MH$BLDG1(P1,3)*100+RN2+200)/100
  TIME1 VARIABLE    MH$FACT1(1,6)/5
*
```

```
          GENERATE    ,,1,1,120          FOR ENTRY OF NEW DATA OTHERWISE *
          ASSIGN      2,V$TIME1          NUMBER OF TIME INTERVALS
DAA1      ASSIGN      1,MH$FACT1(1,7)    NUMBER OF FLOOR DESTINATIONS
DAA       MSAVEVALUE  BLDG1,P1,P2,V*2,H  PASSENGER DESTINATIONS
          LOOP        1,DAA
          MSAVEVALUE  BLDG1,1,P2,0,H     NOBODY FOR FIRST FLOOR
          LOOP        2,DAA1
*         MSAVEVALUE  FACT1,1,1,10,H     TIME INTERVAL FOR ARRIVALS
*         MSAVEVALUE  FACT1,1,2,1,H      TIME INTERVAL BETWEEN FLOORS UP
*         MSAVEVALUE  FACT1,1,3,3,H      EXIT TIME MEAN
*         MSAVEVALUE  FACT1,1,4,2,H      EXIT TIME SPREAD
*         MSAVEVALUE  FACT1,1,8,5,H      NUMBER OF ELEVATORS
          MSAVEVALUE  FACT1,1,6,30,H     TOTAL TIME TO BE SIMULATED
          ASSIGN      1,MH$FACT1(1,8)    NUMBER OF FLOORS
DBB       MSAVEVALUE  ELEVS,1,*1,MH$FACT1(1,9),H ELEVATOR CAPACITY
          LOOP        1,DBB
          TERMINATE
*         PAGE
*
          GENERATE    ,,1,1,127       INTERACTIVE TRANSACTION
          HELP        TYPLIN,BLANK,1,2
          HELP        TYPLIN,PRMB1,1,52       THIS MODEL
          HELP        TYPLIN,PRMB2,1,37       ELEVATOR SERVICE
          HELP        TYPLIN,PRMB3,1,37       TWENTY FLOORS AND ELEVATORS
          HELP        TYPLIN,PRMB4,1,66       ENTER YOUR BUILDING
          HELP        TYPLIN,PRMB5,1,22       FLOORS, ELEVATORS, CAPACTIY
          HELP        RDVAL,FACT1,7,9,1       NO.,FLOORS,ELEVATORS,CAPACITY
GGG       ADVANCE     100             EVERY TEN MINUTES
          MSAVEVALUE  FACT1,1,5,Q$UPSG1,H     CURRENT SIZE OF QUEUE
          MSAVEVALUE  FACT1,1,10,QM$UPSG1,H   MAX QUEUE SIZE
          HELP        TYPLIN,LINE1,1,47,0,1        NUMBER PEOPLE
          HELP        TYPVAL,FACT1,5,5,1      ACTUAL NUMBER
          HELP        TYPLIN,LINE2,1,28,0,1      MAX NUMBER
          HELP        TYPVAL,FACT1,10,10,1    ACTUAL MAX
          SAVEVALUE   COUNT+,1,H      FOR THIRTY MINUTES ONLY
          TEST NE     XH$COUNT,3,GAZ
          HELP        TYPLIN,LINE3,1,46,0,1      CHANGE QUERY
          HELP        RDLINE,LINE4,1,1        READ REPLY
          TEST NE     MX$LINE4(1,1),MX$YES(1,1),GAA    CHECK ANSWER
          SPLIT       1,GAZ
          TRANSFER    ,GGG            IF NO CONTINUE
GAA       SAVEVALUE   NMSSTMH$FACT1(1,8)    NUMBER OF ELEVATORS
          HELP        EQUATE,LINE5,NMBR,7,2      FILL IN NUMBER
          HELP        TYPLIN,LINE5,1,16,0,1      ENTER VALUES
          HELP        RDVAL,ELEVS,1,X$NMBR,1     READ NEW CAPACITIES
          SPLIT       1,GGG
GAZ       TERMINATE   1
*         PAGE
          SAVE
          START       3,,,1
          MSTORE      MH$FACT1
```

```
DSPLY REPORT      NRG,1,72,14000
LIST OUTPUT

''   COMPILE  FROM COLUMN 1 TO 72, USE 14,000 BYTES TOTAL.

            SKIP 3 LINES

            PRINT 1 LINE AS FOLLOWS
      PASSENGER DESTINATIONS FOR FIVE MINUTE

SKIP 1 LINE

            PRINT THUS
                  INTERVALS

              SKIP 3 LINES

            PRINT THUS
      FIVE MINUTE TIME INTERVALS


      LET LIMIT = MH$FACT1(1,7) + 1
FOR LOOP = 1 TO MH$FACT1(1,7) DO THE FOLLOWING

SKIP 1 LINE
LET FLOOR = LIMIT - LOOP                PRINT WITH FLOOR AS FOLLOWS
          FLOOR #**

FOR COL = 1 TO 6
PRINT 1 LINE     WITH MH$BLDG1(FLOOR,COL) AS FOLLOWS
***   ***   ***   ***   ***   ***

            REPEAT THE ABOVE


              SKIP 3 LINES
              PRINT 1 LINE WITH QM$UPSG1 AS FOLLOWS
THE MAXIMUM NUMBER IN THE LOBBY WAS ***

      START NEW PAGE   '' SKIP TO THE BOTTOM OF THIS PAGE''


      LIST TERMINAL

            FINISH REPORT
            END
```

Job Control Language. Eliminate all need to know and use System 360 Job Control Language. Instead, an interactive question and answer monitor leads the user through the running of the model, and the earlier phases of data library establishment, model construction, debugging, and updating.

To demonstrate the use of the interactive GPSS system a model has been constructed to provide insight into how a building might be loaded during the morning rush half hour. The scenario to load the building is fixed. However, the user determines the size building, number of elevators and each elevator's capacity. The use of the interactive system is illustrated by printouts from the remote printer. These start on page 389.

For the terminal session the lower case represents input and the upper case response. The REQUESTS from the computer are numbered only for human convenience. The model was originally assembled and saved. Now the READ SAVE feature will be used to restart and run for three 10-minute intervals. After each interval different statistics will be reviewed. The interactivity in English was established through a set of generic HELP blocks. The report output is used to detail the fixed scenario of the number of people seeking to enter the building during the 30-minute interval. The listing of the model starts on page 395.

Summary

Models which rely on a statistical inference may also benefit from the availability of a computer. Results do not have to be gathered and later analyzed. Instead, those results subject to statistical variation can be analyzed while the model is being run. A moving average can compare the current length of a queue with its limited prior history, or equally useful results at this instant may be compared with a previous run's results at the same instant. All that is required is the foresight to store the intermediate results in an accessible fashion for comparative purposes.

The major limitation to the use of the printer as a means of communication between computer and design team is its slowness and quantity of output. However, there are numerous areas which cannot be represented simply. Therefore there are occasions when the full output has meaning. In cases where there are production runs an effort should be made to reduce the amount of output by replacing tabular listings with graphs.

The gain from the on-line display device, the cathode ray tube, results from its tie-in with the computer, which provides prompt and adequate amounts of output data. The time-shared terminal also provides this capability. Its limitation is the rate of data display. These devices certainly contribute to making simulation a more practical, useful system design tool. Overall, no one device is instrumental in the success of simulation; instead the advantages are in the use of various devices—printers, terminals, plotters, and cathode ray tube displays—to provide a variety of outputs when the system design team needs them and in form suitable for transportation to where the results are needed and to accomplish this as quickly as possible.

BIBLIOGRAPHY

1. Computer-Assisted Design of Complex Organic Synthesis, E. J. Corey and W. T. Wipke, *Science,* **166,** 3902:178–192, 1969.
2. EASL—A Digital Computer Language for Hands—On Simulation, S. I. Schlesinger and L. Sashkin, *Simulation,* **6,** 2:110–120, 1966.
3. Man-Computer Interaction: A Challange for Human Factor Research, R. S. Nickerson, *Ergonomics,* **12,** 4:501–517, 1969.
4. On Line, Incremental Simulation, M. Greenberger and M. Jones, *Simulation Programming Languages,* J. N. Buxton (ed.), 13–30, North-Holland, Amsterdam, 1968.
5. On Man-Computer Interaction: A Model and Some Related Issues, J. R. Carbonell, *IEEE Trans. Sys. Sci. Cyber.,* **SSC-5,** 1:16–26, 1969.
6. Man-Machine Simulation Experience, M. A. Geisler and A. S. Ginsberg, *Proc. IBM Scientific Computing Symp. Simulation Models and Gaming,* IBM Form 320-1940:225–242, 1966.
7. Sketchpad: A Man-Machine Graphical Communication System, I. E. Southerland, *Proc. SJCC,* 329–346, Spartan, Washington, 1963.
8. Principles of Interactive Systems, C. I. Johnson, *IBM Sys. J.,* **7,** 3 & 4:147–173, 1968.
9. Aspects of Display Technology, A. Appel, T. P. Dankowski, and R. L. Dougherty, *ibid.,* 176–187.
10. GRAIL/GPSS: Graphic On-Line Modeling, J. P. Haverty, *Proc. IBM Seminar Oper. Res. Aerosp. Ind.,* 197–210, 1968.
11. The GPSS On-Line Monitor, E. W. Ziegler, Jr., *IEEE Trans. Sys. Sci. Cyber.,* **SSC-4,** 4:438–441, 1968.
12. GPSS/360-Norden, An Improved System Analysis Tool, W. A. Walde, D. Eig, and S. R. Hunter, *ibid.,* 442–445.
13. GPSS/360-Norden, A Partial Conversational GPSS, S. R. Hunter and J. Reitman, *Proc. 2nd Conf. Appl. Stimul.,* 147–150, New York, 1968.
14. Computer Graphics for Simulation Problem-Solving, T. E. Bell, *Proc. 3rd Conf. Appl. Simul.,* 47–56, Los Angeles, 1969.
15. Job Shop Scheduling Simulations for Interactive Use in Computer Graphics, J. W. O'Leary, *4th Conf. Appl. Simul.,* 213–218, New York, 1970.
16. A Complete Interactive Simulation Environment, GPSS/360-Norden, J. Reitman, D. Ingerman, J. Katzke, J. Shapiro, K. Simon, and B. Smith, *Proc. 4th Conf. Appl. Simul.,* New York, 1970.
17. *GPSS/Norden Simulation Language Application User's Guide,* National CSS, Inc., Stamford, Conn., 1971.

16

Simulation and Tomorrow— This is only The Beginning

The current state of the simulation art is one of rapid progress and of great achievement in limited areas. We see simulation being applied to problems associated with the design, analysis, and prediction of system behavior, for limited size systems. Next we expect successful applications of simulation to involve very large systems. More significant to the user is the growth to interaction of real-time operation and simulation. This is the logical outgrowth of experience, transferring information among simulation groups, and the systematic improvement of simulation languages. However, it has been a slow process tightly related to improved capabilities of computer systems. On this basis the improvements forecast for simulation can be divided into three areas:

1. Language extension due to evolution and the experiences of numerous applications.

2. Improved system design ability from having used simulation.

3. Improved and extended computer system hardware, software, and data management.

Language Improvement

Simulation needs and requires convenient languages. These languages exist today. The difficulty with the current languages is in obtaining tighter coupling between the people attempting to solve the problems and the people who are putting the problems on the computer. The concept of activity-oriented languages, wherein the natural language of the problem becomes the means of stating the problem for the computer and for simulation, will come closer to realization. This improvement will occur either through the development of a special-purpose language for a problem area or through the superposition of the activity language converter before the general-purpose simulation language. Both techniques undoubtedly will be developed.

402

In the computer system field where the number of potential users is as large, the cost of an activity-oriented language could be spread out over a large number of users. The advantages in this bounded area arise from a reduction in the effort to simulate a computer system with a computer and to obtain minimum running time. One means of avoiding the tedious details characteristic of simulation of computer systems would be to develop within the language a macro structure of the instruction reportory of the computer. This permanent library would be used to provide simulated behavior of instructions, either elemental or macro. The simulation would have to consider the effects of cycle stealing, channel capabilities, and instruction look ahead without the penalty of lengthy running times.

Activity-oriented languages for traffic analysis and similar fields should relate in a general sense to the natural language of the field. The traffic simulation example, used in Chapter 12, shows how the natural language can be superimposed—in this instance, on GPSS II. This result was certainly much less expensive and faster than developing a new language, especially since running time will become even less significant than it is now while labor costs will continue to rise.

Concentration of effort to improve general-purpose simulation languages should begin to reach basic standards through the process of adding features from one language to another. Since each language is a dynamic evolving simulation tool, the individual groups refining each language will, as they have in the past, borrow good ideas from each other. Under these circumstances, any group developing a new language from the beginning must provide the equivalent performance of existing languages. This factor may mitigate against new languages as such, but not against new implementations of old languages. GPSS, for example, is implemented in assembly language in all its IBM versions, while the Univac GPSS II version is in FORTRAN. How others would develop GPSS-like languages would be up to them, but a PL/I version of GPSS or only some parts of GPSS is conceivable. It is realistic to consider for future implementations a multiplicity of higher-order languages being applied to develop one simulation language. The greatest value of assembly language is to generate the most efficient code for running the model, then perhaps that is the only part of the language in assembly language. Other parts of the language could more appropriately utilize higher-order languages according to the function required: list-processing flexibility, LISP, or SNOBOL; mathematics, PL/I or FORTRAN; and output, COBOL. Advantages of this approach might be to keep an interpretive structure similar to the execution portion of GPSS while gaining in flexibility to create new syntax, add symbols, provide a more user-oriented vocabulary and be more compatible with a time-sharing environment. These gains should not slow down running time. Rather, faster arithmetic routines or a more specific form of a general report generator could be added. One final point, changes to develop this structure would tend to reduce language differences because the language would be a subset of a more user-oriented simulation language.

The growth of simulation language use will make it harder for manufacturers to produce incompatible versions of the same language. It can be expected that major computer system manufacturers will have at least one, and probably several, simulation languages. Basic versions of the languages will have to be available from time-sharing services. The most sophisticated versions will eventually also be available, through both typewriter and display. Since the direction of language development is toward vastly greater capability, there is the associated problem of ease of training. As the tool becomes more versatile, more able to mirror the nuance of the problem, must we also expect the penalty of greater complexity when learning to use this tool? Here the answer must be no. There must be retained an elementary subset of the language which requires minimum training *and produces immediate results.* Beyond this ability, there must be a teaching machine approach to permit the elementary user to upgrade his ability. We are on the threshold of this development.

The critical area in which simulation languages need improvement is interaction between man and computer. At present, conversational interaction between the man with the problem and the computer is possible; it needs to become convenient. The role of the computer must change from that of a hostile assistant attempting and succeeding in showing the human to be a relatively inept provider of instructions. Instead, the computer must become a patient teacher. The problems or their subareas will be structured in languages attuned to user knowledge and problem requirements. The system designer will sit down at a console, introduce what he thinks the problem is, and maintain a dialogue with a computer program to establish system characteristics, logical rules, equations, input data, and output presentation until he is able to reach an initial result. From that point on, as shown in Figure 16.1, an iterative procedure becomes established. The system designer obtains insight, modifies input data, obtains more insight, makes new decisions, creates new ideas and new rules, and obtains new answers. In this way, the roles of the system designer and the computer become tightly intertwined. The computer, operating through a simulation language, becomes a very responsive tool for the system designer.

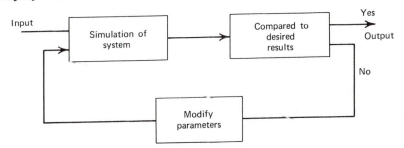

Figure 16.1 Iteration may become automatic.

However, the system designer will make errors both in syntax and concept; the close interaction between man and machine will reduce the time and effort to eliminate the errors. Syntax errors would be corrected by an immediate request from the computer for clarification. Conceptual errors would be more apparent with graphic displays and the comparison of results with expected values. The concept of conversational interaction between man and computer promises an ability to cope with larger, more realistic problems in a considerably shortened time frame.

The present languages handle the technical aspects of simulation satisfactorily except for two concepts: the representation of simultaneous events and the interaction between continuous and discrete event processes. Simultaneous events can not, in our present computer systems, be handled as simultaneous events except by the artifice of placing all these events on a current events list and repeatedly processing the list until it is reduced to zero. This may be adequate; but as we undertake to simulate larger problems with extensive parallelism—a city street traffic pattern for example, there will be needs for computer systems that are able to handle parallel events by duplicate computational facilities operating in parallel. The computer system may have several arithmetical units, each able to process an intersection.

The other aspect of parallel actions involves the interaction of phenomena represented by both discrete event and continuous simulations. These need to be married to provide the ability to simulate systems in which we have the differential equations and understand the analytical structure of a dynamic system. Superimposed on the analog continuous simulation solutions must be the characteristics of the external world imposed by the logic of discrete event simulation. The controlling discrete event framework provides logic, control, and overall timing through the overview scenario. The continuous simulation runs independently of the discrete event scenario until specific times when data are interchanged. The constants for the continuous simulation change according to the scenario, which, in turn, changes from new data it has received. One example might be the effectiveness of braking action depending on road surface, weather, and vehicle characteristics. This concept of hybrid computation, when effectively implemented, will be able to handle economically many problems currently too awkward to solve even by our current hybrid computer systems. The characteristics of computer hardware and software will determine the extent to which these sophisticated system concepts develop.

One possible implementation consists of the control of the overall system in a GPSS model. The discrete event scenario would provide the route the car takes and the driving strategy. From this GPSS structure a CSMP model obtains through HELP routines the coefficients to derive the acceleration and deceleration profiles and the responses of the suspension to potholes and road surfaces. If there were independent systems in the vehicle or when several vehicles are interacting in the simulation, the GPSS structure would feed several CSMP programs, each representing one vehicle. Obviously this concept could be implemented in a variety of languages. GPSS might be easier since the interpretive aspect of the language provides a simple reference base for data locations.

The Changing System Design Environment

The system designer has until now been forced to rely too heavily on experience, ability to quickly learn the problem area, discover simplified analytical expressions, and finally develop a reliable intuitive judgment. This situation need no longer be true. Various tools are available depending on the system characteristics. For example, one does not *design* solutions to global problems. There are numerous divergent schools of economic theory. Each model of an economic system is different. No one model has universal acceptance. Fortunately, system designers face less than global problems and, moreover, can agree on a representation of the system. The constraints and boundaries both assumed and accepted can provide an unambiguous system scope and definition. Undoubtably, these tactical systems are the ones most likely to benefit from simulation.

Tactical problems lend themselves to the set of definitions that can be manipulated with computer programs. A set of limited resources must be managed, scheduled, priced out, and their performance compared. This, the tactical environment rather than the global one, is where simulation will provide the tool for better system designs, analysis, and implementations.

The variety of systems that lend themselves to useful analysis via simulation is growing. The opportunity will emerge to tie the simulated system directly to the real system. While this will benefit systems now being investigated—communications, factories, traffic, and resource management—interesting technical progress will occur through the extension of the technique to social areas. The use of simulation to predict interactions within and among social groups will radically extend the disciplines employing simulation. These new applications represent more emphasis on the deterministic structure of the system design rather than on the statistical. Again this fits in with the ease with which complex interactions can be simulated. Where analytical expressions exist and are regarded as accurate, they will be used. The combination of both deterministic and statistical methods will determine the need for simulation.

The use of simulation for system synthesis depends on establishing confidence in the tool. This comes only with experience. Once the confusions between the global and tactical goals of the system design are resolved, simulation can help the tactical design. Only with simulation does the system designer have the opportunity to make mistakes, obtain insight, and get the feel of a system, before it is operational. An iterative design process can succeed, evaluating one scheme after another. Then, when the insight points in a particular direction, the system can be explored, modified, explored further, and one possible solution accepted, until there is further information. But the structure of the evaluation is quantitative and relative, rather than intuitive or arbitrary.

This is a radical change in the environment of system synthesis: a meaningful prediction of the system performance without simplifications that could lead to absurdity. Later, when the scope of a system is enlarged to represent the real world more accurately, the system design can be updated quickly within the time frame of the system design effort, and at reasonable cost. The cost may be greater than the budget, but realistic budgets are difficult to construct for system analysis. More experience is necessary. However, if there is the desire to adequately synthesize a system, construct the simulation, and learn the characteristics of the system, then the significant conclusion is that simulation adds little either to the system synthesis cost or time frame. This is the breakthrough.

As this powerful tool is used more routinely, the problems will grow, become more complex, less constrained, and more closely represent the system scope. Environmental and ecological problems such as air and water pollution can be studied. Life-cycle costs for a system can be calculated. Dynamic growth can be part of the pattern of change for a system. Worst case conditions can be represented to a useful degree. And, despite these gains, the overall system cost will be less. There will emerge a strong probability the system is designed right the first time; or the system may not be designed at all. It is better to know during system synthesis what is the expected performance and for what estimated cost.

This does not imply that all the problems are solved, merely that there will be an opportunity for better system synthesis for those systems susceptible to the use of simulation. Furthermore, the system designer will have the opportunity to increase the level of detail during the system design process in order to increase the degree of confidence. Specific complications can be included. The system design can be tuned for particular applications. Various scenarios can be used to explore system behavior under infrequent conditions. And these studies need not be lengthy or unduly complex. Unfortunately all is not optimistic. There is the consideration that if the wrong factors are entered into the model the wrong results will most likely emerge. There is no substitute for the understanding and competence of the design team.

Besides the system designers, there are others who can make use of the simulation. How should a manager react to a problem? Use his intuition? Rely on previous similar experiences? Compare notes with others? Now he has an alternative. He can have the simulation prepared in anticipation. He can use it in hypothetical situations. But the manager has to be conditioned to the existence of simulation. He cannot be expected to have found this out for himself. It is not yet part of his training. Conceptually, it is an engineering rather than an operations research or linear programming approach. But nothing succeeds like success, and if the cliche is correct, then low cost, flexibility, and ability to mirror his real world will make simulation one of his accepted tools.

This is not to suggest that simulation will replace management games. There are different purposes for each. Broad policy is one thing, day-to-day management is another. Again there is the distinction between strategic implications of policy and the tactics employed to carry out policy. Simulation belongs to the latter class. Of course, simulation could influence policy. Imagine the influence of more accurate cost projections or of performance figures that match the goal.

The gulf between systems of the real world and those of simulation will narrow. A traffic-control system ties sensors, communications and computers. The computers could also be used for a simulation. Under normal conditions, the computer is used with a deterministic controlling program to control traffic. Then, when something unexpected happens or there is an intimation that something will happen, the computer can also be coupled to the simulation program to evaluate alternative approaches to the problem. Obviously, if the situation were frequent, the program would grow to include this case, but where there is a wide variety of unexpected problems, the man must be in the loop. Let the operating system use his pattern-recognition abilities, provide him with information, and let him choose the solution, based on quantitative projections. This would change an abstract tool into a day-to-day helper for the decision maker.

The practical environment for the user of simulation is rapidly improving. Conferences are being held, providing a meeting place for the exchange of techniques, experience, and tricks. The nucleus of simulation proponents now exceeds the critical number. Dissemination of information is becoming easier and the audience is becoming identified. Simulation has been measured for the insight provided and found both helpful and wanting. All these are signs of acceptance and a degree of maturity.

The system designer is in line to become one of the earliest beneficiaries of improved computer hardware. He will be able to use improved man-machine communication to a much greater extent than the seeker of a single solution to an analytical expression. The system designer has many questions to ask, and one answer suggests another question. The display console can present more adequately the numerous conditions which provide the insight into the expected system performance. Hardware and software still make it possible to have significant degrees of interaction even while the simulation is going on. In a dynamic sense, the problem unfolds under the eye of the model developer, enabling him to have insight to the behavior of the system in simulated time. He is able to see the lengths of queues grow and diminish, and he can intercede to provide different resources which may control the queue length.

Simulation activity will also grow from unilateral simulation control to multiple control, the development of games. The system model will be used in a competitive manner analogous to the way it is expected to perform in the real world. Different operating tactics would be considered and the effectiveness of one policy against another evaluated. Opposing teams would be able to both manipulate the model and have access to joint and private data banks. Under these circumstances, the model would operate in a truly competitive manner.

Improvements in Computer Hardware and Software

The rapid improvements in both hardware and software indicate benefits where they are needed, better responsive systems, wider range of suitable computers, standardization, more versatile software, and reduced costs.

Simulations require large-sized computers. However, the use of peripheral random access storage devices will allow small machines to process large problems, but slowly. For really big simulations, the multiprocessor concept of several computers tied together will raise the upper limit of capability and reduce the running-time limitations. More important for the simulation user, the multitask computer, shown in Figure 16.2, is able to produce simulation results along with other programs and display the results. While the human being is determining the next step, the computer is processing an assortment of tasks. This feature markedly reduces the cost per hour of display equipment, since the hourly rates only have to reflect the cost of the display equipment and that part of the computer core being used rather than the entire machine. Dedicated computer systems cost too much to permit man-machine interaction. However, time-shared central processing units and partitioned large core storage systems prorate the costs. This makes use of the system economically feasible.

At the other extreme, future simulation languages will be able to run on smaller machines. Conceivably, a different instruction set would be used for simulation. Since there are so many small machines, this

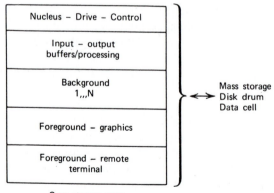

Figure 16.2

means that the problem of accessibility to a suitable computer will be reduced. The accessible computer is the machine to use; and if many more can perform simulation, this improves the practicability of using simulation. On the hardware horizon is the fast low-cost remote display terminal providing a cathode ray tube display for data input and its correction, high-speed temporary display of output, and hard copy. This device requires an ordinary telephone line for connection to the computer. Perfecting the use of these existing devices in simulation systems will provide a widely available tool for the system designer.

Simulation will be one of the uses of the computer that will benefit from the use of remote time-sharing equipments. The need for large quantities of output data make the usual time-shared terminal too slow. Two developments help this situation. First, the development of low-cost, higher-speed hard copy devices make it practical to obtain the output data. Second, the improvements in the simulation languages reduce the need for output by permitting a selection of which output to receive and still allow another opportunity to request further output details.

Standardization of languages so that models in one of the common languages will run on different manufacturer's machines will improve the opportunities to use simulation. This practice is becoming more the rule instead of the exception, which is most encouraging.

The general improvements in the software for the computer systems, as distinguished from simulation software, has considerable value for system designers. The biggest single gain lies in new data management techniques. This laborious area needs accessible, organized data banks or libraries which may be interrogated for different purposes by numerous different models. In addition, the files may be manipulated, reorganized, modified, and kept up to date with minimal effort on the part of the systems designer. His housekeeping practices might not be exemplary, but under the data management system he no longer has the active responsibility to control the data and the errors therein. Permanent data libraries suggest that what otherwise might be a theoretical model would instead be based on the utilization of the latest available information.

New Applications

The principal new areas lie in the ability to couple simulation with reality. Simulation based on actual data can provide new transportation-activity schedules and control job shops, and indicate the adequacy of computer and telephone facilities. These areas are the next logical evolution, held back primarily because of the difficulty of tying real-world data and models together. Since this limitation is dissolving under the development of large permanent data libraries, the new areas which will provide greater challenges lie where the data are present, yet obscure.

One such new area would be a combination of sensors, computers, and displays which could provide civil or industrial management information equivalent to the military command and control "war room." The purpose could be to control any critical resources capable of modification and reevaluation, such as vehicular traffic control, servicing routes, air traffic control, assembly lines, or even dynamic school room assignment. In each case, there is a system and although it is operational, the question is how to make it operate better. The war room approach is an educational one. The simulation and display become a user-oriented teaching machine. Why make mistakes in real time when you can try out the "solution," determine how effective it is and then continue by either progressing to another trial or by implementing the simulated solution? This approach trains the human; it does not require elaborate special programming and can improve performance. The additional cost beyond the cost of simulation should be slight. The difficulty is in developing meaningful displays that are user-oriented, operational, flexible, and able to educate the user.

The use of simulation to guide the process of urban renewal may seem unnecessary, useless, or visionary, depending on the scope of the problem to be studied. When urban renewal is considered in terms of the gradual transformation of an area from one form of land use to another, it may be appropriate to consider, in addition to costs and tax returns, the extent of dislocation and inconvenience. This broader scope is within the range of feasibility for simulation, especially when directed toward minimizing the disruption and inconvenience to the individuals and firms in one specific area.

Urban renewal is probably a more difficult system to model than city planning, evaluation of potential land uses, or the system plans for new educational facilities, but it shares the advantages for simulation of having some available data and the construction of the model through the development of partitioned submodels, which eventually are aggregated together. It is certainly a series of applications as far as possible from conventional system analysis, but obviously in areas that need help.

Thus we see that some of our most difficult system problems, those concerned with how we live and how we can improve our environment, may be approached and possibly successfully handled through a series of models. Models used to tell us the possible results of tax policies, air pollution, traffic congestion, and population growth.

Simulation will not solve the problems or make the decisions. It will provide more and better information to the decision makers for their use in arriving at a conclusion.

BIBLIOGRAPHY

1. SMPS—A Toolbox For Military Communications Staffs, K. Jacoby, D. Fackenthal, and A. Cassel, *AFIPS SJCC*, 127–139 (1966).
2. Digital Computer Simulation of Sampled Data Communication Systems Using the Block Diagram Compiler: BLODIB, R. M. Golden, *Bell Sys. Tech. J.*, **45**, 3:345–358 (1966).
3. Aiding the Decision Maker—A Decision Process Model, L. P. Schrenk, *Ergonomics*, **12**, 4:542–557 (1969).
4. Simulation of Geologic System: An Overview, C. Bonham-Carter and J. W. Harbaugh, *Simulation*, **12**, 2:81–86 (1969).
5. Modeling a Weather Environment, C. J. Green, *Digest 2nd Conf. Applic. Simul.*, 337–345, New York, 1968.
6. *Simulation Gaming for Management Development*, J. L. McKenney, Division of Research, Harvard Business School, Boston, 1967.
7. Some Comments on Gaming for Teaching and Research Purposes, M. Shubik, *Proc. IBM Sci. Comput. Symp., Simul. Models and Gaming*, IBM Form 320-1940, 243–248, 1966.
8. *Simulation of Information and Decision System in the Firm*, C. Bonini, Prentice-Hall, Englewood Cliffs, N.J., 1963.
9. *A Microanalysis of Socioeconomic Systems*, G. Orcutt, M. Greenberger, J. Korbel, and A. Rivlin, Harper and Row, New York, 1961.
10. *Candidates, Issues, and Strategies: A Computer Simulation of the 1960 Presidential Election*, I. deS. Pool, R. P. Abelson, and S. L. Popkin, MIT Press, Cambridge, Mass., 1964.
11. *Computer Applications in the Behavioral Sciences*, H. Borko (ed.), Prentice-Hall, Englewood Cliffs, N.J., 1962.
12. Experimental Simulation of a Social System, J. V. Rosenhead, *Oper. Res. Quart.*, **19**, 4:389–407 (1968).
13. On Parameter Identification and Quantification in Social System Simulation, J. D. Palmer, *Rec. IEEE Sys. Sci. Cyber. Conf.*, 10–19, 1969.

Index